Praise for *The Old Iron Road*

"Highly recommended . . . Historic accounts of particular people and places along the way make for lively and interesting reading, along with Bain's entertaining descriptions of and reflections on modern events and sights. The overall effect is a modern exploration of the American West and its development of a sense of place in the tradition of Charles Kuralt and Bill Moyers."
—*Library Journal*

A richly interesting book of travel and history that also has the virtue of being a beautiful and original portrait of a family together definitely *not* going to Disneyland. Bain has produced an excellent and inspiring book. It so inspired this reader that he made Kool-Aid (Black Cherry—10 cents a packet!) for the first time since the Nixon administration, rented *Butch Cassidy and the Sundance Kid*, made plans to visit his local historical museum and for the first time ever considered wanting a family so he could take them on an American road trip like the one Bain took with his."
—*The Oregonian* (Portland)

Praise for *Empire Express*

"One of the greatest of all American stories has finally found a chronicler up to the task of telling it. David Haward Bain has managed to encompass it all—genuine heroism and brutal dispossession, utopian vision and rampant corruption, technological wonders and war with the elements—in a vivid narrative that no one interested in the American character will want to miss."
—Geoffrey C. Ward, author of *The West: An Illustrated History* and coauthor of *The Civil War*

"A drama of conflict, adventure, excitement, and suspense . . . *Empire Express* promises to endure as the standard history of the Pacific Railroad."
—*The New York Times Book Review*

"The definitive story of heroism and heartbreak that produced a railroad, changed the landscape of the country, and altered the horizon of the nation."
—*The Wall Street Journal*

"A breathtaking tale enthusiastically told—of vision, greed, adventure, courage betrayal, accomplishment . . . *Empire Express* is a spirited telling of a complicated tale." —*Chicago Sun-Times*

"*Empire Express* is more than a study of the building of a railroad. It encompasses the range of nineteenth-century American life as it swept up Native Americans, women, settlers, con men, and speculators in one of man's greatest accomplishments." —*The Denver Post*

ABOUT THE AUTHOR

David Haward Bain is the author of four previous works of nonfiction, including *Empire Express* and *Sitting in Darkness*, which received a Robert F. Kennedy Memorial Book Award honorable mention. His articles and essays have appeared in *Smithsonian, American Heritage, The Kenyon Review*, and *Prairie Schooner*, and he reviews regularly for *The New York Times Book Review, The Washington Post*, and *Newsday*. He coproduced the documentary, "Transcontinental Railroad" for the PBS *American Experience* series and has appeared widely in venues as varied as Bill Moyers's "Becoming American: The Chinese Experience" and C-SPAN's *Booknotes* with Brian Lamb. A teacher at Middlebury College and the Bread Loaf Writers' Conference, Bain lives in Orwell, Vermont. Visit the author on his Web site at www.davidhbain.com.

THE OLD IRON ROAD

An Epic of Rails, Roads,
and the Urge to Go West

DAVID HAWARD BAIN

PENGUIN BOOKS

For Mary Smyth Duffy
and for G. J. "Chris" Graves

PENGUIN BOOKS
Published by the Penguin Group
Penguin Group (USA) Inc., 375 Hudson Street, New York, New York 10014, U.S.A.
Penguin Group (Canada), 10 Alcorn Avenue, Toronto, Ontario, Canada M4V 3B2
     (a division of Pearson Penguin Canada Inc.)
Penguin Books Ltd, 80 Strand, London WC2R 0RL, England
Penguin Ireland, 25 St Stephen's Green, Dublin 2, Ireland (a division of Penguin Books Ltd)
Penguin Group (Australia), 250 Camberwell Road, Camberwell, Victoria 3124, Australia
     (a division of Pearson Australia Group Pty Ltd)
Penguin Books India Pvt Ltd, 11 Community Centre, Panchsheel Park, New Delhi - 110 017, India
Penguin Group (NZ), cnr Airborne and Rosedale Roads, Albany, Auckland 1310, New Zealand
     (a division of Pearson New Zealand Ltd)
Penguin Books (South Africa) (Pty) Ltd, 24 Sturdee Avenue, Rosebank,
     Johannesburg 2196, South Africa

Penguin Books Ltd, Registered Offices:
80 Strand, London WC2R 0RL, England

First published in the United States of America Viking Penguin,
a member of Penguin Group (USA) Inc. 2004
Published in Penguin Books 2005

10   9   8   7   6   5   4   3   2   1

Copyright © David Haward Bain, 2004
All rights reserved

Map on pages xii–xiii drawn by David Lindroth

THE LIBRARY OF CONGRESS HAS CATALOGED THE HARDCOVER EDITION AS FOLLOWS:
Bain, David Haward.
The old iron road : an epic of rails, roads, and the urge to go West / David Haward Bain.
p.   cm.
Includes bibliographical references and index.
ISBN 0-670-03308-1 (hc.)
ISBN 0 14 30.3526 6 (pbk.)
1. Railroads—United States—History.   2. West (U.S.)—History.   I. Title.
TF22.B35   2004
385'.0973—dc22   2003071066

Printed in the United States of America

Acknowledgments

I have many people to thank, but my deepest go to my wife, Mary Smyth Duffy, for her endless support, encouragement, and good humor—especially during our little jaunt across the country, but also for many valuable suggestions for research and editing. For more than two decades of my writing life I treasured her faith. My ever-resourceful literary agent and dear friend Ellen Levine has my warmest thanks for seeing our plans for a family expedition and encouraging the book project it swiftly became. Ellen and three family members—Rosemary Haward Bain, my mother, and Katherine and John Duffy, my in-laws—were my most careful and helpful readers, and will always have my gratitude.

I owe a special debt to my good friends Chris and Carol Graves. From the moment he learned of our plans, Chris was extraordinarily generous with his time and expertise in things historical and geographical. Thanks to his presence on the journey between Ogden and Sacramento in July and August 2000, an enormous, multilooping itinerary, and to his always resourceful research and personal contacts, the project grew exponentially. I'm grateful for his editorial help as well as to Carol's many affirmations. In a number of visits to their home in Auburn, California, Chris and Carol's hospitality always made me feel at home. Our friendship grew out of this book project but will continue, I trust, for a long time.

My presumption of long associations extend to my Ogden friends Charles Sweet and Robert Chugg, who gave so much of their time and expertise as our itinerary took us across Utah and eastern Nevada, and who have welcomed me so warmly during numerous return engagements to the Salt Lake Valley. They have been the most steadfast of friends. During our family expedition, my wife's California relatives James and Elizabeth Smyth of Stockton, and Neil Smyth and Maria Castro of Santa Cruz, were so generous with their time and accommodations. I treasure my connections to them. My deep thanks also go to my friends Bob Geier, Director of the Ogden Union Station, and William Allison, Professor of History at Weber State University, for their encouragement and expertise as I composed my manuscript, and for their warm hospitality arranging my forays to Utah over the last several years.

Thanks, also, to my friend Mark Zwonitzer, a terrific writer of books and documentary films; our work together on his Hidden Hill production of "Transcontinental Railroad," for PBS's *American Experience*, coincided with my heaviest loads on *The Old Iron Road* and in my personal life, and his support kept me going in many ways. I also want to thank Kathy and Bruce Hornsby, wishing them well, and Robert Utley, for his belief and enthusiasm. Arthur King Peters, whose beautiful book, *Seven Trails West*, is a proud possession, gave me great inspiration and is dearly missed.

In Nebraska, my fulsome thanks go to Don Snoddy and Bill Kratville for their longstanding help with the Union Pacific photographic archives as well as for their affirmations about this book project; Chad Wall, Nebraska State Historical Society, Lincoln; Steve Gough, Great Platte River Road Archway Monument, Kearney; Barb Vondras; Dawson County Museum, Lexington; Dick Witt, Hastings County Historical Society, Hastings; Valerie Naylor, Robert Hamann, and Dean Knudsen, Scotts Bluff National Monument, Gering.

In Wyoming, my deep thanks go to the staff, board, and national membership of the Union Pacific Historical Society, Cheyenne, particularly Mary Nystrom, James L. Ehernberger, Lou Schmitz, Jim Hanna, Chris Plummer, Richard Rice, and Glenn Diehl; Ron LeVene and the Cheyenne Gunslingers Association; Ellen Stump, Wyoming State Museum, Cheyenne; Brent H. Breithaupt, University of Wyoming Geological Museum, Laramie; Joyce Kelley, Carbon County Museum, Rawlins; Robert Nelson and Audra Oliver, Rock Springs Museum; Diane Butler, New Studio, Rock Springs; Ruth Lauritzen, Sweetwater County Historical Museum, Green River.

In Utah, my deep thanks to Mary Risser, Superintendent, Golden Spike National Historic Site, Promontory, and to all the site staff and volunteers, with great affection; Professors Richard Sadler, Richard Ulibari, Gene Sessions, and Russell Burrows, Weber State University, Ogden, for many courtesies; William Critchlow and family; John Eldredge; Ken Verdoia, KUED-TV, Salt Lake City, for encouragement and for his excellent PBS documentary, "Promontory"; Daniel B. Kuhn; Stewart Johnson; Terry Tempest Williams; DeLone Bradford-Glover; Norman Nelson and the membership of the Golden Spike Association; and the Pole Patch Irregulars, a motley group if ever there was one.

In Nevada, special thanks to my friend Andria Daley-Taylor and her lovely family; Juanita Colvin, Justice of the Peace, Esmeralda County Courthouse, Goldfield; Michael A. Bedeau; and Professor Sue-Fawn Chung, University of Nevada-Las Vegas.

In California, great thanks to J. S. Holliday, friend and author of two great books on the Gold Rush, *The World Rushed In* and *Rush for Riches*; Gary Kurutz,

Curator, Special Collections, California State Library, and KD Kurutz, California Consortium for Arts Education, Sacramento, for their ideas and enthusiasm and for their delightful book, *California Calls You: The Art of Promoting the Golden State, 1870 to 1940*; Ken Yeo; Lucky J. Owyang; Members of the Auburn Parlor, Native Sons of the Golden West; Dana Scanlan; Mead Kibbey, for his *The Railroad Photographs of Alfred A. Hart, Artist*; Ellen L. Halteman, Librarian, California State Railroad Museum, Sacramento. David Lavender, a superb historian, was very encouraging, and I'm sorry he is no longer here to see the finished book.

Brian Lamb and Robin Scullin at C-SPAN were both wonderfully hospitable and heartening as I first began musing about this book. Arlynn Greenbaum, Authors Unlimited, was also encouraging.

At Middlebury College in Vermont, where I have taught since 1987, I'm grateful for support, enthusiasm, and the commerce of ideas. Special thanks go to Alison Byerly for her help through the Faculty Professional Development Fund, but also for her many expressions of belief in my work. Similarly, I'm lucky to have the companionship and advice of John Elder, David Price, John Bertolini; Julia Alvarez, Rob Cohen, Jay Parini; Brett Millier, Karl Lindholm, Bob Buckeye, Robert Schine, and fellow Western travelers Cates Baldridge, Jim Ralph, and Keith Conkin. At Middlebury's Starr Library, Fleur Laslocky of Interlibrary Loan; Michael Knapp and Danielle Rougeau of the Archives and Special Collections; Ellen Waagen of Circulation; and Hans Raum of Government Documents helped in many ways. President John McCardell and his wife, Bonnie McCardell, were unfailingly supportive of this project. They, and all my colleagues, rallied during my family's loss, and they will forever have my gratitude. At Middlebury's Bread Loaf Writers' Conference, I have been also fortunate in friends and colleagues, particularly Michael Collier and Katherine Branch; Devon Jersild; Noreen Cargill; Andrea Barrett; and Thomas Mallon. My friendship with Mary Pope Osborne began at Bread Loaf in 1982, and our consultations during the length of this book project came on top of her abiding deep feelings for my wife and children.

At Viking, I am as grateful as ever to my editor, Kathryn Court, for her many insights, patience, and fellowship over many years; thanks also to Stephen Morrison and Sarah Mangso, for their expert editorial counsel, to Fred Chase, for his careful copyediting; to Stephanie Curci and Ali Bothwell, for their sundry efficiencies and good humor. Thanks, also, to Patricia Bozza, Nancy Resnick, and Cohen & Carruth, Inc.

I was glad to find the gifted cartographer David Lindroth (thanks to Ann Close for that introduction). It is, moreover, an honor to have on the cover the photographs of Jim Richardson, whose *National Geographic* work has captivated me for years.

I have been lucky with my large book-loving family, all of whose enthusiasm about this project, and good ideas, meant much to me. In particular, thanks to Terry Bain and Marc Santiago; Christopher and Andrea Bain; Lisa Bain and William Schwarz; Sarah Duffy; and John and Kris Duffy.

Mimi Bain and David M. Bain, my extraordinary children, have never flagged, and they are my light. I don't think, however, that I should dare to ask them to climb aboard another caboose—at least for a few more years.

Finally, my thanks to KarenLise Bjerring, for her surprise reappearance in our lives.

Contents

PART V

PART VI

THE OLD
IRON ROAD

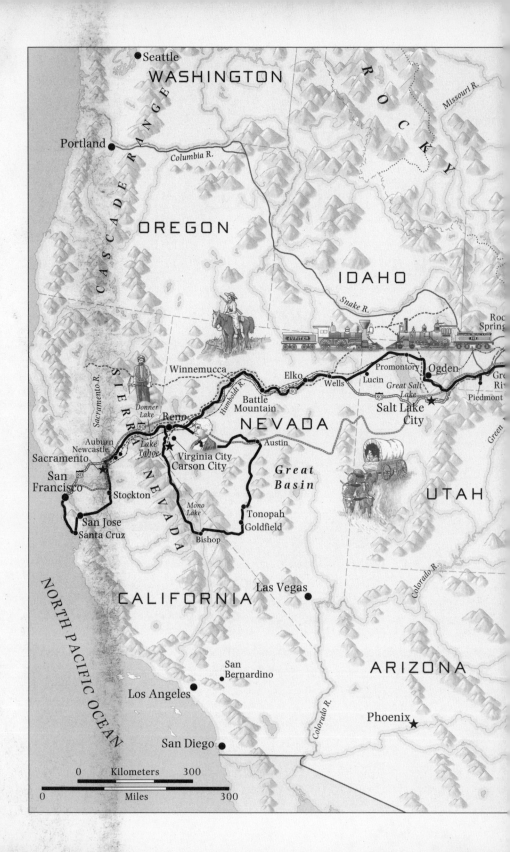

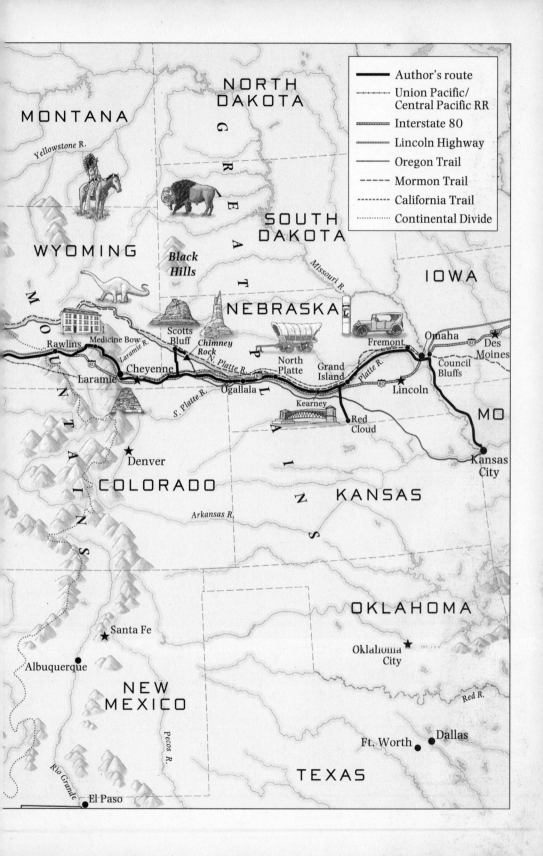

MONTANA

NORTH DAKOTA

Yellowstone R.

WYOMING

SOUTH DAKOTA

Black Hills

G R E A T

NEBRASKA

IOWA

Missouri R.

Rawlins
Medicine Bow
Laramie R.
Cheyenne
Laramie
Scotts Bluff
Chimney Rock
N. Platte R.
North Platte
Grand Island
Fremont
Omaha
Des Moines

Council Bluffs

Ogallala
S. Platte R.
Platte R.
Lincoln

MO

Kearney
Red Cloud

P L A I N S

Denver
COLORADO
Kansas City

Arkansas R.
KANSAS

OKLAHOMA

Santa Fe
Oklahoma City

Albuquerque

NEW MEXICO

Pecos R.

Red R.

Ft. Worth
Dallas

Rio Grande
El Paso

TEXAS

	Author's route
	Union Pacific/Central Pacific RR
	Interstate 80
	Lincoln Highway
	Oregon Trail
	Mormon Trail
	California Trail
	Continental Divide

PART I

1

The Odyssey Begins

We had been driving north on the old Leavenworth to Fort Laramie military road, now designated Kansas Highway 7/73, concrete and strips of softening tar winding through attractive wooded hills, wild trumpet vines and daylilies sprouting at roadside, willows and poplars alternating with pastures, hayfields, and stands of corn. Hawks rode warm air currents far overhead. Wood thrushes darted in and out of the shade trees. I pulled our car over to the roadside and shut off the engine. It ticked in the early summer breeze. "This is as good a place as any," I said to my wife and children.

Somewhere on this road between Atchison and Leavenworth in eastern Kansas my grandmother Rose Donahue Haward had been born in a covered wagon in the year 1889.

We were on the first leg of a summer exploring expedition, eight days out from our home in the Champlain Valley in Vermont, following half-forgotten footsteps, barely discernible wheel ruts, and vanished iron rails across the width of our continent for two months. Here, on the Kansas side of the Missouri River, seemed a fitting spiritual start to our odyssey—with my grandmother's humble beginning in that canvas-covered wagon.

We might have gotten to the Missouri River, to Kansas City, and to Leavenworth by car, but we had really been conveyed by a train called *Empire Express*, my book about the building of the first Pacific railroad.

On July 4, 1999, I signed my name to the preface of *Empire Express: Building the First Transcontinental Railroad* and closed an era in my life that had begun more than fourteen years before. When I started work on the book in the spring of 1985, I was living with my wife in a five-room ground-floor apartment, with a tiny garden, in Brooklyn, New York. Mary and I had been together for five years and our household consisted of us and two cats, Fred and Ginger. By the time the research and writing was done, we had moved twice—first, to Shoreham, Vermont, where we fixed up a dilapidated, 140-year-old farmhouse and tended sheep, and then down Route 22A to Orwell, where we bought an equally old house that had once been the village's Methodist parsonage, a solid Greek Revival–style house with a lovely wraparound porch with trumpet vines. By then we were raising our two children, Mimi and David.

Despite our bucolic surroundings, I will admit that those intervening years were hard. During that time I went around the country for research, depended on the kindness of many strangers, wore out the interlibrary loan staff at the Middlebury College library, filled up a four-drawer filing cabinet with photocopied handwritten documents and official reports and a floor-to-ceiling bookcase with books, and wrote a 1,100-page manuscript, which translated to 800 book pages. A publisher's advance in 1985 stretched out pretty thin over thirteen years, on top of which was my wife's small salary and mine as a part-time writing instructor at Middlebury College. There were no grants, and as part-time faculty I was not eligible for paid leaves. Fortunately we had access to health insurance, for we did have periods of serious illnesses.

What sustained us through most of these hard times was the warm, bright light our children brought into our lives, and also the fact that pursuing such a project as *Empire Express* was the greatest gift I could be given as a writer and historian. The research covered three decades of tumultuous, absolutely pivotal American history; the greatest single construction project our nation ever faced; and an extraordinary cast of characters. No previous chronicler, in my opinion, had done this historical narrative justice. There were myths to be shattered, new information to be unearthed, new characters to finally be given their due, and—perhaps most important of all— connecting lines to be drawn between the central drama of the railroad and the larger, enveloping national context, between long-accepted isolated events that were actually integrally part of a whole. It may have been a challenge for our family to get to the end of each succeeding month over fourteen years, but I seldom sat down at my desk in the morning without a rising sense of excitement and curiosity about the people and their stories, and how they all fit together, and how the narrative was going to be built.

Back in November 1997, Mary was recovering from her last session of chemotherapy after a mastectomy, and I was at the college on a teaching day. An English Department colleague, Cates Baldridge, came up to me in the faculty lunchroom. "David," he told me, "I've got terrible news for you." *Terrible news?* I thought bleakly of the past six months. *What could be more terrible?* He continued: "I was watching the Charlie Rose show," he said, "and Charlie was interviewing Stephen Ambrose, and Charlie asked Stephen what was his next book, and Ambrose said, 'I'm going to do a book on the first transcontinental railroad.'"

Well, yes, that was terrible news, but not as bad as a cancer diagnosis.

Standing there in the faculty lunchroom, I had the most eerie cinematic moment. My surroundings melted away, and I was out somewhere on the Forty-Mile Desert in Nevada, pumping away on a railroad handcar, sweating like an animal, making slow progress but definitely making progress. Then there was a rumbling vibration I could feel through the pump handle of my handcar, and I looked around to see a great, gleaming, gold-plated leviathan, belching black smoke and white steam, thundering down the tracks and gaining on me every moment—the *Ambrose Limited.* I could even see his face on the front of the locomotive, like Thomas the Tank Engine only his was emphatically not a warm, happy face: it was cold, expressionless, predatory. The vision faded away and I was back in the lunchroom. I staggered back to my office and made some calls, thinking that thirteen years of work were about to be blown away.

Well, again, our luck held. Three of my most faithful book friends in the world, my agent, editor, and publisher, all of whom I've been close to for more than twenty-five years, concocted a formula, to which my college superiors instantly acquiesced. Viking bought me out of my teaching contract for thirteen months, and Middlebury College promised that my untenured job would resume when I finished. All of my book research was already done. More than a decade of following leads like a detective had filled a filing cabinet. Chronologically I was in the spring of 1867—two years to go in the narrative before the Golden Spike, and then the epilogue with all its railroad scandal. All I needed was a year to finish the narrative, a matter of connecting the dots. For the next thirteen months, I wrote seven days a week, all of it in my memory a glorious blur. Usually it's very enjoyable to live with one foot in the past and one in the present, but for expediency's sake I just mentally stepped back into the 1860s and stayed there, twenty-four hours a day. I drove the Golden Spike in early December 1998. I finished the epilogue on January 5, 1999, a beautiful, white winter Vermont day. After the long editorial period of winter and spring, I was freed to finish the preface on July 4, with the sound of snare and bass drums booming down on Main Street and the town green.

Empire Express was released with a ten-month lead on its indefatigably popular competitor, and on my own and my publisher's terms it did extremely well. As to my vision of being out on Nevada's Forty-Mile Desert, the leviathan may have caught up with me, but he didn't knock me off the tracks, at least not entirely, and I pumped myself a pretty good distance, or, as railroaders would say, made the grade.

After some of the excitement abated, I began to ask myself, "How can I repay my spouse for fourteen years of belief and support in this project? How can I reward my good children for not getting complexes because their dad always had a faraway look in his eyes, and was always tired, their entire lives?"

I did not take them to Disney World—not yet, at least. My mind and my imagination had been out West, in the 1860s, since before Mimi and David were born. Flying over the prairie, the Plains, the Rockies, the Wasatch, and the Sierra during the book tour, taking all these little puddle-jumper flights from city to city, with the wide-open land always visible below, I was seized with the idea of taking them out West to see the sights I'd seen during research, of showing them places of history. It would take up the whole summer. I began creating an itinerary for two months of cross-country driving, staying for the most part off the interstates and on state, county, or town roads. Or, in many cases, on no roads at all.

My goal was to trace portions of many old emigrant routes between the Missouri River and the Golden Gate—parts of the Oregon, California, Mormon Pioneer, and Overland trails; the Pony Express; the first transcontinental railroad; and several exploring expeditions, all the way up through our most recent century's old Lincoln Highway. Most of them were along one extraordinarily resonant and historical corridor. I also wanted to draw an impressionistic narrative line from the Indians, trappers, and traders; the explorers, engineers, and emigrants; to those who actually found what they were looking for and settled into the tiny, isolated pioneer communities that grew up, spread out, and transformed the West, confiscating one kind of life and implanting another. What, after all, had the trails and the train wrought?

A modern-day travel narrative from this would have certain kinds of literary antecedents. I had enjoyed works by William Least Heat-Moon, Colin Thubron, Jonathan Raban, and Bruce Chatwin, to name only a few, and had passionately admired John McPhee's three-book odyssey across the length of Interstate 80 as a way to illuminate the historical geography of North America. A previous book of mine, *Sitting in Darkness*, had been a two-level narrative set during the turn-of-the-century Philippine-American War and during colonialism's logical extension in the Marcos era, these different levels linked by my expedition up the isolated northeast coast of Luzon, following old footsteps and once famous but now obscure doings. But this new work would contain many more historical yarns—I've always been convinced that a travel narrative could be supple but strong, that it could indeed support a generous amount of linked historical digressions.

The journey, then, would be many digressions—as I told my children in our driveway as we were pulling out onto our first road, "This isn't about the destination, so it's no use asking, 'Are we there yet?' This is about the journey, about hearing voices and discovering stories. We'll be *there* every mile of the trip."

At the start of this journey in the Vermont village of Orwell, my passengers and I knew that the continent was 3,000 miles wide. As it would turn out, we'd log more than 7,000 miles—*one way*. I stopped noting the odometer two months after we began, at a truck terminal in the Pacific slope town of

Watsonville, California, as I consigned the battered, scratched, and dusty car for flatbed transport home—we had to fly back East in time for my mother's eightieth birthday party. Back at the spiritual start of the journey in the Missouri Valley, I would tell my mother, I'd paused to stand on the Kansas soil near where her mother, Rose Donahue Haward, had been born in her emigrant parents' covered wagon, in the year 1889.

Her life spoke to me. Rose's father was named Peter Donahue and he was born in County Galway, Ireland. He had fled to America during the Great Famine; he was eight years old, traveling with his older brother, Thomas, who was fourteen, and his seven-year-old brother, Patrick. They landed at Baltimore. The year was 1856, and we can only guess how those three boys lived, part of the despised Irish hordes who took the bottom-rung jobs when they could rise that far. We know that Peter and Patrick finally joined the U.S. Army in January 1866, nine months after Appomattox. They lied about their ages; Peter was seventeen but said he was twenty-one, and Patrick said he was twenty though he was sixteen. Three square meals a day and clothing provided were quite an enticement. In the year after the war the 5th Cavalry was garrisoned at Washington, D.C., and then was dispatched to the South for Reconstruction duties, probably police and civil rebuilding, around Atlanta.

When my great-grandfather was mustered out in January 1869 in Atlanta, he disappeared into the obscurity of the very poor, who often eluded census takers and weren't important enough to be found in city directories. He emerged in 1880 in South English, Keokuk County, southeast Iowa, having acquired a wife, Catherine Coughlin, born in Wisconsin Territory (now Minnesota). That year they had a one-year-old daughter and there was a son on the way. Peter Donahue was a railroad laborer with the Chicago & Rock Island. He had grown to his full height, five feet five inches tall, and he had brown hair and brown eyes; his Galway bog accent could not have gone far. According to family lore, the Donahues moved on, taking root in the northeastern part of the state.

The hilly pasture country around Cresco, Iowa, seat of Howard County, was already full of Lutheran German, Norwegian, and Dutch landowners, and as elsewhere there was bad blood between them and Irish Catholics. Moreover, there in the 1880s crop prices were stagnating and population was leveling off. The Donahues had produced five children by then, and were probably tenant farmers—one in every four Iowa farmers was a tenant farmer—since they left no land records behind. Tenant farming was a miserable, hardscrabble life, cursed by circumstance and impermanence. Constantly it was a wandering—always that search for slightly better conditions and the chance to get an inch or two ahead. February was always moving time for tenants so that they could be ready for spring planting in March; in February the rutted, icy roads were always filled with sojourners of this type, a depressing sight for those who had means and sympathy. In late winter 1889 the Donahues hit bottom, so once again, with Catherine in advanced pregnancy with their sixth child, they piled their few possessions into a canvas-topped wagon. They headed west and south across Iowa and Missouri, toward eastern Kansas and Catherine's brother Jack Coughlin, who had a farm in Shawnee township and who, they hoped, might help them get a new start.

The roads would have taken them down to the redbrick city of St. Joseph, on the Missouri River, and they would have crossed there by ferry or even by toll bridge a little south, opposite Atchison. Perhaps ex-soldier Peter Donahue had an old army buddy stationed at Fort Leavenworth over in Kansas. Perhaps he hoped to work a little for cash at the post. It was getting late for planting in March, but they had their eyes set on Jack Coughlin's farm, still two or three days distant, when Catherine felt the unmistakable pains and knew it was time. The baby, my grandmother Rose Elizabeth Donahue, was born on the road on March 25, 1889.

Looking at the farms and hills now green in early summer, I thought of what my wife, Mary, had endured in bearing our children, Mimi and David, now eight and eleven and sitting behind us in the truck looking out at the rolling terrain—two difficult, complicated, emergency births that in Catherine Donahue's time would have left the husband a widower. And Catherine had survived nine childbirths, with one in early spring on the road in the back of a cold, canvas-topped wagon. No wonder that Rose's patron saint was Saint Christopher, comforter of all travelers. She used to send me and my sister Terry tiny blue and silver Saint Christopher medals taped onto index cards, which my mother, the last Catholic in our family, sometimes permitted us to wear. I remember the baby blue enamel against my chest; I was probably five or six years old and certainly soon lost the medals. With such a heritage it's no wonder my family has always identified with sojourners, particularly those chasing a dream into the unknown.

Peter and Catherine continued their journey down the road as soon as they could to her brother's farm in Shawnee, a few miles west of Kansas City. But most of their hopes were unmet. Jack Coughlin had a good-sized spread, some 200 acres, which he had cleared back in 1862. Jack loved his sister but had little use for her husband, and the summer and fall of 1889 were hardly a time for generosity. Peter Donahue took his team of horses and contracted as a road grader in and around Kansas City. He was never a success at it. And the children kept coming. Rose remembered little but privation, and told one story about how excited the family was once when Peter came home on a Saturday night and dragged a 100-pound bag of potatoes in the back door, guaranteeing them weeks of filling meals. Always tired and always distant, he left childrearing entirely in his wife's care. But once when she was five or six, Rose remembered, her father called her out on the front porch to ask her what time it was. "I don't know," she replied. "I don't know how to tell time." "You don't know how to tell time?" he exclaimed as if seeing her for the first time in her life. "Rose, go get the clock from the mantel and bring it out on the porch." She did, and then climbed up on his lap and he taught her how to tell time. "That was one of the few things he ever did for me," she said much later.

By 1900, living on Hunter Avenue in Kansas City, Missouri, all eight children were residing in the house and one more was on the way. The oldest, Mamie, did piecework sewing in a factory; Joseph, the next, was following in his father's footsteps as a day laborer; Sarah and Maggie were recorded in the census as "nurse girls," whether for toddler Martin or out of the house we don't know. My grandmother and two siblings, John and Tom, were still in school. Peter's older brother, Thomas, now fifty-six, boarded with the family and worked as a railroad day laborer, but Peter reported that he had been unemployed for six months.

At some point Peter Donahue just decided he'd had enough of teamstering and family life. He was worn out at fifty-six. So one day he announced to his surprised wife and children that he was moving away to the Old Soldier's Home up in Leavenworth, which is what he did. There had been an old injury to his right side in Washington while in the army, and this disability bought him entrance. It can't have helped his family, since most of whatever army pension he received was deducted for room and board

He spent his declining years in the veterans' barracks, looking out from his rocking chair on the second-floor front porch across the slope of the veterans' cemetery toward the brown Missouri. Peter Donahue died on February 1, 1907, of lobar pneumonia, aged fifty-nine. The hospital inventory noted he had $1.95 in cash and $2.80 in effects. He was buried with military efficiency the next day, about 300 yards away from his barracks.

Catherine was left to supervise the brimming household, as indeed she had been doing for some years. Rose, then pushing eighteen and out of school, had long since handed over child care of the youngest siblings to the next sister and was out in the labor force, working at a variety of jobs until she found a place in a printing plant owned by Robert E. Haward on Union Avenue in Kansas City. She was a sheet feeder, inserting paper into the big, loud presses. She was nice-looking and knew it; when Rose was twenty-one or twenty-two and had just had her hair done in an elaborate swept-up bun, she slid into a starched and pleated, high-necked blouse and locket and paid a photographer to capture front, side, and rear views of the hairdo. Sometime after this endearingly prideful act, she attracted the attention of the plant owner's son, Charles William Haward, also born in 1889. The Hawards, who had emigrated from Suffolk, England, to America after the Civil War, were stalwart

Baptists and frowned on their pride and joy's relationship with an Irish girl, but Charles and Rose were married in May 1914. They had a son, Charles Jr., in 1915 at their house on Troost Avenue in Kansas City, but didn't stay there long. During the Great War and for a few years more, Charles left the printing business to set up a small chain of stationery and supply stores with his brother John, situating them near army camps in Kansas, Arkansas, and Iowa. While managing a store at Fort Riley, Kansas, Charles and Rose had a daughter, Rosemary, my mother, in August 1920 in nearby Junction City. Around this time the stores stopped turning profits and Charles went back to printing in Kansas City.

My mother recalls her childhood in the 1920s and early 1930s as a long series of rented houses—a pump in the kitchen, chickens out in the yard, and strangers in the neighborhood—with occasional trips out to the permanency of Uncle Jack Coughlin's farm, where young cousins would play in the big hayloft and ride on workhorses. It was a big, sunlit, rambling house with a front and back staircase. One time she was invited to spend a few days at Christmastime and slept in a feather bed, the acme of luxury to the little girl. There was a notable contrast between the Coughlin homestead, settled and elaborated for sixty years, and all the temporary quarters of Charles and Rose Haward. Then things changed. On her own, not telling her husband, Rose marched out one day and with minuscule savings made a down payment on a modest stucco house southeast of downtown Kansas City on 92nd Street. This was at the height of the Depression, when many were walking away from their homes, but now, Rose declared, they could begin to put down roots. The house sat back from the road on four acres with views eastward over wooded bottomland. Rose lovingly enlarged the house, beautified the property, and commissioned a little sign to swing from a post at the head of the driveway: "Circle H Ranch." She ordered ink pads and rubber stamps from the stationery store with that name, branding iron figure and all, and the address, "92nd and Harrison." I have one still, as well as lush memories of summers there.

After college and the war my mother had gotten a job with a Kansas City radio station, and there, at a company picnic, she met David Bain, then a rep from the RCA broadcast equipment division out of Chicago. He bore a remarkable resemblance to Fred Astaire and, working in radio stations all over the South in the 1930s, he had assiduously erased all traces of a drawl from boyhood homes in Wilmington, North Carolina, and Jacksonville, Florida. During the war he had worked on a top secret radar project for the navy, being recruited after peace with many fellow engineers into the RCA family. "Remember me?" he wrote Rosemary on a postcard from Chicago's Hotel Eastgate, where he lived. "Please do, 'cause I'll be calling you one of these days." He did, and in Kansas City they married in 1948, the year before I was born out East in the RCA world headquarters city of Camden, New Jersey, to which my father had been transferred.

I think we spent part of nearly every summer at the Hawards' "ranch" in Kansas City during the 1950s. I know that we lived in at least three apartments and five houses, virtually all of them in different towns, before I was twelve as my father was frequently transferred between RCA's South Jersey and Washington, D.C., offices. Much later, in college, I met army brats and we understood each other with our constantly uprooted childhoods. It seems as if the Kansas City "ranch" was as comparatively solid to me as it had been to my mother and grandparents, much as the Shawnee farm of Uncle Jack Coughlin had been to Rosemary and Rose. As a boy I remember trudging up the 200-foot driveway bordered with petunias to get my grandfather's mail, turning around at the Circle H Ranch sign to go back toward the neat stucco house with its screened-in side porch. Inside it seemed dim and of another century. The living room had an oriental carpet and heavy overstuffed old furniture, with a print of Gainsborough's *Blue Boy* framed above. But other places fired my imagination and memory: there was a bunk bed nook upstairs off the master bedroom in which my grandfather read to me and my sister Terry, and a little silver-floored balcony outside we could visit, and there were 7-Up bottles kept cool down in wooden cases in the stone cellar, and back outside there was a little bunkhouse that always smelled of mown grass and charcoal from an ancient fire. Beyond the back door and a grape trellis (in which was wired a little plastic bird), my grandmother had fashioned and landscaped a grotto out of a half-tumbled building foundation, where there stood her statue of Our Lady of Fatima, and beyond this and a garage was a lower yard with a dank cyclone cellar and a tenant house that was usually empty, where we played.

I know there were summer gatherings of the Haward family there at 92nd and Harrison, but the ones I really remember are of the Donahues—two or three picnic tables set out in the cicada-serenaded yard near a hammock, attended by card tables with great plattered piles of barbecued chicken, corn on the cob, pickles, and potato salad—with cousins and their parents, some of Rose's many siblings, Aunts Agnes, Sarah, and Elizabeth, and Uncles John and Tom, the latter of whom was most distinguishable because of his old-fashioned metal spectacles, suspenders, and missing right leg, which he had lost as a boy when swinging onto a slow-moving freight train to hitch a ride into the city as all the kids did, but then falling under the boxcar wheel. Another family connection with trains, I was told, came through my grandfather, who when a boy had been minding his little brother Oliver Haward out on the sidewalk and the boy ran out in front of a streetcar and was killed. And there were other tragedies, some of which we young ones didn't learn about until much later, like the fact that Charles and Rose's only son, Charles, had shot himself in 1940 in the American embassy in Peking, where he worked, a victim of the clinical depression passed down through his mother and grand-father. My mother had been summoned home from college in St. Louis only to read of the suicide in the paper on the way to Kansas City. But tragedies went unaddressed at my grandparents' in the 1950s.

Terry and I were Charles and Rose's only grandchildren and the focus of great attention—Rose encouraged me through my mother into piano lessons, and Charles saw my interest in the Civil War and the Old West and sent me a subscription to a new magazine called *American Heritage*. Then two more sib-lings arrived, Christopher in 1955 and Lisa in 1957, but for them there are no memories of the Circle H Ranch. On November 6, 1958, Rose woke sometime before dawn and moved to sit in a bedroom rocking chair, where, quietly, her heart stopped. Charles stayed on in Kansas City for a few years but as his health declined he closed up the house and moved in with us in the New York suburbs. He followed Rose in August 1968. Someone bought their old place.

Two days before we stopped on the Leavenworth–Atchison road to approxi-mate the place of my grandmother's birth, I led my family southeast of down-town Kansas City to 92nd Street. I had heard from my mother in New York that the house was no longer there, but I wanted to walk on the open land, even if it meant hopping fences and ignoring NO TRESPASSING signs. Just off Holmes, on the eastward way to Troost, 92nd was a forgotten waste, weeds growing in pavement cracks, the street lined with exhausted or abandoned bungalows and cottages. From the 1920s through the 1950s this had been a modest, genteel outskirts-of-the-city kind of road with summertime radio broadcasts of the Kansas City A's, the Athletics, pouring from every house, laundry on lines strung at the back screen door, pitchers of iced tea on the counters, and friendly waves from side porches when one walked or drove by. Now it was difficult to tell whether anyone still lived on the street, although a few houses still struggled to look current. At the end of the street stood two metal barriers, then 100 feet more of cracked pavement, then a final dead-end barrier and a sumac thicket. Despite the decay, in my mind's eye a drive-way stretched down alongside flower beds and honeysuckle to the graceful stucco house, and I smelled grass clippings from my grandfather's electric mower, and there was the carbonated tang of a swallow of 7-Up in my mouth, but what I really took in, where ranch and house once were, was a great gray steel prefab warehouse, that took up virtually all the square acreage of the homestead. Even the topsoil had been bulldozed down to the underlying shale and carted away. There were no markings on the building. We walked down along the back of the warehouse until thickets blocked the way at roughly the place where I used to catapult out the side porch screen door into the yard. I could see old neighbors' forgotten sheds and Forties-era car hulks off in the trees. Behind us the houses across the street were also razed, overtaken by weeds and saplings, but we found a neglected little family cemetery with a dozen or so stones visible in the poison ivy—Douglases, who worked the land there in the nineteenth century and broke the farm up into building lots in the 1920s so that people like my grandparents could own a little place. Blue-jays and cardinals flew through the grove. Memories I have still not completely catalogued overwhelmed me. All that was there was gone, except what was in the heads of a middle-aged brother and sister and their eighty-year-old mother.

A few miles away at the high-walled Mount St. Mary's Cemetery, 23rd and Cleveland, with its large, well-kept lots, with gardeners out pulling away felled limbs from the previous night's seventy-mile-per-hour winds, we found the Donahue family plot—great-aunts, great-uncles; my sad Uncle Charles, who when he shot his own heart broke my grandmother's; and finally Rose Elizabeth Donahue Haward herself.

Two days later we were in Leavenworth, Kansas, at the Old Soldier's Home and the veterans' burial ground, now called the Dwight David Eisenhower Veterans' Administration Medical Center and the National Cemetery. Standing beneath a spreading maple tree near the bottom of a hill was Peter Donahue's gravestone, his name elaborated only by his military unit, Company C, 5th U.S. Cavalry. Uphill was the Old Soldier's Home barracks, now empty— the large modern facility that had replaced it was away on the other side of a placid reflecting pool decorated by ducks. The two old buildings, one brick, one clapboard, had massive columns reaching up past second-story screened-in porches toward austere eaves and silver tin hip roofs. Here taciturn Peter, having dumped his family, waited for an end to a lifetime of wandering, bitter prejudice, and hard labor, thinking occasionally, I suppose, about the green of County Galway, an Atlantic voyage, the hard regularity of army life, the procession westward with wife and children. I tried to shrug him off. I could hear songbirds, and even more comforting, from out of the northeast, a train whistle.

Frequently during the writing of *Empire Express*, my mother and I had talked about how Peter Donahue had drifted west like unchronicled thousands in the nineteenth century, particularly after the Civil War. This was the same time when the Union Pacific Railroad had papered the Eastern and Midwestern cities and towns with handbills hoping to recruit thousands of laborers, many happening to be Irish, to the great work west of the Missouri. Now we know that as a seventeen-year-old Irish orphan, my great-grandfather had not answered that locomotive's whistle but had lied about his age and gotten the only steady job of his life, a three-year hitch in the army, being discharged in Georgia just four months before the driving of the Golden Spike in Utah in May 1869. And then he had drifted westward and disappeared for a time. Meanwhile thousands of his countrymen had labored out on that historic trackside. Where had he gone when he vanished? What might he have done if, instead of enlisting in the army, he had found a Union Pacific handbill and followed the call out past the Missouri? Maybe on some level this journey was to see what might have happened out there, for him, through my eyes.

And so, on the old Leavenworth to Fort Laramie military road, now designated Kansas Highway 7/73, we paused on a gravel shoulder to watch an imperturbable red-tailed hawk get bullied away from raiding a pasture oak nest by two tiny, swooping, protective parent birds, who gave him the bum's rush off across stretching acres of cornstalks toward northeastern hills, with the Missouri River beyond. I thought about the infant Rose seeing her first light of day filtered through white canvas—and also about all those who hadn't stopped westering here but kept on going until they dropped or found shreds of what they were seeking.

I pulled us back onto the concrete and we headed north toward the river, which bent to meet us.

2

Jumping Off

T he day's journey lay up the Missouri River Valley from the first big el-
bow of the river, where metropolitan Kansas City sprawled and the
Kansas River flowed in, to where Council Bluffs, Iowa, and Omaha,
Nebraska, faced each other on opposing riverbanks.

The drive was about 200 miles but in history, an even longer span of years.
Far back into prehistory, settlements of the Missouri, Osage, Kansa, Iowa, Oto,
Ponca, and Pawnee occupied permanent wintertime lodge communities in
the lower valley and well up the tributary streams, going out on extended buffalo hunts on the Great Plains in summer, forever jostling one another. In
1682 the French colonizer La Salle had claimed possession of the vast water-
sheds of the Mississippi and the Missouri for the Sun King, Louis XIV, and
named it Louisiana. Forts and missions were planted along the Mississippi,
which was then established as the eastern boundary of Louisiana, but French
penetration of the Missouri Valley was slow until the 1720s.

Then, the adventurer Etienne de Bourgmont built a fort near the mouth
of the Osage River and ascended the Missouri at least as far as the Platte, in
present-day Nebraska. He established trade relationships with the tribal peoples that began to take on more importance into the next era, of Spanish
ownership, between 1762 and 1801, and the following, when Napoleonic
France regained Louisiana and so alarmed President Thomas Jefferson that
he engineered the purchase of those extraordinary lands in April 1803. Four
months before the transaction, Jefferson persuaded Congress to appropriate
funds for an exploring expedition to cultivate relations with the native tribes
and extend internal commerce, to be organized under the command of Captain Meriwether Lewis and Lieutenant William Clark.

The Corps of Discovery began its descent of the Ohio on August 31, 1803,
and left St. Louis in their ascent of the Missouri on May 14, 1804, some forty
members rowing, paddling, poling, or pulling one keelboat and two piroques
against the current, menaced by snags, floating debris, and trick currents.
They landed at the confluence with the Kansas River on June 26, some 390
miles above the mouth of the Missouri. It was, Captain Clark wrote, "verry
fine."

Today, threading through gentle hills, following concrete roads that knit the Kansas farming communities to the Missouri River Valley, and driving along country roads that slope eastward down to the tawny bluffs, railroad tracks, and the river, one has to subtract a high proportion of the trees one sees on the terrain to approximate what the earliest explorers saw when they reconnoitered from riverside camps. Most woods here had been discouraged by repeated prairie fires, sometimes engendered by lightning storms but often by Indians deliberately setting fires to encourage new grass growth and hungry foraging game animals. Cultivation by prairie fires up the valley was crude but effective, as William Clark noted on June 30, 1804. "Deer to be Seen in every direction and their tracks ar as plenty as Hogs about a *farm*," he marveled.

All wayfarers in this early era of exploration commented on the richness of the lower valley and its suitability for colonization. After Lewis and Clark had driven a link between the headwaters of the Missouri and the Columbia, gazed at the Pacific, and returned to the East in early autumn 1806, it became inevitable that traders and trappers, not just travelers, would begin to follow in their wake. Parties aiming for the upper Missouri and the Yellowstone in 1807 ushered in the Rocky Mountain fur trade, active for more than three decades. Scientists such as John Bradbury, an English botanist from Liverpool who had been sent to investigate plant life in the United States, recorded minute observations up the length of the Missouri in 1811, greatly expanding the knowledge imparted by the Lewis and Clark accounts.

For his account Bradbury must be noted. Soon after he arrived in America he had visited Thomas Jefferson in his retirement at Monticello, and the former president suggested Bradbury make St. Louis the center of his work, from which he made several short trips in the Ohio River Valley. Then in St. Louis in early 1811 he met Wilson P. Hunt, an agent for John Jacob Astor. Hunt was planning a trading expedition up the Missouri River some 1,800 miles, and invited Bradbury to join him. The eager botanist signed on for the duration, and even continued his exploration another 200 miles upriver past his host's final destination before returning downriver all the way to the Mississippi and New Orleans. John Bradbury's absorbing book, *Travels in the Interior of America, in the Years 1809, 1810, and 1811*, published in Liverpool in 1817, was popular on both sides of the Atlantic and gave readers the first detailed observations of the Missouri River Valley since the voyage of Lewis and Clark. It was also a rarity for the time, showing an Englishman's positive feelings about Americans in the antagonistic wake of the War of 1812.

Bradbury left St. Louis on March 13, 1811, and joined the expedition at St. Charles. The party had a boat powered by ten Canadian oarsmen, who sang call-and-response songs in French as they worked. When they could pick up a breeze they would quickly hoist a sail, but they went upstream mostly by muscle power. Bradbury saw an old man on the bank at La Charette, to whom he was introduced: it was Daniel Boone, "the discoverer of Kentucky," who thumped his chest and proudly told Bradbury that he was eighty-four and had just returned from his spring hunt with nearly sixty beaver skins. Farther upriver, Bradbury related exciting encounters with charging bears and being chased by a vengeful skunk. They discovered a large colony of honeybees in a hollow tree, from which they extracted three gallons of honey.

The log-palisaded Fort Osage, built some eleven miles east of the present day Kansas City, represented the outer limit of permanent white settlement on the Missouri. On April 8, 1811, the expedition landed there, and Bradbury recorded detailed observations of the Osage nation while visiting a nearby village. "The Osage are so tall and robust as almost to warrant the application of the term gigantic," he said. "Few of them appear to be under six feet, and many are above it." With a dispassionate eye he watched warriors perform a scalp dance with trophies recently taken in raids on distant tribal villages, and he described their lamentations of their own dead, but Bradbury was taken aback by the deportment of some Osage women, who gathered on the riverbank where traders' boats landed, he was assured, "for the same purpose as females of a certain class in the maritime towns of Europe crowd around vessels lately arrived from a long voyage, and it must be admitted with the same success."

Some miles north of the mouth of the Kansas River, the voyagers began to notice a great number of drowned (or killed) buffalo floating down the Missouri, with vast numbers already thrown ashore by the current, to which were attracted "an immense number of turkey buzzards, and as the preceding night had been rainy, multitudes of them were sitting on the trees, with their backs towards the sun, and their wings spread out to dry, a common practice with these birds, after rain." To this macabre but poetic image, young Bradbury offered a postscript with no discernible trace of irony, when, a day or so later, he discovered during a walk in some woods that there were many pigeons feeding. He watched these "gregarious animals" and their feeding habits for some time, taking detailed notes. Then, with his fowling piece, he began shooting them—in a few hours he killed 271, "when I desisted." Having provided the local buzzard population a sudden rich change in their accustomed diet, John Bradbury moved blithely, unconsciously, on.

As natural habitat as well as Native American hunting territory, most of the trans-Missouri West obtained a twenty-year lease on life thanks to the explorer Stephen Long, whose men ascended the Missouri and traced the Platte River westward until the Rocky Mountains were in sight; they returned roundabout via the Arkansas and Canadian rivers. The Long report was published in two volumes in 1822 and 1823, and it held a bleak, exaggerated view of the territory east of the Rockies and west of the Missouri: it was "an unfit residence for any but a nomad population ... forever [to] remain the unmolested haunt of the native hunter, the bison, and the jackall."

Already maps of the West had imaginatively endorsed this impression by labeling the trans-Missouri West as "The Great American Desert," using the mythic ideas of Zebulon Pike, government explorer of the Southwest: "This area in time might become as celebrated as the African deserts," he wrote. "In various places [there were] tracts of many leagues, where the wind had thrown up the sand in all the fanciful forms of the ocean's rolling wave, and on which not a spear of vegetable matter existed."

Dry became dead. Openness became waste and sterility. The received notion of the Great American Desert west of the Missouri River took hold, stemming interlopers for nigh unto two decades. "What do we want with this vast and worthless area, of this region of savages and wild beasts, of deserts, of shifting sands and whirlwinds, of dust, of cactus and prairie dogs," thundered Daniel Webster on the floor of the House in 1824. "To what use could we ever hope to put these great deserts, or those endless mountain ranges, impenetrable and covered to their very base with eternal snow? What can we ever hope to do with the western coast, a coast of 3,000 miles, rockbound, cheerless, uninviting and not a harbor in it? Mr. President, I will never vote one cent from the public treasury to place the Pacific Coast one inch nearer Boston than it is now."

By the early 1840s, though, sentiment had shifted, thanks to the allure of American enclaves in Texas and Oregon, and the new American lands of California, taken with the Mexican War in 1846. Published and widely circulated reports of the dauntless explorer and self-promoter John C. Frémont spurred the popular imagination, too—he had led one expedition to the Wind River Range of the Rockies in 1842, and a second (1843–44) from the Missouri River to Oregon country, Nevada, California, and Utah. The first wave of "the great migration" got underway. In 1841 the California Trail opened, which followed old fur trappers' routes and ran all the way from the Missouri across the Plains, through South Pass (a low place in the Rockies' Wind River Range) and the Wasatch Range, before striking the Humboldt River corridor across Nevada and, beyond, the looming Sierra Nevada of California. Within a year or two settlers were also choosing the Oregon Trail, which branched northwesterly from South Pass, aiming to reach green and pleasant Astoria, at the mouth of the Columbia River. Only a few years later, with the discovery of gold on the western slope of the California Sierra in 1848, there was even less heed given to overland fears.

That once mysterious and unknown watercourse of the Missouri was now a frontier to be crossed. Settlements along its banks at ferry crossings became known as "jumping-off places"—sojourners were jumping off from civilization, jumping off into the harsh uncertainty of the Great Plains, the mountains, the unknown. And the Missouri town of Independence, founded in 1827, became the principal jumping-off place, the eastern terminus of the California and Oregon trails as it had been for the Santa Fe Trail, the commercial trade route to the Southwest begun in the 1820s. The town, a few miles inland from the river with a permanent population of several hundred, thrived. Regular overland mail stagecoaches—the first line into the Far West—began rolling out of Independence in 1846 for Santa Fe. Thousands of emigrants with their rattling, canvas-topped wagons, bawling ox teams and livestock, thronged the streets and jammed the squares, congregating at campsites near a large free public spring, doing business at outfitters and blacksmith shops. The excitement and anticipation must have been extraordinary. One observer described it in 1846 as "a great Babel of African slaves, indolent dark-skinned Spaniards, profane and dust-laden bullwhackers going to and from Santa Fe with their immense wagons, and emigrant families bound for the Pacific, all cheerful and intent on their embarkation upon the great prairie wilderness."

Today, Independence proudly advertises itself as the "Queen City of the Trails." Since 1990 this has been underscored by the National Frontier Trails Center, which we visited for a day while staying in nearby Kansas City. Inside is one of the best collections of literature on the Western migration between the Missouri River and the Pacific Ocean. The core of the library, 1,350 volumes, was bequeathed by the late Merrill J. Mattes, a longtime National Parks historian and co-founder of the Oregon-California Trails Association, which also has its headquarters at the center. Since the library's founding in 1991 it has grown to include 1,200 more books on the West, some fifty cubic feet of manuscripts, 2,000 maps, hundreds of periodicals, together with drawings, photographs, and microfilm. There are more than 2,300 trail diaries, letters, and memoirs, many in their original state, and these are the source of the great power of the exhibits.

After viewing an award-winning introductory film, *West*, one chooses in turn from three winding exhibit corridors that guide one along the Oregon, California, and Santa Fe trails. Vast color maps and greatly enlarged prints of sketches, drawings, lithographs, paintings, and photographs densely cover the walls; everywhere there are portraits of people who made the great crossing, alongside generous extracts from their letters and diaries: one looks into the eyes of those sojourners and reads their words, and the effect is haunting and indelible upon the mind, like experiencing the rigor, even the horror, of the Western trails through personal experience. Here, among many accounts, begins that of James Frazier Reed, born in Ireland and emigrant to America when a child, who became a merchant and farmer near Springfield, Illinois, but decided to move his family to California. With three wagons including an elephantine, double-decker coach for sitting and sleeping, Reed's party set off from Independence on May 12, 1846, joined by a wagon driven by a black-smith, Hiram Miller, and those of two neighbors, George Donner and Jacob Donner, whose names would one day be connected to the most awful emigrant tragedy in American history.

One can immerse oneself in their story at Independence; one can push buttons next to wall portraits and hear actors read portions of the testimony of Reed and his wife, and some of the thousands of others who undertook the journey. Interspersed are exhibit cases of artifacts—oxen yokes, wagons, parts, and tools, personal effects, and even castoffs from trailside, such as an heirloom rocker tossed into the dust to lighten the load.

In a central place is a children's interactive exhibit, one that we were to see at a number of historical centers in the West: a wagon box of exact dimensions, and, scattered in piles on the floor, blocks of varying sizes and shapes and labeled with the "contents" and "weight." When one is driving a wagon 2,000 miles across the plains and mountains, what does one take? What proportion of food, and what kind, and what of private possessions? What is the relative value of fifty pounds of bacon versus fifty pounds of books? A mirror and a porcelain pitcher to remind one of home? Or an extra sack of grain? My children were occupied at this process of picking, choosing, weighing, comparing, stacking, and replacing, for well over an hour (in fact, they had to be pulled away from the efforts), all the while trying to avoid setting off the red light and buzzer that signified the wagon's limit had been reached. One can think about the words of a correspondent to the *Missouri Republican*, who wrote in 1849 under the nom de plume of "Pawnee" from Fort Kearney in Indian Territory (present-day central Nebraska).

"The great California caravan has at length swept by this point, and the prairies are beginning to resume their wonted state of quiet and loneliness," he said.

> Occasionally, however, a solitary wagon may be seen hurrying on like buffalo on the outskirts of a band, but all the organized, as well as disorganized, companies have cut loose from civilization, and are pushing towards the Pacific. Five thousand five hundred and sixteen wagons up to the present time, have passed here, on this bank of the river, while, on the other, from the best information that can be obtained, about six hundred have gone along. These two roads unite at the base of the mountains, and the whole emigration will then roll along over the same road. At a moderate calculation, there are 20,000 persons and 60,000 animals now upon the road between this point and Fort Hall. This is below the actual number, as the numerous trains of pack mules are thrown in. The question naturally suggests itself, can this vast crowd suc-

ceed in crossing the mountains in safety? It cannot. The leading trains will doubtless succeed, but those behind, will find the grass gone, and their heavy teams must then fail.

Many are but scantily supplied with provisions, and any little detention, which will throw men behind their time, will bring famine upon them. There is one thing, however, in favor of the emigration, which will be of vast advantage. The grass is better this season than it has been for years. The heavy rains, although they have made the roads bad, have made most ample amends in pushing forward an unusual amount of growing grass. Had some of the sawmill, blacksmith shops, gold diggers, grind-stones and gold workers been left at home and lighter wagons provided, a large number would have made much better progress than they are now making.

Much sickness has prevailed against the emigrants, and many have died. The different roads leading to the frontiers are lined with graves.

Ascending the Missouri Valley—first on the Kansas side, as if in parallel to Lewis and Clark and John Bradbury, crossing long-erased ruts once imprinted across the dark prairie sod by wagons climbing up from the Missouri ferries— the feeling of having our front wheels in the present and our rear wheels in the past was palpable. Impressions came sailing in from many different eras, as if we were unstuck in time like Kurt Vonnegut's Billy Pilgrim. We passed Fort Leavenworth, in continuous operation since its founding in 1827, then threaded through the brick-fronted streets of Leavenworth, organized in 1854 by land-hungry, pro-slavery Missourians who came across the river to squat on land owned by the Delaware Indians. Driving northward just a couple of miles we passed the foundation of Buffalo Bill's Log Cabin at Salt Creek, where William Frederick Cody spent part of his boyhood before taking a job working for the Pony Express. Nearby was Atchison, established as a pro-slavery town in 1854 after agents of the New England Emigrant Aid Company founded the town of Lawrence and threatened to send 20,000 Free Soilers into Kansas each year. John Brown operated in Atchison in 1857. In 1859 during a speaking tour to attentive, even warm audiences, Abraham Lincoln warned that secessionists would be hanged surely as would John Brown. Lincoln spoke from the same theater stage that, in four years, would suffer the boot heels of John Wilkes Booth, doing *Richard III*. Atchison is also known as birthplace and childhood home of the spirited tomboy and future aviatrix Amelia Earhart, who climbed trees, rode horses bareback, and explored local Indian caves, all to the general bafflement of the Atchison gentry, in which she had a place but absolutely no interest.

The feeling of being unstuck in time persisted as we crossed railroad tracks laid down originally for the Chicago, Burlington & Quincy line, which we would parallel as we rumbled onto a high steel bridge that carried us in mere seconds over the great river and down onto Missouri soil again, heading northward alongside the railroad, now the domain of the Burlington Northern Santa Fe, toward the distant Iowa border. A freight train with more than 100 hopper cars passed us, heading down to Kansas City.

The road took us into the proletarian outskirts of St. Joseph with its ancient stockyards and packing mills, weathered redbrick wholesale houses, grain mills, and factories giving way to bungalows, shotgun shacks, and alleyways. It was hard to tell the alleyways from the streets. Then, the main town: brick warehouses and office buildings, while draped above over the hillside were turn-of-the-century Victorians, many still elegant and show-offy with their gingerbread, scalloped shingles, turrets, and dormers under bright paint.

St. Joseph was redolent with historical associations, being sited at a good ferrying point on the Missouri and therefore another principal jumping-off place for westering thousands. It was founded by Joseph Robidoux, who first saw the area in 1799 while on a trading mission, and in 1826 he returned to stay for the American Fur Company, soon buying the place from his employers and establishing a virtual barony with vast cornfields and pasture lands. Northwestern Missouri was still Indian country, but in 1836 some two million acres were bought from the Indians by the federal government, and settlers poured into the area. Many of them were slaveholders, and Robidoux's Post, as the settlement was called, was an important commercial center. In 1843 Robidoux platted a town site there, naming it after his patron saint, and watched as St. Joseph pulled in the customers. In 1844, a large wagon train, led by General Cornelius Gilliam, crossed the river just north of the new village. Among some 800 immigrants in the wagons was James W. Marshall, an itinerant wheelwright and carpenter who, while working on a sawmill at Coloma, California, for a Swiss named Sutter in 1848, noticed flakes of gold in the riverbed and went to tell his boss.

Before the Gold Rush sent many tens of thousands crowding through St. Joseph straining for the California dream, Robidoux's settlement had won title to the county seat and boasted of 350 houses, 2 churches, a city hall, and a jail. With the emigrant tide swelling the wagon roads and the river itself, St. Joseph became the place where furs and buffalo hides of the Far West were exchanged for cash and goods sent upriver from St. Louis. Mountain men in their fringed buckskin and battered big hats, probably complaining about the falling price of pelts, elbowed droves of anxious wayfarers getting outfitted for their long summer journeys before they joined the ponderously moving queues of wagon trains bottlenecked down toward the ferries. Stockyards corralled thousands of beef cattle waiting either for a long walk across the Western plains and mountains or a short one to the slaughterhouses south of town. In 1847 the state chartered the Hannibal & St. Joseph Railroad (one of its incorporators was John Marshall Clemens, the father of Mark Twain), which finally forged an iron link between the Mississippi and Missouri rivers in 1859; the last spike of that line was driven by none other than Joseph Robidoux. A year later, on April 3, 1860, scrawny young men working for the local shipping firm of Russell, Majors & Waddell leaped onto steeds in St. Joseph and Sacramento, California, and began the short but glorious chapter called the Pony Express. With Western mail arriving in St. Joseph in only eleven days, it then stopped in town long enough to be sorted—until the local postmaster proposed that they speed the mail on by having it sorted in a special railroad car hurtling eastward. That practice on the Hannibal & St. Joseph, approved by the postmaster general, soon became national, wherever trains rolled.

Being a slave city in a slave state, a short boat ride from "Free Kansas," St. Joseph was no stranger to conflict long before the firing on Fort Sumter, South Carolina, in 1861, nor were many towns in Missouri and Kansas. Slaves were "stolen" by abolitionists and liberated across the river, enflaming the tempers in pro-slavery Missouri. Mob action and skirmishes were commonplace. John Brown and family were just the tip of the iceberg. St. Joseph became the base for pro-slavery guerrilla bands who terrorized the countryside once war was declared. One of these Confederate raiders, in the band led by Charles Quantrill, was Jesse James, who operated with his brother Frank and his friends Cole and James Younger, all of whom turned to a different occupation after the end of the war. Years after their outlaw spree began, Jesse James sought some rest in St. Joseph under the pseudonym of "Mr. Howard," living quietly at 1318 Lafayette Street in a modest one-story cottage with his family. Here, on April 5, 1882, he was shot in the back by his former lieutenant, Bob Ford, who desired the $10,000 reward.

Ninety-five miles north of St. Joseph, and fifty-five miles south of Council Bluffs, we crossed the Missouri line into the Hawkeye State. Near here, back in frontier days, a trader named Hitchcock built his cabin right on the border. He lived there for several years with a number of wives. I do not know whether Hitchcock was Mormon or not; in the early 1830s there was a large community of Mormons in Independence, Missouri, under the leadership of Joseph Smith, which erected a temple and published a newspaper, but the settlement was burned out by mobs and driven away. Straddling the border, though, Hitchcock kept his harem, which upset some neighbors and the authorities. Sheriffs from Missouri would periodically knock on his door to ask questions about his marital situation, but he would retreat to the Iowa side of his cabin; peace officers from Iowa, when they went calling, were frustrated when

Hitchcock stepped over to the Missouri side. This went on for several years—until the authorities combined forces and left him with no retreat, and he was brought to justice. I have heard of similar stories about justice thwarted by geography in the Prohibition era on the Vermont-Canada border.

Finally we were closing in on the city named by Lewis and Clark for the steep clay banks overlooking that stretch of Missouri River bottomland. Council Bluffs! Since I was a boy reading about explorers and railroaders, the name has sent a thrill through me. It did this day, too, and I tried to ignore the dull gray condos creeping up the flanks of the bluff as I remembered the first time I had come here.

It was August 1973, I was twenty-four, and I had been working at a high-piled editorial desk at Alfred A. Knopf for eight months. I'd been surprised and pleased to learn that the liberal policies of RCA, the parent company, gave me a week of vacation time. I took two, one unpaid, and decided to go to Berkeley and San Francisco to stay with friends. I was still enough of a hippie to sign up with a bus company advertising in *The Village Voice* called something like "White Rabbit Ltd.," harkening back, I suppose, to the song of the same name by Jefferson Airplane, if not to Lewis Carroll.

An absurdly small amount of money bought me a seat in a cross-country van, to be driven by the passengers themselves. Behind the two front seats was a sea of sleeping bags and duffels. There were ten adults and one dog. The only windows behind the driver's seat were in the back double doors. We felt like illegal immigrants, and in some way we were. We drove night and day, and covered the distance between the George Washington Bridge and the Oakland Bay Bridge in fifty-eight hours. To do this I believe we exceeded the speed limit nearly all the time, fueled by cheap thirty-nine-cents-a-gallon gasoline and an ocean of coffee. The atmosphere in back, particularly with the dog and the fatty road food everyone ate, was so thick that I contrived to be in a front seat with an open window as much as possible, and I did most of the driving. It seems like every mile is still imprinted in my mind's eye, nearly three decades later. And when Interstate 80 swooped us down across the farmland of Iowa's Pottawattamie County and I saw the first green road signs for Council Bluffs, a heavy energy filled my chest and my head, and all I could do was blabber about Lewis and Clark, mountain men and fur traders, Indians, and the railroad, to the indifferent raggedy sojourners with me. The feeling propelled me far into the rolling prairie farmland of Nebraska, and I felt something very much like it twenty-seven years later as I woke my children from their naps and my wife steered us through town and across the Douglas Street Bridge into Omaha, a town that I greatly like. There was much to experience there and across the river in Council Bluffs, and several days' stay would reward us.

PART II

3

Rails and the River

Omaha is the birthplace of Marlon Brando, Malcolm X, and Fred Astaire, an unlikelier trio one could not hope to find but one that in a decidedly contrary way reflects on the city itself. Omaha? They are identified with *anything* but Omaha. Malcolm X, a minister's son, born Malcolm Little in 1925, has associations more with the Boston of his young manhood or the Harlem of his firebrand days with the Nation of Islam. Marlon Brando Jr., whose father was a salesman and mother an amateur actress, was born in 1924; it seems more likely he was born in the Actors Studio in New York with greater associations with the jagged Hudson piers of *On the Waterfront*, the menacing, motorcycle-exhaust-filled townscape of *The Wild One*, the sultry New Orleans of *A Streetcar Named Desire*, or even the murderous Cambodian stronghold of *Apocalypse Now*. But the biggest surprise is Fred Astaire, born Frederic Austerlitz Jr. in 1899, son of an Austrian brewery worker. *Omaha. Fred Astaire.* The names resist being placed in the same sentence. Epitome of the glamorous Manhattan and Hollywood smart set, or the giddy nightclub whirl of Paris, Rio, or Venice, his name should be illuminated in a myriad of marquee bulbs, never plebeian neon. His usual costume must always be a top hat, white tie, and tails, not a humble Nebraskan store-bought brown suit or, perish the thought, overalls. Tap shoes, never sturdy brogans.

This trio's brief residencies are interesting. Malcolm X was born in an Omaha hospital shortly after hooded night riders of the Ku Klux Klan surrounded his parents' house, breaking windows, shouting threats, and calling out his father, a Baptist minister who was an organizer for Marcus Garvey's Universal Negro Improvement Association. The 1920s had seen a big resurgence of the Klan across the Midwest. "My mother went to the front door and opened it," Malcolm recalled. "Standing where they could see her pregnant condition, she told them that she was alone with her three small children, and that my father was away, preaching, in Milwaukee. The Klansmen shouted threats and warnings at her that we had better get out of town because 'the good Christian white people' were not going to stand for my father's 'spreading trouble' among the 'good' Negroes of Omaha with the 'back to Africa' preachings of Marcus Garvey." When Malcolm's father returned he was livid and vowed they would move as soon as the baby was born.

Many of Marlon Brando's Omaha memories were pleasant by contrast—his primarily Irish family had lived for generations in Nebraska—but both his parents were alcoholics and spent most of their time avoiding each other in favor of other partners. For a while the boy was unaware. "Until I was seven," he wrote, "we lived in a big wood-shingled house on a broad street in Omaha lined on both sides with houses much like our own, and with leafy elm trees that at the time seemed taller than anything that a young boy could imagine." "Bud," as he was called, remembered

> the sweet aroma of fresh-cut hay, the fragrance of burning leaves and the redolence of leaf dust as I scuffed through them. I remember the fragrance of the lilies of the valley in the garden where I often slept on the hot afternoons in Omaha, and I suppose the fragrance will always be with me. I don't think I'll ever forget the smell of lilacs or wild roses or the almost chic appearance of the trees in our neighborhood dressed in the silver lamé of a spring ice storm. Or the unforgettable sound that grates on me even today, the squeak of midwestern snow beneath my boots when the temperature was fifteen below.

Fred Astaire remembered the small-town feeling of Omaha in the early years of the twentieth century despite its booming size and population. "We lived in a wooden frame house on North Nineteenth Street, about twenty minutes' walk to the business district and not much faster by buggy," he recalled. "My father used to take me to town on Sunday afternoons, riding beside him in the two-seater to visit his friends at Saks' Cigar Store." He also wrote of "the rumbling of locomotives in the distance as engines switched freight cars in the evening when we sat on the front porch and also after I went to bed. That and the railroad whistles in the night, going someplace. I used to imagine that I was riding on a train."

If this trio is known for full lives lived *away* from Omaha (Malcolm was taken away in diapers; Brando, at seven; Astaire, at four), it matters not to those who keep track of such things there. It counts for something that they were born here and imbibed some ineffable but permanent essence, and one cannot disagree—it has to do with the hallmark quality of the West, that of possibility, and of always having the ability to remake oneself. "The city has the small town's interest in local boys who made good," the Nebraska writers admitted in the 1939 WPA guide; "the front page always has space for the doings of any 'former Omahan,' whether he wrote a script for Hollywood or was arrested for theft in Denver." Boosterism in all of its forms helped build the

Midwest and the Far West, sometimes for better, sometimes for worse. It was as vibrant at the town's settlement in 1854, when the resident Omaha tribe agreed to relocate to reserved lands some sixty miles upriver to make way for eager whites from Iowa and points east. It would stay vibrant during its explosive growth as a place for trade as well as a jumping-off place. It would continue into its heyday as a railroading, manufacturing, and processing center.

Fifty-one weeks and 145 years before I took my family into Omaha, on July 4, 1854, a picnic party came over on a ferry from Council Bluffs to celebrate Independence Day as well as the bright future of this new town. At that point the "town" was merely verdant bottomland rising to prairie, rising to a pretty, enclosing chain of low round hills. On one, later called Capitol Hill for its short-lived hosting of the territorial government until the capital was moved to the town of Lincoln, the picnickers made themselves comfortable on the prairie grass, enjoying, no doubt, the breeze and the unimpeded view. There were ample toasts and speeches and an "anvil salute," a homey form of fireworks display in which two big, heavy blacksmith's anvils were turned upside down on the ground after the central holes were tamped full of black powder and fuses attached. Someone lit the fuses and the blast sent the anvils more than 100 feet in the air. I suppose the small crowd scattered and evaded the iron as it returned to earth, but the resumption of speechifying was interrupted when a band of Indians who were camping nearby came over to see what all the noise was about. This alarmed the women and children. The celebrants all leaped aboard their wagons and lashed their horses into a mad gallop down to the ferry landing and a quick boat ride back to the settled Iowa riverbank, leaving the Indians to nudge the iron anvils with their toes and puzzle out what mysterious rites had been afoot.

Especially in the earlier days, boosters employed something close to a religious faith to overlook certain drawbacks. The bottomland, for instance, was perennially subject to river floods and it was full of sloughs with their pesky wildlife. Even back in 1804, Lewis and Clark noted this when they camped near the site of today's Douglas Street Bridge, where Interstate 680 swoops over from Iowa. The Corps of Expedition killed and roasted a deer and enjoyed a beautiful breeze from the northwest, "which would have been verry agreeable," wrote Meriwether Lewis, "had the Misquiters been tolerably [peaceable]," but they "were rageing all night, Some about the Si[z]e of house [flies]." John Bradbury remarked on the river valley's mosquitoes in 1811:

> The Missouri had overflowed its banks some time before our arrival, and on receding it had left numberless pools in the alluvion. In these the mosquitoes had been generated in numbers inconceivably great. In walking it was necessary to have one hand constantly employed to keep them out of the eyes; and although a

person killed hundreds, thousands were ready to take their place. At evening the horses collected in a body round the Fort, waiting until fires were made, to produce smoke, in which they might stand for protection. This was regularly done, and a quantity of green weeds thrown on each fire to increase the smoke. These fires caused much quarrelling and fighting, each horse contending for the centre of the smoke and the place nearest the fire.

Another great impediment to perfection also came on the wind. As the English journalist Henry Morton Stanley noted in 1867,

No town on the Missouri River is more annoyed, even afflicted, by moving clouds of dust and sand—when the wind is up—than Omaha. It is absolutely terrific. The lower terrace along the river is a waste of fine sand, which is blown about in drifts, and banked up against the houses, like snow in a wintry storm. For two or three days people have been obliged to shut themselves up in their houses for protection from the sand. But Omaha is a wide-awake, energetic town, and is beautifully located, on the second high terrace from the river, with a cordon of hills which are mantled with country residences.

The solution for these problems was simple. In time, Omahans filled in the sloughs with the dust.

A good portion of that lowland along the riverbank was overspread in the 1860s by railroad yards and shops of the Union Pacific, which was instrumental in powering the first boom in land speculation, settlement, and trade, when Council Bluffs and Omaha were designated to be the Missouri River terminus of the first transcontinental railroad. That decision had its root in an encounter between the Illinois attorney, Abraham Lincoln, and an Iowan civil engineer named Grenville Dodge, in Council Bluffs in 1859.

Lincoln was devoting much of the latter half of the year to the speaking tour around the Midwestern states that aimed to build coalitions within the fledgling Republican Party and pull people away from nativist and pro-slavery sentiments, anticipating a hot and crowded presidential race in 1860. He had agreed to lend $3,000 to a political associate, who had offered some Council Bluffs real estate as security, so Lincoln decided to look over the property; supporters organized a reception and a public talk. Dodge, originally from Massachusetts, had relocated to Council Bluffs where he had a number of small commercial enterprises between engineering assignments for the struggling

Mississippi & Missouri Railroad, which was inching across central Iowa from Illinois lines, and which, Dodge fervently hoped, would become a link in the Pacific railroad. Dodge wanted to have a major role in that enterprise, and as it would turn out, he would. Already he had done reconnaissance work up the Platte River Valley toward Colorado and Wyoming.

Lincoln heard about the young engineer, who happened to be staying at the same hotel, and found him relaxing out on the front porch. He asked him what he knew about the country on the far side of the Missouri. Dodge, who was a Republican and who had heard Lincoln's talk, enthusiastically told him about the superiority of the 42nd Parallel route up the Platte Valley—it was far better than competing routes westward from St. Paul and Kansas City. As Dodge would later explain, "the Lord had so constructed the country that any engineer who failed to take advantage of the great open road from here west to Salt Lake would not have been fit to belong to the profession; 600 miles of it up a single valley without a grade to exceed fifteen feet; the natural pass over the Rocky mountains, the lowest in all the range; and the divide of the continent, instead of being a mountain summit, has a basin 500 feet below the general level." Lincoln skillfully kept Dodge talking for two hours, saying that nothing was more important than the transcontinental railroad. Later the engineer ruefully realized that he had been thoroughly emptied of much proprietary information; Lincoln was then well known as the stump debater with fellow Illinoisan Stephen Douglas, but he was also a railroad lawyer who had won high-profile cases for the Rock Island line. But Dodge decided it was for a good cause—and he would come to count on Lincoln's political support in years ahead. Before Lincoln left town, he was escorted by supporters to the edge of the western bluffs, where from that height he could view the Missouri, the growing town of Omaha, the rolling plains, and the Platte Valley. His hosts predicted that the Pacific railroad would someday run through Council Bluffs. "Not one, but many roads," replied Lincoln, "will someday center here."

The Union Pacific Railroad, created by the 1862 Pacific Railroad Act and signed by President Lincoln, broke ground on December 2, 1863, at the intersection of Seventh and Davenport streets in Omaha. A thousand citizens cheered as territorial governor Alvin Saunders thrust his shovel into frosty ground, accompanied by booming cannons and a brass band. President Lincoln and many dignitaries sent congratulatory telegrams.

The keynote speaker was the eccentric financier George Francis Train, attired in his trademark white suit. "America possesses one-half the common sense, three-fourths the enterprise, and seven-eighths the beauty of the world," he bellowed out to the crowd—and with the Pacific railroad, he promised, it would control the world's commerce. A local newspaper editorial pronounced it "the raciest, liveliest, best natured, and most tip-top speech ever delivered west of the Missouri." Finally the celebrants retired to the Herndon House, the premier hotel of the time, for their gala banquet and ball. The blowout would all but empty the corporate treasury. It would require eighteen months of financial scrambling and political backroom maneuvering before the Union Pacific could actually lay any tracks.

It seemed like everyone we encountered in Omaha and Council Bluffs had "iron in the blood," or an ancestor connected with the railroads. By the heyday of rail travel in the twentieth century—the 1930s and 1940s—the roster of passenger and freight lines servicing the two cities on their opposite Missouri banks illustrated the strong, seemingly indestructible steel spiderweb of America's transportation network that had been flung across the Midwest and West and anchored at distant cities and towns: the Union Pacific; the Chicago, Burlington & Quincy; the Chicago & North Western; the Chicago, Great Western; the Chicago, Milwaukee, St. Paul & Pacific; the Chicago, Rock Island & Pacific; the Illinois Central; the Missouri Pacific; and the Wabash.

The railroad was central to American life. Everyone had a personal connection. And untold legions had familial or ancestral ties, as true today as it was long ago. All during the years of research for *Empire Express*, especially in the West, I met librarians and archivists who brightened when I told them my mission, and then they related a story about a father, a grandfather, a great-grandfather who worked for one line or another. When the book was finished and the publisher enterprisingly sent me off on a publicity tour from New York and Washington out to Western cities with old and sometimes vanished railroad connections, people lined up to tell me of their relatives.

Some were legendary. There were two tiny elderly ladies at the Smithsonian stage steps in Washington, each of them tottering under the weight of my book, who were great-granddaughters of Charles Crocker, one of the founders of the Central Pacific Railroad, which had vied against the Union Pacific in the first transcontinental race. Another direct descendant of Charles Crocker found me at a Library of Congress talk. There was a beaming woman in Ogden, Utah, who was a descendant of Thomas Durant, the machiavellian vice president of the Union Pacific who had schemed many millions out of the company, repeatedly almost killing it off; I had to use some inventive and diplomatic language to congratulate the woman for such a forebear. There was a frail, pale-eyed, and soft-spoken man in a Salt Lake City bookstore whose great-grandfather was Chauncey West, a major Mormon figure in early Utah history and a principal contractor for the railroads (who cheated him out of his payment for the work, which killed him). There was, in California, a great-grandnephew of James Harvey Strobridge, the colorful construction boss of the Central Pacific.

Of course there were many other people whose ancestors had played smaller roles in the legend as one of the force of more than 10,000 working for the U.P. during the 1860s race—the mostly Irish tracklayers, the graders, tie contractors, and bridge builders—or whose forebears had, in following decades, shoveled coal, tended switches, chopped ice, lit lamps, oiled wheels, strung wires, braked cars, patrolled yards, stocked freight, sorted mail, punched tickets, tapped telegrams, chased hoboes, searched no-shows, painted stock, crushed rock, flipped hash, hauled ash. There was a fellow at the library in Council Bluffs who could point to a branch in his family tree and a forebear who had chopped cottonwoods for ties, went on to become a baggage handler and then a conductor on the U.P., and had seen a rustler lynched out somewhere in Wyoming. There was a chambermaid in one motel near the Omaha airport, on whose paternal line a great-great-great- had fought to push the Cheyenne away from the tracks out on the Plains as an army bluecoat, and on whose maternal side was a member, far back, around the same time, of that tribe. There was a man working in an Omaha steakhouse whose father's grandfather had lost eight of his ten fingers and all but one of his toes when lost in his own railyard during a blizzard. There was a customer in a used-book store who said that his uncle had fallen off the observation deck of a speeding train while drunk and lived to ride again. There were plenty of people who were still working for the railroad, or recently retired, all seemingly with stories to spare.

The Union Pacific corporate headquarters is at 15th and Dodge, and there are many satellite locations. The yards stretch out across many downtown city blocks west and south of the grand Art Deco–style Union Terminal down on 10th Street, a true palace full of echoes and memories.

Opened in 1931 to replace the time-worn 1899 brick and stone station, the Union Passenger Terminal was designed by Los Angeles architect Gilbert Stanley Underwood and constructed on a twenty-three-acre plot by a workforce of six hundred men. It looked like a massive frosted butter cake with its exterior walls of cream-colored, glazed terra-cotta braced by buttresses and pylons. At night it was spotlit. Over the northern passenger entrance were sculptured figures of a brakeman and a locomotive engineer; over the western entrance on 10th Street stood figures of a civil engineer and a railroad mechanic. Inside there was a cathedral-like waiting room, sixty feet from the colored terrazzo floor to the silver- and gold-leafed ceiling, lit by ten windows of rose, amber, and green translucent plate glass and six massive crystal and bronze chandeliers, each weighing 2,000 pounds. The interior walls were Black Belgian marble wainscoting and Caen stone upper walls, with blue marble columnettes flanking each window. Passengers stood in line at ten ticket windows against the north wall and waited for trains in the long, high-backed and solid wooden benches just opposite. They could grab a sandwich at the stylized lunch counter at the east end of the waiting room or enter Omaha's most sumptuous dining room, with oak wainscoting and silver-accented Italian travertine walls, and with window borders of patent leather. Up near the ceiling were six murals depicting dramatic developments in Western history.

Thanks to a $24 million renovation, the entire building but especially the main-level waiting room looks just like photographs taken in the station in the 1930s (although the ticket window area now encompasses a gift shop), minus the crowds. I've seen pictures of the great room in April 1939 when it was swarming with people who spilled outside into the parking lot and down 10th Street, on the occasion of the world premiere of the feature film *Union Pacific*. They had all turned out to meet the Hollywood train with stars Joel McCrea and Barbara Stanwyck.

Back in those glory days thirty to forty passenger trains came in per day, especially in the late 1930s when fares were lowered to encourage traffic and the railroads introduced the slick, fast new streamliners. Across the railyard, farther south on 10th Street, was the graceful Burlington Station, faced with Indiana limestone and with a waiting room decorated with gold medallion bas-relief borders and heavy bronze chandeliers hanging from the rose and gold ceiling. "The room is completely equipped," said a guidebook, "with overstuffed furniture."

I think it is the humble but solicitous touch of "overstuffed furniture" that nearly takes me over the edge. Every time I read these descriptions I sigh, especially knowing that in Omaha nowadays the solitary daily Amtrak passenger train comes in on the Burlington tracks and pauses impatiently at a nondescript brick building built in 1984 nearby at Ninth and Pacific—a tenth of the size of Union Station and a hundredth of its character. One can disembark from a Chicago or Denver or San Francisco train, dislodge with one's tongue the toothsome residue of a plastic-wrapped tunafish sandwich or microwaved cheeseburger, glance witheringly at the little brick Amtrak station, and then raise one's eyes to the holy tabernacles of transportation, bereft of their congregations, on the right side of the tracks.

But, mirabile dictu, the buildings still stand, and in a section of town that is a powerful magnet for those with time on their hands and money in their pockets. Across and down the street is Old Market, teeming with window shoppers and people watchers—the old wholesale fruit and vegetable district, now given over to blocks upon blocks of upscale shops, boutiques, galleries, brew pubs, coffeehouses, and more than thirty restaurants in those old brick warehouses with their wide, sidewalk-sheltering metal awnings. And Union Station is shining and shrouded with reverence.

Omaha's Union Terminal was, in the summer of 2000, home to the Durham Western Heritage Museum and the Union Pacific Railroad Museum, both located in the lower concourse. With the spotless, gleaming waiting room, looking like its heyday in the 1930s and 1940s, and with the downstairs exhibits on Omaha and Nebraska history alongside a small procession of climb-aboard Union Pacific rolling stock and an assortment of display cases and framed photos, maps, and posters, it is understandable that one might want to spend a good amount of time there. This is true for Bob Fahey, a friendly and fit-looking septuagenarian who volunteered his time in the U.P. exhibit to answer questions. "My father was a signalman in the Omaha yards," he told me proudly. "He was posted at an employee crossway over all the tracks, keeping people from getting in the way of the trains, which, if they did, tended to slow operations down a bit." He grinned. "I myself had just about every job in the place. I started here as baggage handler in 1941. Then I worked as a redcap, getting passengers and their baggage from the terminal entrance down onto the trains, and vice versa. That was backbreaking work— no soft-sided duffel bags and little wheely-things. Trunks, more trunks, hardshell or leather suitcases, makeup cases, and even hat boxes were more like it. Hard on the back, even when you're young. I rose up to Omaha stationmaster." I was very impressed and glad for my luck, striking up a conversation with a cheerful docent and finding that he had presided over this entire edifice for decades. "I met my wife here in Union Station," he went on. "She was a government travel dispatcher, creating itineraries for soldiers during and after the war. We've been together fifty-three years now!"

Bob Fahey did the station announcements for so many years that "they'll always be with me," he said. "And you know, you had to pay attention to the correct way to announce stations. You had to adjust for the size and acoustics of the station interior—so you wouldn't catch up with your own echo. You paused at places, you used a certain cadence, and above all you had to speak clearly." He straightened up and raised his head so his eyeglasses gleamed in the light, and stretched his throat a little. "Union Pacific San Francisco Overland Limited . . . with connections for Los Angeles Limited, Portland Rose, Sun Valley, and Boise, 10:25 A.M. on Track Number Two," he intoned evenly and authoritatively. "The Union Pacific Challenger for San Francisco with connections for Sun Valley and Portland, departing 10:40 A.M. on Track Number Three." I could have listened to him all day and it pleased him to be told so. Upstairs in the passenger terminal, loudspeakers broadcast similar bulletins on nonexistent trains for station atmosphere. Unfortunately it was someone else's voice and unintelligible, though that might be changing soon if Bob's voice was given another chance. "I recently recorded all of the stationmaster's calls for a certain day in 1947," he said, "and I guess they're going to do something with that up there. It'll be nice to hear my voice, but a little strange after all these years. I retired in 1971."

He looked nostalgically across the exhibit area to the U.P. Pullman, canary yellow with merry red trim, and the Southern Pacific lounge car, and the brawny black engine surrounded by little boys. "It was really something back then." He sighed. "I'll tell you what was hard. As stationmaster I watched a sad process beginning in the 1960s when the number of station employees began to dwindle. First there were sixty people working here. Then there were forty, then thirty, and then twenty, until finally it was just down to me on the four-to-twelve shift." He shook his head sadly. "Well, things change, and you got to change with the times." He stopped talking with me long enough to answer the questions of a middle-aged man pushing a stroller, and then a stout young woman in a T-shirt advertising Omaha's Henry Doorly Zoo, with three toddlers harnessed together with straps. Then he returned to me. "You know," he said, "my wife and I met here. We spent our lives in this terminal. Maybe we should be buried here, too!—some kind of urn, or maybe behind one of those marble wall plaques." He gave a wide grin. I told him if something like that were ever possible in a place like this, he and his wife would have a lot of company.

Upstairs, in the quarters once occupied by the station barber shop, was the office and library of the Union Pacific collection, presided over by the genial Don Snoddy, a Kansas native whose previous port of call was at the Nebraska State Historical Society in Lincoln. The collection had been founded in 1921. In the main-floor barber shop quarters, Don Snoddy had the large library and the U.P.'s extraordinary photograph collection, at present well catalogued and accessible and comprising some 590,000 images from the Union Pacific, Southern Pacific, Chicago & North Western, and Missouri Pacific Railroads. I once spent many giddy days inside this treasure trove, looking at hundreds of rare stereographic cards from the 1860s, by the pioneer photographers Andrew Jackson Russell, Alfred A. Hart, and others, who chronicled the great work on the first transcontinental railroad across (respectively) the dangerous plains and mountains of Nebraska and Wyoming; the unforgiving dry stretches across Utah and Nevada; and the harsh mountain slopes of the Sierra Nevada. Staring into those stereo photographs using a handheld viewer, as the optical illusion of three-dimensional space snaps in, one can almost smell the sagebrush and the blasting powder and almost see the workers, their teams of mules, and the magnificent old rolling stock begin to move. To take a visual break during my hours of viewing, I would stray over to other storage cabinets full of publicity photos of Hollywood stars taken aboard trains and in stations of the Southern Pacific and Union Pacific.

Strangely enough, the massive U.P. collection office had a staff of one—director Don Snoddy, assisted part-time by docent and fellow rail historian Bill Kratville, author of a number of books about routes and rolling stock. In two

years, having supervised the collection's removal to larger space in the former Council Bluffs public library building, Don was to be rewarded by the corporation with a pink slip, not only a heartless move but one that forecast a
stony indifference to its rich history. This of course was unknown to us in
June 2000. "Come back in a year," he said happily, "or maybe Labor Day to be
safe. We'll have a lot of material out that hasn't been seen in a century.

"Hey," he said, changing the subject by looking over to my eight-year-old
son and eleven-year-old daughter. "Want to see some neat stuff?" He turned
and rummaged under his desk, which groaned under the weight of computer
equipment, stacks of books, and great piles of papers. "I'm taking these to a
talk at a local school this afternoon," he explained as he emerged. To my
wife's and children's delight he came up with an 1860s buffalo gun—a huge
and heavy rifle—and a large hank of auburn hair—the tanned scalp of one
William Thompson, a Union Pacific employee who had been scalped in the
summer of 1867 and lived to tell about it. Mary and Mimi and David had
heard Thompson's story from me, and at their urging we took a photograph
of gleeful Don Snoddy with his treasures, not the most dignified of portraits of
a renowned curator and historian, but one certainly in keeping with the man
and with the moment.

In 1869, after the Golden Spike Ceremony at Promontory, Utah, had welded
the wide nation with iron rails—during which event the Union Pacific chief
engineer had told the assembled throng, "This is the way to India"—Grenville
Dodge built a handsome mansion high on the rock-walled, terraced slope in
Council Bluffs at 605 Third Street, with a cliff at its back. Thanks to his railroad investments, speculations, and political connections, he was one of the
richest and most powerful men in Iowa. The three-story, fourteen-room brick
and stone house, designed in the popular Second Empire style with mansard
roof, elaborate bay windows, and a wraparound colonnaded verandah, cost
$35,000 and commanded an excellent view of the city, the river, and the Ne-
braska lands he had subdued and scored for the road. From the library Dodge
would be able to see each of the eight railroads that were to build into Council
Bluffs. He resigned from his post at the U.P. in January 1870, writing to his
friend General William Tecumseh Sherman that he had "no taste" for administrative duties. "Like all roads, the Union Pacific was managed a thousand
miles away—the mere play thing for Wall Street," he complained, "to be set
up or down as a circus may dictate and in such hands will make no effort to
repay the Government for what it has done, or make good the stock of those
who went in good faith."

The engineer was right on target, though he had no cause to complain
as far as his personal finances were concerned; in a few years when his own

culpability in shady business and political practices (such as bribing congressmen) became known, he went underground to evade a summons to a congressional hearing. The committee chairman, Jeremiah Wilson, vowed that federal agents would track him down. A former associate of Dodge's told Wilson he was dreaming: Dodge was "a man of wonderful resources, and can live in Texas all winter, out of doors, if he wants to, and where none of your marshals can go, and if he don't want to come he will not come." If he was cornered Dodge would offer a $10,000 bribe, "and you have not got a marshall in the state of Texas that will stand that sort of treatment. . . . Money is no object to him." Grenville Dodge would weather that scandal, and outlive every one of the important figures in the transcontinental railroad story, working on new railroads in the Southwest and becoming more wealthy, more powerful, and of course more admired.

His mansion in Council Bluffs, where he died in January 1916, was maintained by the family in approximate semblance as it had been during his life—in the 1930s, according to the WPA guide, a resident caretaker would in special cases admit visitors—and in 1963 the city board of parks and the county historical society acquired the place and it was designated a National Historic Landmark. Since then, the Dodge House trustees have worked to restore the house strictly to the period of Dodge's residency, and it is open to visitors six days a week. Inside, the house reminds me of Sagamore Hill, Theodore Roosevelt's home in Oyster Bay, New York. It is a lavish Victorian home well used by a man of action, decorated with animal trophy heads, weaponry, souvenirs of extensive travels and legions of worshippers, glass bookcases full of histories, biographies, and memoirs, and a writing desk suitably massive for a "great man." For years I struggled to read documents with Dodge's bad spelling and truly horrible handwriting; much had been scrawled out on the plains and deserts with blunt pencils, but some were composed there at that desk, so ironic in its clean lines and elegance.

At the mansion, whispering docents, some of whom seem old enough to have seen General Dodge in the living flesh, escort small tours past roped-off plush settees and couches, remarking on the walnut, cherry, and butternut woodwork, parquet floors, Italian marble fireplaces, and paintings. Between his wife, Anne's, desire to live as elegantly as possible, Dodge's curiosity about newfangled domestic appointments, and his inventive engineering eye, the house has many modern conveniences quite forward for the time, such as servant signal systems, water storage and supply, sanitation, and heating. Even outside, in the garage, Dodge installed an ingenious metal turntable so that an automobile driver would never have to back out or in, just like the larger versions employed in railroad roundhouses; I have seen only one like it, at a 1920s-era Vanderbilt mansion in Centerport, Long Island. It must, I will confess ruefully, save on replacement taillights.

As much as Omaha, Council Bluffs is still at its spiritual heart a railroad town, and displays an active appreciation of its own history though its economy seems to have been shifted to malls and other service sectors and to three garish "riverboat" casinos plunked down in clear view of Interstates 80 and 29. The downtown area—two- and three-story brick and stone buildings showing the town's architectural and economic surge to have been from the Victorian to the Art Deco eras—is undergoing a revival, with many restorations of storefronts and buildings, pocket parks and fountains. Sidewalks have been redone in warm brick with a nice touch, inset enameled plaques of local historical scenes—Lewis and Clark; the Jesuit missionary Pierre Jean DeSmet, who worked with the Pottawattamie tribe in the late 1830s; the large Mormon settlement here when it was called Kanesville; and of course, Abraham Lincoln, Grenville Dodge, and the Union Pacific. Some of the new storefronts not yet rented carried signs of the N. P. Dodge Real Estate firm, founded by Grenville Dodge's brother Nathan in the 1850s. With the Union Pacific museum installed in the old public library, the downtown will have a fine balance.

There is a heroic-sized portrait of General Dodge in his Civil War uniform in the restaurant of the old Kiehl Hotel on South Main Street. Called Duncan's Café, the place is a step back in time; it has dark walls, a black and white checkerboard tile floor, big wideboard booths, and the perennial buzz of a milk shake stirrer. At Duncan's, everyone smoked, patrons and employees alike including waiters and the cook, and everyone—all ages, both sexes—had tattoos. We took one look at the big menus and realized the cuisine was also a step back in time. Of course I ordered the special. It was a pork tenderloin sandwich with jacket fries and coleslaw. When it came it was the size of a watermelon slice, with the fries as big as cantaloupe pieces, and the slaw made fresh that morning. After such a meal—it was truly memorable, and I did it justice, though it was about as medically incorrect as one could find out in the heartland—I stumbled as best I could up to the cashier to pay our bill. He squinted at the waiter's scrawls, took a puff, and exhaled toward the ceiling. "Next time," he rasped, "get it with homemade hash and fried eggs on top—but don't plan to do any work that afternoon." Even with the streamlined version I was nearly comatose, like a bear overdue for hibernation.

A few miles south of town, not far from the Missouri levee, there was the modern Western Historic Trails Center, an attractive one-story stone and ramada-timbered building built by the National Park Service and local partners in 1997 and since maintained by the State Historical Society of Iowa. It is reached by driving down a mile-long road winding through prairie wildflowers, and there are walking paths down to the riverbank. As much as the wonderful center down in Independence, it is tailor-made for today's travelers following the Lewis and Clark route and those tracing routes of the Oregon,

California, and Mormon trails. Interesting, well-crafted dioramas and sculptures stand above interpretive exhibits, and small video terminals show, at the press of a button, footage of the trails as they are today. There are illuminated maps and both period and contemporary photographs and art, and an excellent film by John Allen in a comfortably appointed little theater.

Omaha rises from the riverside Heartland of America Park, with its elaborately lit fountain display and walking paths, to the most immediate neighborhood—that which encompasses the busy "shop till you drop, eat till you're beat" Old Market. Nearby one will find Union Station, many towering brick warehouses where remnants of ancient painted advertisements are still visible, and two excellent hotels (one from the 1930s, an Art Deco palace, and one big upscale tourist haven). New glassy office buildings intrude from place to place in the otherwise old neighborhood. Westward, toward the ring of hills, modern Omaha takes over and one can point one's car past the rose-colored Georgian Joslyn Art Museum, stately and severe, on the rim of a hill. Farther along are beautiful, tree-shaded old neighborhoods of turn-of-the-century houses, the campus of the Jesuit Creighton University, and pockets of business activity such as the swoopingly modern tower of the Mutual of Omaha headquarters.

We went directly to the Joslyn Art Museum, named for the powerful newspaper publisher George A. Joslyn. Opened in 1931 and featuring exceptional public spaces, exhibition galleries, a 1,000-seat concert hall, and an art reference library, the museum dramatically expanded its space in 1994 with a modern pavilion, glass atrium, and upper-story bridge. It has a wonderful European collection ranging from Old Masters to the Impressionists: from Titian and Veronese, El Greco and Lorrain, to Delacroix, Courbet, Degas, Monet, Pissarro, and Renoir. The American collection runs from early American portraiture through the nineteenth-century Hudson River School; the twentieth-century collection features Modernist, Regionalist, Abstract Expressionist, and even Pop Art styles. There are also very good Asian and Ancient Greek collections, but our main attention during this Western pilgrimage was on the excellent Western American collection—the panoramic canvases of Albert Bierstadt and the watercolors of Thomas Moran, the brawny romanticism of Frederic Remington, and the stately Native American portraits of Edward S. Curtis. We went with greatest interest to admire the watercolors and prints of the Swiss artist Karl Bodmer.

Bodmer (1809–1893) was born in Zurich and showed great talent at a young age; while on a visit to the Rhineland his work was noticed by the German naturalist Maximilian, prince of Wied-Neuwied, then planning a survey of America and its Western territories. Maximilian had attracted great scientific and popular attention with his ethnological work among the South

American Indians; he intended this trip as a comparative study among the tribes of the American West. The prince hired Karl Bodmer as his illustrator. In 1832 the party made its way to St. Louis where it was booked on the *Yellowstone*, a steamboat of the American Fur Company, to its trading post at Fort Union in present-day North Dakota. From there they went upriver to Fort McKenzie, somewhat north of today's Great Falls, Montana, and thence down to Fort Clark, near present-day Bismarck, North Dakota.

These trading posts were always attended by many Native American tribes, which kept the scientist Maximilian busy with his notebook and Karl Bodmer with sketching and painting. The young artist captured extraordinary likenesses of the Sioux, Omaha, Blackfoot, Assinboin, Minatarre, and Mandan people and the stark, beautiful landscapes in which they lived. I remember as a boy reading about the journey of Maximilian and Karl Bodmer, and seeing reproductions of the Indian portraits and landscapes in *American Heritage* magazine, thanks to the subscription sent me by my grandfather, and many years later reading the account of Bernard DeVoto in his narrative of the fur trade, *Across the Wide Missouri*. Bodmer's paintings, he said, "have the focus and selectivity of medical art, the illustration of anatomy or surgical technique by drawing, which permits a clarity, emphasis and separation of parts and planes beyond the capacity of the camera lens."

It was quite an event to go to the Joslyn, which owns most of the watercolors and the hand-colored aquatints, along with Maximilian's handwritten journals used to illustrate Maximilian's book, *Travels in the Interior of North America, 1832–34*. The collection also includes Bodmer's *Kiäsax, Piegan Blackfeet Man*, a beautiful three-quarter portrait of a proud tribesman clad in a striped blanket, and *Chan-Chä-Uiá-Teüin, Teton Woman*, a full-length view of a woman in a colorfully figured buckskin dress with shawl, fringed breeches, and beaded moccasins. It was breathtaking to see the product of Bodmer's wilderness sojourn up so close that one could read the delicate brushstrokes and even light pencil lines. My wife, who is an artist, could have stayed looking at those watercolors for days. When we had to pull her away to get the children fed, her eyes were far away, in a different time, thinking not only of the world seen and rendered by Karl Bodmer but of her childhood, filled with books about horses, Indians, beads, buckskin, and romance.

There is no memorial in Omaha to Fred Astaire—in fact, the nearest site of the Fred Astaire Dance Academy is in the capital city, Lincoln, forty-five minutes away, which is good for all the government workers and university students, but what about the ballroom king's birthplace? No statue stands to commemorate Marlon Brando, either. The house as well as the entire neighborhood briefly called home by the Little family is bulldozed into memory, though on maps and tourist brochures the eleven-acre space fronted at 3448 Pinkney Street in North Omaha's African-American neighborhood is called the Malcolm X Birthsite and contains a subtle marker. There are dim hopes of funding some kind of meditative park, according to a volunteer reached by phone at the nearby Great Plains Black History Museum, located in an old brick building on Lake Street. We called at the latter several times during its posted hours, but were disappointed to find it closed. Not so, however, for the Mormon Pioneer Winter Quarters and Mormon Pioneer Cemetery, at North Ridge Drive and State Street, which marks the home of some 4,000 Latter-day Saints who had fled from their violent ouster from Nauvoo, Illinois, in 1846 only to endure an extraordinarily bad winter and several disease plagues, killing more than 600 people. It is a sad and stirring place.

There are many more sites of historical interest in Omaha and Council Bluffs—more, in fact, than we could contemplate for the three days I allotted for the sister cities. If we had put up for a family vote to choose between the sprawling and superbly rated Henry Doorly Zoo, which boasts of 1.6 million visitors annually, and the humble President Gerald R. Ford Birthsite (rose garden, a booth with White House memorabilia, and a replica of part of the house, which burned in 1971), the zoo would most certainly have won.

But the Plains beckoned. And our route was far more enticing than the multilane, characterless interstate that slices up the old Platte River Valley route across Nebraska for 450 miles without once giving its millions of sojourners a chance to see anything. We would avoid I-80 unless absolutely necessary. For us, rolling along the original route of the Union Pacific Railroad, and following the ruts of the wagon pioneers, we would hug old Route 30—the path of the Lincoln Highway—relishing the extra time it would take, for the multitude of impressions ahead. And that would make all the difference.

4

The Lincoln Highway

B ack when I was a little boy in the early 1950s, our summer expeditions out to my grandparents' farm in Missouri would have begun either in South Jersey or the Maryland suburbs outside Washington, depending on where we were living at the time. The trips would usually siphon after Philadelphia onto the Pennsylvania Turnpike, which had been conveying cars and trucks on smooth, divided concrete roads since around 1940. It still felt new to everyone, and the gentle mountain grades (the turnpike had been built on an old railroad grade) were always worthy of approving comment by my father, whose natural Southern garrulousness would be heightened by the No-Doz caffeine pills he swallowed to keep himself awake behind the wheel and us on the road. The single-bore tunnels on the turnpike mystified and fascinated me; many were originally used by the railroads, and of course now they always occurred in pairs. Inside the tunnels my sister Terry and I would try to hold our breath the whole way. I remember the thrill of seeing an orange-tiled roof on the horizon, knowing that it was a Howard Johnson's restaurant and that we were going to stop for hot dogs on heavily buttered split rolls, palate-searing french fries, a foul, chemical-tasting HJ cola I often forgot to avoid, and the Eternal Question: which of the twenty-eight flavors of ice cream would win out today?

In some of these early journeys, though, I have dim memories of hewing at least partway to the old Lincoln Highway. It was called "the old Lincoln Highway" even then in those antiquated times, for by our standards a turnpike or a toll road was a highway, not a route like Route 30 that ran through cities, towns, and hamlets like a string through pearls. This was not what we would call today "limited access"—access to the Lincoln Highway was limited only to where the curbs were broken by driveways, lanes, roads, avenues, and boulevards. Now, for us, taking the Lincoln Highway in those days would only make sense from a directional standpoint in Pennsylvania and for about half the width of Ohio—after this we would have tended southwestward toward Kansas City while the Lincoln Highway would have borne northwestward toward Chicago.

But the memories I have are auditory (I hear my father telling someone,

"Out by way of the Pennsylvania pike, the old Lincoln Highway," and so on) and they are visual, because I remember peering out the window looking for Abraham Lincoln's bronze profile on the irregularly spaced, battered old concrete marker posts as we sped by, and seeing, less frequently, the faded-almost-to-invisible paint stripes of red, white, and blue on poles. I remember "collecting" these markers like I would "collect" license plates.

We played a lot of games on the road: collecting, counting, guessing, and spelling games, especially before 1955, when my parents began the second half of their childbearing with the younger two, Christopher and Lisa, and the increasing numbers made drives more of a managerial problem. After that, the main tool of control and justice was a rolled-up copy of *Life* magazine wielded by my mother like a policeman's baton. We graduated, too, after about 1960, from Mercury sedans to Ford station wagons, and even with three rows of seats and a cargo area we'd run out of storage capacity, with luggage piled high on the top of the car and covered with a brown Sears, Roebuck tarpaulin.

A generation later, I was herding my own family into a three-rowed vehicle piled high with duffel bags stuffed into a rooftop carrier, and we were pulling out of our motel parking lot in Omaha and pointing west, over the hills on a local road, Route 6. On my highly detailed DeLorme map book of Nebraska I had spied the legend "Old Lincoln Highway" out west of town. I badly wanted to see a metallic Lincoln profile marking a roadway again. Past Boys Town, the residence and school for homeless and troubled boys started in Omaha by Father Flanagan in 1917, we came to a sign pole with red, white, and blue stripes and the identifying "L." We turned aside, bumping onto a wonderful narrow old roadway. It was paved entirely with red bricks, with a painted double center line, back in 1919–20.

Weeks after the Union Pacific and Central Pacific railroads joined at Promontory Summit in Utah on May 10, 1869, the adventurous *New York Tribune* journalist Albert Richardson, who had done time in a Confederate military prison camp on spying charges, took a transcontinental railroad trip and wrote about it for the newspaper. In several popular books and many serialized newspaper accounts Richardson had chronicled life in the far-flung Western settlements—encounters with the celebrated and the obscure, and picturesque descriptions of the landscape. In 1869, he commented wryly during a pause in Omaha, "The unpaved soil of the Omaha streets is jet-black and would yield splendid corn or wheat. When the weather—which may change a dozen times a day—is wet, the mud is incredible in depth and stickiness; but a few hours of sunshine dry it up completely." What is interesting is how far into the so-called modern era this was the case for most roads, outside the towns and cities, particularly in the Midwest and West.

The great railway-building era in the United States, roughly between 1840 and 1910, had monopolized the public and government attention as iron and then steel rails spiderwebbed across the East, Midwest, and West, connecting the important cities and towns and creating subsidiary settlements along the way. Most of the old wagon trails were abandoned with joy, relief, and disdain. Few resources were expended on roads—why do so when one could ride or send freight on the train? There were, by 1910, some two million miles of unconnected roads radiating out from railway depots and market towns to the nearer farms and dependent hamlets, some of them graded or graveled, a very few of them actually paved. In wet weather all hopes of getting anywhere would sink far below the level of mud. Road names and directional signs were exceptionally rare. But gasoline-powered automobiles had become more affordable after the turn of the century, and opportunities for leisure time had increased; now there was something called "pleasure driving." The problem badly needed to be solved.

The story of Carl G. Fisher is a perfect example of the situation as it existed around 1910. Fisher was an automobile promoter and inventor who had perfected the carbide lamp process for car headlamps and founded the Prest-O-Lite Corporation, which reigned supreme for decades until the invention of the electrical light system. "Three of us drove out nine miles from Indianapolis," he recalled,

> and being delayed, were overtaken by darkness on the return trip. To complicate matters, it began to rain pretty hard, and you know automobiles didn't have any tops on them in those days, so we all three got pretty wet. We guessed our way along as well as we could, until we came to a place where the road forked three ways. It was black as the inside of your pocket. We couldn't see any light from the city, and none of us could remember which of the three roads we had followed in driving out; if, indeed, we had come that way at all.
>
> So we stopped and held a consultation. Presently, by the light of our headlamps, reflected up in the rain, one of us thought he saw a sign on a pole. It was too high up to read and we had no means of throwing a light on it, so there was nothing to be done but climb the pole in the wet and darkness and see if we could make out some road direction on the sign.
>
> We matched to see who should climb. I lost. I was halfway up the pole when I remembered that my matches were inside my overcoat and I couldn't reach them. So down I had to come, dig out the matches, put them in my hat, and climb up again.

Eventually, by hard climbing, I got up to the sign. I scratched a match and before the wind blew it out I read the sign.

It said: "Chew Battle-Ax Plug."

Confronted by few route markers and directional signs, unpaved roads, and other hardships, a man like Horatio Nelson Jackson, of Burlington, Vermont, became a pioneer of long-distance tourism. Answering a gentlemen's club bet of $50, Jackson set out from San Francisco in May 1903 with a chauffeur named Sewell K. Crocker, driving a Winton touring car with a 20-horsepower gasoline engine and a chain drive. The roundabout route, which bypassed the Sierra Nevada range and difficult desert stretches in Nevada and Utah, was fraught with wrong turns and serious mechanical mishaps, but Jackson's muddy and battered car finally rolled into New York City some sixty-three days, twelve hours, and thirty minutes later. He won his $50 bet—the trip set him back about $8,000, including the $3,000 for the car—but Jackson was widely credited with the first truly transcontinental auto trip. The coast-to-coast record was soon improved by the Packard Motor Company, which commissioned a team later that same summer and reached its goal in fifty-one days.

One traveler, a writer named Bellamy Partridge, was inspired to try it himself and wrote to the newly created American Automobile Association for advice. The reply came from A. L. Westgard, the group's pathfinder. "He wrote that he thought I would find the trip a very enjoyable adventure," recalled Partridge, "if I was fond of changing tires and digging a car out of sand pits and mud holes. He warned me, however, that west of Omaha I would find the roads very informal, signboards about 1,000 miles apart, water as scarce as gasoline, and no bridges across streams under 50 feet in width. He also called my attention to the fact that there was no such thing as a road map for the country west of Chicago."

My favorite story concerns a Packard bigwig, President Henry B. Joy, an enthusiastic proponent of long-distance travel. During one journey he got himself to Omaha, and asked his local distributor for directions to the road west. "There isn't any," the Omaha man replied. "Then how do I go?" asked Mr. Joy. "Follow me and I'll show you." They drove westward away from Omaha until they came to a wire fence. "Just take down the fence and drive on and when you come to the next fence, take that down and go on again." "A little farther," said Mr. Joy, "and there were no fences, no fields, nothing but two ruts across the prairie."

"But," finished Mr. Joy's interviewer, "some distance farther there were plenty of ruts, deep, grass-grown ones, marked by rotted bits of broken wagons, rusted tires and occasional relics of a grimmer sort, mementos of the thousands who had struggled westward on the Overland Route in 1849 and '50, breaking trail for the railroad, pioneering the highway of today."

Sometime after his road-hunting adventure in the wilderness of rural Indiana, auto lamp manufacturer Carl Fisher conceived a program to promote an all-weather, hard-surfaced, clear-signed roadway. He called it "The Coast-to-Coast Rock Highway." Financing such a scheme would not depend on federal, state, or local government: he preferred to apply modest voluntary tithes to the automobile industry, whether the contributors were manufacturers, suppliers, or dealers, and to create an automobile owners association with annual dues. Fisher thought he could raise $25 million in this fashion, all of it to go to construction materials; these would be donated to government road authorities, whose task it would be to do the building. Nearly every business leader pledged large amounts—the main exception being the cantankerous Henry Ford, who believed government should do it all. Nevertheless, the project gained momentum into the spring of 1913, when Fisher and his fellow industrialists decided to change the name of the project to appeal to patriotic sentiments. "The Lincoln Highway" won out over all other contenders, and so the Lincoln Highway Association was born, with Packard's Henry Joy as president. The by-laws of the new association announced its purpose: "To Procure the establishment of a continuous improved highway from the Atlantic to the Pacific, open to lawful traffic of all description without toll charges: such highway to be known, in memory of Abraham Lincoln, as 'The Lincoln Highway.'"

No route was announced in press releases, though publicity for the project was enormous. The association already had several hundred thousand dollars in cash on hand and corporate millions pledged for subsequent stages of the project. After the stories hit the newspapers letters poured in to the office in Detroit from everywhere in the nation, offering donations—though many pledges were dependent on the Lincoln Highway passing through the donor's locality. Suspense grew as the directors met, equipped with maps, guides, logs, journals, and even railroad mileage tables, to evaluate geography, road conditions, weather, and firsthand experiences; the last must have generated some heat, since most of these executives had done the difficult trip themselves and were invested in their own discoveries and conclusions. The debate cannot have been easy, but the results were finished in time to announce at the annual National Governors' Conference, held in Colorado Springs in August 1913.

The map distributed to the governors brought satisfied grins to twelve officials and scowls to the rest: the highway ran through New York, New Jersey, Pennsylvania, Ohio, Indiana, Illinois, Iowa, Nebraska, Wyoming, Utah, Nevada, and California. It would be 3,389 miles long, beginning in Manhattan at Times Square, and ending in San Francisco at Lincoln Park, in view of the Pacific. From New York City it ran down New Jersey through Trenton to Philadelphia and west through Pittsburgh into Ohio; it went from Canton, Fort Wayne, and South Bend to Chicago. It crossed Illinois and Iowa down to Council Bluffs and Omaha, at which point the Lincoln Highway joined forces with the Union Pacific Railroad route and ran westward up the Platte Valley to Cheyenne, up the Laramie Plains to Rawlins, and across the easy Continental Divide. It passed through the eastern Utah mountains to Salt Lake City and then, departing from the railroad's trail blazing, went around the southern end of Great Salt Lake and through the deserts to washboard Nevada, tracing the old Pony Express route through Ely, Eureka, and Austin to Reno, before rejoining the first transcontinental railroad route through the Sierras' Donner Pass over to Sacramento, Stockton, and Oakland. Bay ferryboats and the streets of San Francisco triumphantly completed the line.

Of course, complaints from the disappointed route candidates rose, never greater than the howl of outrage from Colorado, which the Lincoln Highway board could not ignore. To placate the state an extra dogleg route was drawn in, from Nebraska straight down to Denver and thence northward to Cheyenne. Another compromise was an alternate route to California from western Nevada, which ran through Carson City before crossing the Sierra at Echo Summit and descending through Placerville to Sacramento. With these added options the roar to bend the route this way or that grew deafening. But the Lincoln Highway Association moved forward to its next phase with a massive new fall publicity campaign, including full-page ads with an "Appeal to Patriots." October 31 was earmarked as a national celebration for this memorial to Abraham Lincoln. Ministers addressed their flocks; politicians climbed onto the bandwagon; parades and rallies got the word out. Subscriptions poured in.

With all the tumult, as Lincoln Highway scholar Drake Hokanson points out, "As yet this great road was nothing more than a line on Henry Joy's map; nothing had changed just because the name Lincoln had been applied to this loose collection of more or less end-to-end roads. They had been ordinary dirt roads one day and dirt roads called the Lincoln Highway the next."

While the fund-raisers worked on finding the means, lobbying legislatures, buttonholing business interests, and cashing checks from small donors, the association encouraged local members to mark the new route. There was also a program to build demonstration sections, or "seedling miles," to show what driving on a hard-surfaced road was like. Still, the effort would take a number of years. Often, as the Lincoln Highway Association ran into the red, wealthy executives of the auto manufacturers and suppliers dipped into their own pockets to keep it all going. They would be paid back handsomely.

In early 1915 one curious motorist decided to assess the state of the Lincoln Highway during a transcontinental drive between New York and San Francisco. It was the society doyenne of correct behavior, Emily Post, who embarked in an open touring car with a woman companion and with Post's son driving them. Stocked with picnic hampers and a silver tea service, her trip began in Manhattan in April 1915 and traced a route through Albany, Buffalo, Cleveland, and Chicago, west of which they joined the Lincoln Highway. She complained,

> As the most important, advertised, and lauded road in our country, its first appearance was not engaging. If it were called the cross continent *trail* you would expect little, and be philosophical about less, but the very word "highway" suggests macadam at the least. And with such titles as "Transcontinental" and "Lincoln" put before it, you dream of a wide straight road like the Route Nationale of France, or state roads in the East, and you wake rather unhappily to the actuality of a meandering dirt road that becomes mud half a foot deep after a day or two of rain. . . . The only "highway" attributes left were the painted red, white and blue signs decorating the telegraph poles along the way. The highway itself disappeared into a wallow of mud. The center of the road was slightly turtle-backed; the sides were of thick, black ooze and ungaugeably deep, and the car was *possessed*, as though it were alive, to pivot around and slide backward into it.

One could wax poetic about that mud—there was so much of it. "Illinois mud is slippery and slyly eager to push unstable tourists into the ditch," Mrs. Post commented, "but in Iowa it lurks in unfathomable treachery, loath to let anything ever get out again that once ventures into it." There were frequent tire punctures; in those days motorists removed, patched, reinflated, and mounted tires by themselves, often several times a day.

Finally they reached Council Bluffs and Omaha and were mightily impressed. "In nearly all Eastern cities automobiles are treated as though they were loitering tramps; continually ordered by the police to 'keep moving along,'" Mrs. Post said. "In Omaha the avenues are so splendidly wide that they can afford chalked-off parking places in the center of the streets where motors can stand unmolested and indefinitely. If only New York and Boston had the space to follow their example!"

They looked forward to dry Nebraska roads, straight and true, so they could make up for lost time. "'We must be sure that everything is in perfect running order,' you exclaim excitedly as you picture your car leaping out of Omaha and shooting to Denver while scarcely turning over its engine," she noted. But a garage man in Omaha disabused them of the fantasy: "'Fr'm here t' Denver is about thutty-five hours' straight travelin',' he warned. 'You gutta slow down t'eight miles through towns and y'can't go over twenty miles an hour nowheres! Road's fast enough! But the law'll have you if you drive over it faster'n twenty miles an hour.'"

There being no traffic behind or in front of us on this remnant of the old road, we could slow to the once required speed of twenty miles an hour to look for the road markers. Brick gave way to concrete, then brick again, and then we were sailing under magnificent cottonwoods in the bungalow town of Elkhorn, population 6,200, named for the river that flows southward to the Platte, draining the northeastern part of the state.

Back around the middle of the nineteenth century, this was a significant mark on the map of the emigrant route called the Council Bluffs Western Trail, or the Mormon Trail. Merrill Mattes, the National Parks historian who made the Platte Valley route his life work, estimated that more than 100,000 covered wagon emigrants used this road prior to 1867, 30,000 of them in 1849 alone. Hurrying to make time on the Plains in the soggy spring, ever mindful that the high Sierra Nevada and its early snows awaited them at the end of the traveling season, they would ride ridges out from Omaha to their first serious obstruction, the melt-swollen, tumbling, and unbridged Elkhorn River. It was a hazardous place to ford, with calamitous spills and drownings. Not until 1850 did anyone provide ferry service, and a bridge did not appear until the 1860s. Ahead, after the Platte confluence, the wagons followed the Platte's north bank toward the site of a large, permanent Pawnee village, knowing nervously that these tribesmen exacted tolls, handouts, and blackmail from the thousands of emigrants, from coins to coffee, cocoa, tobacco, or liquor.

Elkhorn advertises itself as combining the best elements of suburb and country. It was a short drive to the glass skyscrapers of Omaha, but it was blessedly rural and unspoiled, and on that warm late-June day the corn in the fields was already chest-high and tasseled. We paused long enough in town to look at a humble monument to the Lincoln Highway—two of the concrete "L" posts and an explanatory sign mounted on a modest brick platform. The inclusion of Elkhorn on the revered national route in 1913 caused a small real estate boom that reached the middle three figures and lasted several days before the excitement died down. Somewhere in town we lost the path of the highway, but it beckoned ahead of us at Route 30, the officially designated memorial route across the width of Nebraska. The corn and the cottonwoods stirred in the light breeze as we found ourselves on Route 275, heading northwest alongside the Union Pacific tracks toward the junction with Route 30 in the town of Fremont, a town named for one of the strangest heroes of the nineteenth century.

The town was staked out on the north bank of the Platte on nice, nearly level ground in view of the mostly bare hills that marked the intersection of the Platte and Elkhorn river valleys. It was August 25, 1856, and a month earlier the famous explorer John C. Frémont (nicknamed "the Pathfinder") had been nominated the presidential candidate at the Republican National Convention in Philadelphia; it was the first national convention for the party, organized two years before to oppose slavery. Just south of Nebraska territory smoldered Kansas, Bleeding Kansas—where because of the slavery issue guerrilla raids, assassinations, massacres, and burnings had been raging for more than a year and government had all but broken down. In a welling of party pride, the Nebraska free-state settlers surveying their lots named their new town Fremont. The Pathfinder lost in November—Democrat James Buchanan won by half a million popular votes in a three-way contest that included the spoiler, former president Millard Fillmore, fronting the nativist, anti-Catholic Know-Nothing Party—but there was no effort in Nebraska to repudiate the town name of Fremont. If anything, it tightened the settlers' resolve. In the first three years of existence, Fremont citizens underwent scary tensions with the local Pawnee tribe and terrible deprivation from failed crops. They subsisted on bread and grease, using a little buffalo meat (there wasn't much hunting nearby) to flavor the gravy.

Given the hero for whom their town was named, this near-starvation had an ironic symmetry. Born in Georgia in 1813 and seeing service in the Army Topographical Corps, John C. Frémont took expeditions into the Wind River branch of the Rockies (1842), the great basin between the Rockies and the Sierras (1843–44), and the mountain-choked transit between the eastern Colorado plains and the fertile, beckoning California valleys (1845–46), assisted by the intrepid scout Kit Carson. Frémont's published reports of these travels, with their highly detailed descriptions of terrain, became best-sellers. They inspired thousands, for better or worse, to make the perilous overland journey to Oregon and California. He was given (and adopted proudly for himself) the nickname of the Pathfinder despite the fact that he followed others and broke no significant new ground. An assiduous self-promoter, Frémont kept up his high profile—partially thanks to his marriage to the daughter of Missouri senator Thomas Hart Benton, one of the most powerful men in Washington; Jessie Benton Frémont would ghost write all of her husband's bestselling travelogues.

He has popped up in my research and assignments for many years. The Pathfinder has always seemed to me, during some twenty-five years of writing,

editing, and reading, like an annoying neighbor whose footfalls clump overhead or whose bass-register music clings right on the threshold of consciousness. His egotism, recklessness, and penchant for blaming others for his
own mistakes are like a line of rusty old disabled cars in the country landscape outside my kitchen window. But certain other human qualities are like
toxic substances leaching down into my well. As is known, Frémont had a
major role in what was called "the conquest of California" from the Mexican
government in 1846—many think he was acting under secret federal orders
to start a war and seize the territory. But what is often missed is that he also
permitted a massacre of Indians in the northern Sacramento Valley that was
similar to massacres at Wounded Knee, Sand Creek, and Bear Creek. "The
number I killed I cannot say," his scout, Kit Carson, recalled. "It was a perfect
butchery."

Being under his supposed protection and guidance, Frémont's men put
their lives at peril in every expedition he led. The worst was Frémont's disastrous expedition into the Rockies in the winter of 1848–49—a sumptuously
wrongheaded notion to follow father-in-law Benton's armchair order to find
a transcontinental railroad route from his home in St. Louis across the impossible, mountainous 38th Parallel, the highest part of the Colorado Rockies, to
California. The Pathfinder's stubborn ego resulted in the deaths (by cold or
starvation) of ten men, the certain cannibalism of victims, and the impossible to ignore charges that Frémont left stragglers behind to die. Despite this
calamity Frémont actually undertook to do the journey again, in 1853; having been passed over for command of an army reconnaissance looking for a
transcontinental railroad route, he mounted his own expedition. They got
two thirds of the distance before bogging down in winter, crossing the Colorado Rockies but failing in the snowy Wasatch Range of southern Utah,
where the men froze and starved until saved by Mormon pioneers. The effort
proved nothing in practical terms.

There was also his Civil War service: he may have been at the top of President Lincoln's appointments list for major generals, but Frémont proved to be
one of the worst generals of the Union Army. Earlier, in California, Frémont
had been court-martialed by General Stephen Watts Kearny for mutiny and
disobedience, and he had several times been subject to censure for roaming
hundreds of miles away from his appointed rounds with his exploring expeditions trustfully trailing behind. With the Civil War his every move could be
scrutinized; finally, involving himself in politics, he defied Lincoln's directives
over slaves in occupied Missouri, and he was removed.

Frémont lived twenty-five years after the war, dabbling unsuccessfully in
railroads, writing a little, and serving for a few years as appointive territorial
governor of Arizona. In 1890 the army gave him major general rank for the
pension, but a shabby life caught up with Frémont and he died soon after. For

many years he was lionized. But better scholarship and finer-tuned sensibilities finally began to restore the truth.

There's an old blues song I happen to revere, Sonny Boy Williamson's (Rice Miller) "Don't Start Me Talkin'—I'll Tell Everything I Know." That seems to be the case with the Pathfinder. "Don't get me started," I might have been overheard growling outside the Fremont Academy of Dance, or the Fremont Beef Company, or the Fremont Foot Clinic, or the Fremont Keno Parlor, or (don't say it) Fremont Therapy. But I tried to keep my reservations to myself to avoid alarming or angering the townspeople, and probably thus eluded being lodged in the Fremont jail and exciting my children's dismay and my wife's annoyance.

But on the other side, after all, the town had been named for a candidate running on an anti-slavery platform, and one must give credit for that. Even though his books were written by his talented spouse, they are highly illustrative and finely detailed, always useful to the scholar as well as the interested reader, so he must be credited with good taste in a wife. And the ancestors of today's citizens of Fremont ("It's Just Right!" is the town slogan) were the pioneers who existed on bread and buffalo grease, who lived for years on the barter system. The village finally began to prosper when it became a natural way stop on the journey from Omaha to Fort Kearney, and emigrants left their coins and trade behind, and then later in 1866 when the Union Pacific Railroad laid tracks just south of town and planted a depot. The people who settled this town were best represented by one George Turner, who built a log boarding house in 1859. He cut the logs on a big island in the Platte River, and then swam across the channel with them, one by one.

We pressed onward on the Lincoln Highway, leaving Fremont's grain elevators and power plant (with its mountain of waiting coal) dominating the skyline in my rearview mirror. We passed the Fremont State Recreation Area—with its little sandpit lakes and its acres of campers and clotheslines—and headed into corn and soy fields where rickety watering contraptions looked like towering daddy long-legs. Grayhead coneflowers and asters lined the concrete way, and a marker advertised that the Stephen Long Army expedition had passed here in June 1820.

The party had acquired a new scribe before leaving the Missouri River: it was a Vermont native named Steven James, a botanist and geologist trained at Middlebury College. The soil along the Platte struck him as fertile but the marked lack of timber beyond the thin riverbank layer of cottonwoods made the whole valley seem distinctly unpromising for settlement. A later traveler, Amos Bachelder, agreed, seeing in 1849 "vast naked plains, and barren hills, grand in their loneliness and solitude. The sun . . . scorches us like a hot iron on these sandy roads. This country now seems suitable only for the residence of the wild, wandering savage."

East of Ames, a hamlet under the cottonwoods, we found another granite marker, this one marking the covered wagon trail from Council Bluffs and the site of a long vanished hamlet called Albion. From a distance the routes across the prairie seemed unbroken with the vista, "the most beautiful sight I ever saw," wrote traveler Charles Putnam on May 11, 1846. "They are filled with beautiful flowers, and they cover over a space as far as the eye can reach." But ahead it was far from effortless. Often when the emigrants encountered one of the many creeks pouring tribute into the Platte they would have to cut down quantities of small brush and trees, if any were nearby, to throw down into the muddy ravines so that the wagons would not swamp or bog down. Some banks were so steep that wagons were lowered with ropes and then had to be laboriously raised on the western side with doubled or quadrupled ox teams. The summer of 1846, young Francis Parkman, two years out of Harvard and with the ink on his law degree not yet dry, had gone onto the Oregon Trail seeking something to write a book about; his route had followed the trail from Independence and Fort Leavenworth diagonally across Kansas toward an eventual crossing of the Platte, but his experience on the Kansas prairie was as universal for travelers as it was a strict cautionary. "Let the emigrant be as enthusiastic as he may, he will find enough to damp his ardor," he wrote.

> His wagons will stick in the mud; his horses will break loose; harness will give way; and axletrees prove unsound. His bed will be a soft one, consisting often of black mud of the richest consistency. . . . The wolves will entertain him with a concert at night, and skulk around him by day, just beyond rifleshot; his horse will step into badger holes. . . . A profusion of snakes will glide away from under his horse's feet, or quietly visit him in his tent at night; while the pertinacious humming of unnumbered mosquitoes will banish sleep from his eyelids.
>
> When thirsty with a long ride in the scorching sun over some boundless reach of prairie, he comes at length to a pool of water, and alights to drink, he discovers a troop of young tadpoles sporting in the bottom of his cup. Add to this, that, all morning, the sun beats upon him with a sultry, penetrating heat, and that, with provoking regularity, at about four o'clock in the afternoon, a thunderstorm rises and drenches him to the skin.

Rolling effortlessly along Route 30, passing old farmhouses with ornate stained glass windows, we could catch fleeting glimpses of the 1920 Lincoln Highway, a narrow strip of concrete threading parallel to us. At North Bend, a town founded by Scottish settlers from Illinois in 1856 and now dominated by tractor yards, we pulled in to buy gas at a service station. The man at the next pump, white-haired and open-faced, filling some fuel cans in the bed of his pickup, saw my green Vermont license plates and his eyes widened. "Man, you're from a long way from here!" he exclaimed. "You goin' or comin'?"

"Still goin'," I said. "We're just eleven days out, following some of the old wagon trails, looking for markers and museums, and," I paused, gesturing at the two-lane, goin'-and-comin' road, "we're trying to stay close to the Lincoln Highway. We figured on heading straight till we hit salt water in California."

"Is that a fact?" he replied, brightening even more. He strode over, grabbed my hand, and pumped it vigorously. "Well, in that case, I want to wish you good luck!" He had clear childhood memories of the Lincoln Highway, he said, before the Second World War gas rationing, when there was a steady stream of auto traffic and the little towns along the route saw more of outsiders. "In the Fifties the Eisenhower speed roads came in," he said, "and emptied us all out. But it was pretty full before. There were tourist cabins, auto camps, and convenience stores, all along here," he said. "Mechanics always had more work than they could handle, and they were careful not to charge an arm and a leg, for fear it would get written up in one of those Lincoln guidebooks and give the town a bad name."

Efforts to promote the idea of the coast-to-coast Lincoln Highway continued into the second and third decades of the twentieth century despite national reversals, the onrush of the Great War, and the giddy recovery, helped by strenuous publicity and the public's natural curiosity and enthusiasm for the automobile. Frequent record breaking kept the route in the headlines. In 1916 Bobby Hammond drove an Empire roadster from San Francisco to New York in 6 days, 10 hours, 59 minutes. People began to feel that anyone could do it. "It's a real road," touted the Lincoln Highway Association, "which will permit a traveller to average 21 miles an hour each hour of the 24, day and night, day in and day out, for practically a week running." Then, a Marmon automobile did the same journey in 5 days, 15 hours; a Hudson Super-Six in 5 days, 3 hours, 31 minutes. For four years this record remained until a Hudson Essex whittled the time down to 4 days, 14 hours, 4 minutes. By then there had been real progress on paving sections of the Lincoln Highway. The association donated steel culverts for the demonstration seedling miles, got cement companies all across the country to donate materials, leaned heavily on county governments to dedicate road budgets, and enlisted local Kiwanis and Rotary clubs in the campaign. No wonder the transcontinental time record seemed to be broken every few weeks. One competitor concentrated on breaking his own record; he was Louis B. Miller, a fiftyish San Francisco businessman who seemed to prefer being behind the wheel to anything else: with a Wills Sainte Claire in 1925 he set records of under 102 hours and then a numbing 83 hours, 12 minutes (which included 45 minutes lost when a New Jersey patrolman pulled him over, just 20 miles shy of the Hudson ferries). Miller later clocked a round-trip record in a Plymouth in 1931, returning to San Francisco in 5 days, 12 hours, 9 minutes. How he managed such feats without sleep was a matter, I suppose, between him and his pharmacist.

Of course most travelers went at a more leisurely pace, helped by a series of maps and guidebooks issued by the Lincoln Highway Association and heavily advertised by auto parts and oil companies as well as local hotels, tourist camps, and sandwich shops. Such publications command high prices today for collectors, and replica copies sell briskly; they are as fascinating to scrutinize as the later WPA state guides. In the earliest years the guidebooks contained such directions as "pass in front of saloon buildings and turn left around shearing pens." Eventually the way became clearer, especially with the help of road markers.

These continued to improve beyond the original painted signs erected by local enthusiasts (and sometimes switched by small business owners seeking to divert the highway past their enterprises). In 1916 more than 8,000 uniform enamel-on-steel signs were affixed to poles between New Jersey and the Nebraska-Wyoming border, an average of five to a mile, paid for through donations and the association's efforts. Signage in Wyoming, Utah, Nevada, and California took a few more years to appear with such uniformity and regularity, but was finally accomplished by local automobile club chapters. Mortality, though, was swift, especially in places where there was much hunting or where country schools were nearby. By 1924 the association noted a serious problem. "A check made in October [1924] showed 1,574 new poles and 2,157 signs needed to replace those destroyed or seriously damaged," it reported to members. "Wooden poles rot, are handy for prying cars out of the ditch, and readily gravitate to the farmers' chicken yards. Enameled steel signs are ideal targets; they react with a distinctive ping from either a bullet or a stone and are themselves badly scarred in the encounter. Once chipped, the metal rusts, destroying the remainder of the sign." It being observed that boys were probably the agents of much of the destruction, an association official got the bright idea of enlisting the Boy Scouts in the highway marking efforts, which quickly won approval from the national leadership on down to local scoutmasters. But implementing a new program languished for lack of money until 1927. By then the Lincoln Highway Association had judged its mission to be over, its goals met, and it found that some $66,600 was left in the treasury. Its final official act was to commission 3,000 concrete posts and coordinate a transcontinental digging party for the Boy Scouts. The markers were seven feet long, reinforced with four steel bars, and distinctively cast so that the lower section was octagonally shaped, above which was the square-shaped sign portion, ending in a pyramid. The Lincoln

Highway insignia and a bronze medallion were on the front; directional arrows were on the side. The medallion featured a relief portrait of Abraham Lincoln, with the wording, "This Highway Dedicated to Abraham Lincoln." Enameled steel signs set on metal, not wooden, posts, were employed in large cities. The concrete markers manufactured in Detroit were sent out to strategic places for the Boy Scout efforts, which took six weeks in summer 1927.

Six years after they were set, it was found, fewer than 5 percent of them had been destroyed or removed. As highway departments improved the roads, great care was taken to respect the monuments, keep them safe from construction equipment, and replant them in proper places where they would remain throughout the 1930s and the 1940s, and then into the Eisenhower 1950s when the little boy who was me kept a tally as his father drove the family out toward his grandparents' place in Missouri. We saw only a few out in the natural daylight and air in the summer of 2000, in places like Elkhorn and Fremont, since the ones not now residing in museums are either in private collectors' hands or buried and forgotten by that inexorable force called progress. But when one hove into view there were two excited boys in that car—one eight, one fifty-one—and the girls were equally happy.

The Platte River was obscured by a scrim of cottonwoods west of North Bend when we came upon a Mormon Trail marker and I sadly noticed that farmers were pulling up old windbreaks and readying the uprooted trees for burning in gigantic piles, something we saw with increasing regularity. Red-winged blackbirds, those dependable denizens of such road- and field-side foliage, were swooping around, and it may not have been my imagination that they seemed bereft at the big disruption in their lives, which was done to give tractors an uninterrupted field for plowing, planting, and harvesting. Am I mistaken, or were the windbreaks planted to lessen the harsh effects of wintertime storms and to give the wind less momentum in dry summers to pick up the topsoil and dump it a thousand miles away in the Gulf of Mexico? Will efficiency and factory farming save the land from another dustbowl condition? I think the red-winged blackbirds would say no.

At Schuyler we passed an old stucco gas station with an overhanging portico and red-tiled roofs. The town, seat of Colfax County, was named (as was the county) for Schuyler Colfax, the wildly popular Indiana politician who was the son of a Revolutionary War general and an influential newspaper publisher until he entered Congress, rising to be speaker of the house. In the summer of 1865 with the war over, Colfax had embarked on a cross-country stagecoach trip with two newspaper friends, Albert Richardson and Samuel Bowles, which attracted crowds in every hamlet and city they entered. The number of children, taverns, and municipalities named after the personable and handsome young congressman grew agreeably. However, Nebraska's Schuyler town and Colfax County were founded a little later, in 1869, by which time their namesake was vice president of the United States (under Ulysses S. Grant) and widely predicted to follow Grant to the White House in 1876. Unfortunately for "Smiler" Colfax (the name he was called by his detractors) there was the small matter of his having sought and accepted railroad securities from the Union Pacific Railroad and then having lied about it in the big Crédit Mobilier scandal that blew up during the Grant reelection campaign. The series of official inquiries paralyzed government for half a year; after that Colfax's political career was at an end.

Just a few years before the trading town of Schuyler was platted, back in the days when the Union Pacific was still struggling to get across the Nebraska plains and threats from Indian war parties were still common, one emigrant left a vivid story about an encounter in this vicinity. Two wagons were making their slow progress when suddenly three or four Indians appeared, he said, "riding towards us waving their arms and shouting.

> We slowed down our teams, but kept moving slowly and commenced to get our guns ready in case this really meant war. It was the first time either of us had met an Indian on his own hunting grounds and to say we were frightened is to put it mildly. There was no house in sight. There was no line of retreat. The Platte River cut us off to the east and south and there was nothing to shelter us anywhere, only the plains and the prairie grass to cover us. As they came within speaking distance we could hear that they were yelling, "Taboch, taboch!" In short, they were perfectly harmless and were only begging for chewing tobacco. We shook our heads. "No chew taboch!" And they turned as quickly as they had come. "Heap damn lie" was the limit of their parting salute.

At 65 miles per hour our car kept pace with a mile-long Union Pacific piggyback freight drawn by four engines (a four-header), the bright yellow and red-trimmed paint glowing against green fields. Soon we were crossing the Loup River, a quicksandy hazard to nineteenth-century emigrants, and approaching the town of Columbus, settled ten years before the railroad got there and a popular supply point for the wagon convoys. For many years Columbus was the home of two brothers long identified with the Indian Wars, Major Frank North and Captain Luther North, Ohio boys whose mother moved them out to Nebraska after her husband froze to death in a storm while working as a surveyor on the plains outside Omaha. The Norths lived so close to a peaceful Pawnee village, about ten miles west of Columbus, that the boys picked up the spoken Pawnee language as well as the sign language of the Plains Indians. After carrying the mail the North boys joined the army for brief stints, and by 1867 they were back in commission and charged with organizing a battalion of Pawnee Scouts to do battle against their tribal enemies, the Sioux, Cheyenne, and Arapaho, who had reacted to the despoilment of their hunting lands by emigrants and the railroad—and to bellicose actions by certain army commanders—by making guerrilla war upon the whites.

The four companies of Pawnee Scouts were deadly and effective, and it's likely that they bought the Union Pacific a couple of years to finish its gigantic enterprise. Frank North was the kind of officer who, when his unit was surrounded and in the open, would tell his men that he would shoot the first man who ran. No one did. But once, during a battle, North's horse slipped on some ice and threw its rider, who hit his head and was knocked unconscious; those same Pawnee Scouts stayed with him, shooting at the attackers and rubbing snow in North's face to bring him out of it, until the war party was driven off. "My boys were so well mounted that they could have easily have ridden away if they had been willing to leave me to my fate," North would say, "but with odds of some 15 to 1 against them, they jumped off their horses, formed a circle around me, fought it out and saved my life—which took cast iron nerve. Is it any wonder that I have always stood up for the Pawnee Scouts?"

Frank North expressed himself right out of the dime novels that were so popular back then. He knew most of the legendary characters in the West, including Wild Bill Hickock, with whom he had regular target shooting competitions. Another paperback hero, Colonel William F. "Buffalo Bill" Cody, would say that Frank North was the best revolver shot, whether standing still or on horseback, whether shooting at running men or at animals, that he had ever seen. It was a fitting end to North's career that after his army service he went to work in Buffalo Bill's Wild West Show, working with a number of Pawnee Scouts who played the part of hostile warriors in the shoot-'em-up action shows that played to packed audiences in the United States and Europe. The mock warfare and hard riding were not without hazards, however. In 1884, Frank North took a spill during a show, broke several ribs and sustained internal injuries. He recovered enough to continue with Cody's show into the next year but then the lung damage, combined with a bad cold and long-standing asthma, finished off the old warrior. He died at home in Columbus, where his brother, Luther, lived to a ripe old age, well into the century, keeping store and regaling all who would listen about the brothers' glory days out on the frontier.

They are buried in Columbus. As for the tribe that was so closely connected to the Norths for decades, it has found a memorial of sorts in Pawnee Park with its playgrounds, pools, and displayed Union Pacific engine. The old

Pawnee tribal ground there at the conjunction of the Loup and Platte rivers, where several thousand people lived in permanent earth lodges, farming and riding out to hunt from the extraordinarily large buffalo herds, are now cornfields broken by windbreaks of junipers and cottonwoods, as, on Route 30, local traffic and the occasional refugees from the interstate putter by.

We pressed westward, diverted by the sight of a white Union Pacific inspection pickup truck fitted with special railroad wheels flashing down the tracks near milepost 95, marking the distance from Omaha. At the moment it was hard to think of the old railroad days when, under the Norths' command, the Pawnee did battle against other Native American tribes on behalf of the Iron Horse. Instead, everywhere we looked there were relics from the heyday of the Lincoln Highway in the 1930s: at Silver Creek, an eerie auto graveyard where most of the scores of stacked-up hulks were circa 1935, and looked as if they had been used in a Jimmy Cagney movie like *Public Enemy*; at Clarks, an aquamarine line of tourist cabins right out of *It Happened One Night*; at Grand Island, a compact, attractive little city, a predominance of turreted brickfronts, masonry, and mansard roofs from the turn of century. But in every direction one looked was 1930s-era signage, whether painted or blinking, sputtering neon, where one would not be surprised if a dusty, rickety, running-boarded old Ford piled high with furniture and baggage and Okies coughed its way around a corner and came to rest in a hiss of radiator steam, as in *The Grapes of Wrath*. Such an image was not entirely out of place there; Grand Island was the birthplace of everyone's image of Tom Joad: Henry Fonda.

He was born on May 16, 1905, in a modest cottage at 622 West Division Street to frugal parents of Dutch and Italian extraction whose families had been in America since Colonial times. His father worked as a printer. Fonda was a small boy, shy, awkward, and inarticulate. His parents moved the family from Grand Island to a suburb of Omaha where, in his early teens, he suddenly shot up to six feet one inch, became athletic, and acquired some vague yearnings toward writing. A couple of years of journalism school at the University of Minnesota did nothing to crystallize his talent, but when he drifted back home to Omaha the summer of his twenty-first year, a friend of his mother's, Dorothy Brando (who had a toddler son named Marlon), persuaded Henry to do a part in a play at the Omaha Community Playhouse. Although not especially good, he worked a series of odd jobs and continued to hang out at the playhouse. A year after his debut, he was surprised to be cast as the lead in a play, *Merton of the Movies*. There he clicked—and was put on salary for the rest of the season. Then he headed east, looking for more acting work, and in subsequent years as a mostly penniless actor on Cape Cod, in Boston,

Baltimore, and New York, he worked hard at his craft, and made contacts that would aid his career—such as with the theatrical agent Leland Hayward, who eventually got him to Hollywood. When he became celebrated in the late 1930s, soon being nominated for an Academy Award for his role of Tom Joad in *The Grapes of Wrath* (he lost to his good friend James Stewart for *The Philadelphia Story*), Henry Fonda thanked Omaha's Community Playhouse for giving him a start by donating 250 new theater seats. And much later, in 1966, when his boyhood home in Grand Island was threatened, he bought the house back and donated it to the open-air Stuhr Museum of the Prairie Pioneer a few miles south of Grand Island, paying for its moving and restoration. Fonda died in 1982.

The museum itself, which stands on a tract of old farmland and is maintained by Hall County to reflect the lifestyle of nineteenth-century Nebraskan pioneers, was made possible through the generosity of a successful local farmer and businessman named Leo Stuhr (1878–1961), son of an original settler, who bequeathed a half million in cash and property to develop it. Nearly 80,000 people visit every year. The large main exhibit building, designed by the renowned architect Edward Durrell Stone (designer of the John F. Kennedy Center for the Performing Arts), is starkly incongruous out on the prairie farmland, but contains many interesting displays in a formal museum gallery setting. Out back, though, open between May and October, is the forty-acre Railroad Town, a fully re-created settlement with shops, a blacksmith and livery stable, a mill, bank, and train station, and a respectable-looking little residential neighborhood, where we found the Henry Fonda House. It's a tiny Victorian bungalow, built in 1883, and furnished as it might have appeared in 1905 when Fonda was born: faded wallpaper, dark horsehair upholstered furniture, fringed lamp shades, plates and decorative bric-a-brac adorning every shelf, and what appears to be the upright piano seen in a photograph I once saw of the Fonda family—sober, abashed, seven-year-old Henry; his two demure sisters, Jayne and Harriet; their cheery Christian Scientist mother, Herberta, playing the piano; and the withdrawn-looking father, William. A pleasant docent in period costume was stationed inside to talk about the life people like the Fondas lived long ago, but I wandered away from my wife and children to be out of earshot of the tour, preferring to think about Hank Fonda's films, particularly the Westerns, which enlivened my childhood.

I remember seeing *How the West Was Won*, that episodic, crowded, multi-star epic, and *The Rounders* in the early 1960s in an ornate, large-screened old film palace called the Beacon Theater in Port Washington, New York. But there were many more Fonda Westerns on television, on *The Late Show* and on the dependable *Million Dollar Movie* where each week one film could be seen fourteen times—this decades before the home videotape player—where one could study a film and its players until everything was memorized. On TV I saw *Fort Apache*, *My Darling Clementine*, *The Ox-Bow Incident*, and *Jesse James* and *The Return of Frank James*, among others. If the history was simplistic, Hollywood-style, driving home the old mythology of the West, there was something strong, graceful, and ethical about his portrayals, which I'm sure many boys unconsciously studied as they cobbled together their own characters, in the flickering gray light, struggling to grow in all the right directions. I know I did—studied and struggled and all the rest.

Grand Island was settled by Germans in 1857. The town labored to scratch a hold, helped during the warm months by trade with the constant stream of emigrant parties, but in 1859 a gold hunter returning to the East from Colorado set fire to the prairie grass because he hated Germans. All houses but one were destroyed. After Grand Island was rebuilt, largely on the charity of sympathetic citizens of Omaha, "ill fortune attended a number of the early settlers and promoters," reported the WPA historians: "one died in the poorhouse, one shot himself, another took strychnine, and one was run over by a train." That last would have been after 1866, when the Union Pacific built past the town, prompting a wholesale removal from the original riverbank site two miles up to trackside. From then on the town flourished as a freight division, supply, and car repair point, and mills, plants, and warehouses enlarged the economic base.

But before the town there was the island called Le Grande Isle by French fur trappers, a sandy hump in the middle of the Platte about one hundred miles long. It later marked the place where the Oregon/California Trail from Independence, St. Joseph, or Fort Leavenworth reached the Platte after plodding catty-corner across Kansas. Francis Parkman first beheld the Platte there, "divided into a dozen thread-like sluices," and at that moment "no living thing was moving throughout the vast landscape, except the lizards that darted over the sand and through the rank grass and prickly pears at our feet." Turning westward to trace it, however, he saw that "skulls and whitening bones of buffalo were scattered everywhere; the ground was tracked by myriads of them, and often covered with the circular indentations where the bulls had wallowed in the hot weather. From every gorge and ravine opening from the hills, descended deep, well-worn paths, where the buffalo issue twice a day in regular procession to drink in the Platte. The river itself runs through the midst, a thin sheet of rapid, turbid water, half a mile wide, and scarcely two feet deep."

Guidebooks and advice from both locals and travelers sent us into downtown Grand Island to a place called the Coney Island Cafe, a diner that I don't think has changed its decor since 1933, the year a man named George Katrouzos established it. The Art Deco touches we saw all over town, and the other Lincoln Highway echoes we had picked up, were reflected in plain and unadorned style in the diner. The color scheme was red, white, and pink, and there was a line of scarred old wooden booths opposite the lunch counter. A milk shake mixer provided a soundtrack and a bubbling fryolater in back added atmosphere as we pushed the screen door open, feeling as if we had stepped into a movie set. Thirty or forty blank-expressioned faces turned to study us for a long moment before returning to plates and counter. We took a booth way in the back next to the bathrooms. Coney Island specialized in hot dogs—aptly combined, given its beachfront amusement park antecedents in Brooklyn, New York—and a reportedly pungent style of chili, which we ordered, being curious to compare the Nebraskan style to the City style we happened to know something about. "We're probably the only people here at this moment who've been to Coney Island," my daughter proclaimed.

"Got that right," muttered an indifferent waitress, who seemed to go out of her way to make us feel not at home, but the food, plain as it was, exceeded expectations. After I pulled my wallet out to pay the bill, and my family trooped outside to wait for me, the waitress warmed a little and said she was having a bad day. "I've stood on these feet, in joints like this, for forty years," she explained, "and I don't think these feet will last another hour." She shrugged and turned, snapping her gum like the 1950s carhop she might have once been. I would like to have told her that she was honoring the memory of Jake Eaton, born in Grand Island, known well before mid-century as the champion gum chewer of the world and who was said to be capable of chewing 300 sticks at a time, but I kept my comment to myself as I went back outdoors into the furnace of mid-afternoon, deciding also not to tell my children. Mimi might try to better the record.

5

The Road from Red Cloud

Around the first week of September 1806, a Spanish cavalry unit numbering 300, under the command of Colonel Malgares, climbed the divide between the Republican and the Little Blue rivers after a long northward march from their outpost at Santa Fe, and approached a great Pawnee village. Nearly 2,000 Indians lived there in earth lodges, farming and hunting buffalo. The colonel presented them with many gifts and proposed opening a trade route to Santa Fe. He left the tribe with an attractive blue and gold Spanish flag after showing them how to secure it to a pole so that it would flutter in the breeze above their lodges.

Three or four weeks later an American expedition under the command of Lieutenant Zebulon M. Pike approached the village in a less imposing force of twenty-one men. He, too, was bearing gifts, his party having been out marching since July 15 when they left St. Louis. On September 29 Lieutenant Pike held a grand council with the Pawnee, telling them that their lands were now claimed by the United States; he presented them with a flag of blue, red, and white, and said it had to replace the blue and gold one. "The Pawnee nation cannot have two fathers," Pike is reputed to have said. "They must either be the children of the Spanish king or acknowledge their American father." When translated these words provoked a long, uncomfortable silence, while the Pawnees probably compared the spectacle of the 300 Spaniards with their banners and horses and wagons to the travel-worn handful of men from the east. The silence dragged on. Then, an old Pawnee stood and went over to the base of the pole where the blue and gold flag was hung. He dejectedly took it down, replacing it with the one with the stripes and stars. "This," traditional historians have said, "ended Spanish authority in Nebraska and on the plains of the Middle West."

When I tell this story I cannot resist peering into the Pawnee leader's more likely thoughts. "Now," I bet he really said to himself, "a new color scheme! We're going to have to redesign the whole place!"

As it turned out, the Pawnee days at this large village were numbered, but when their conquerers rode through in the 1830s they were the more aggressive wanderers, the Oglala Sioux, who some years before had come down from the northeast and in the Missouri Valley had easily defeated the Omaha tribe, who had no horses and only wanted to be left alone. Now the Oglala were spoiling for new hunting territory, and the Pawnee were forced to retreat after suffering numerous humiliating raids. They regrouped with other Pawnee villages up between the forks of the Platte, the grand island, and the confluence with the Loup River.

Seventeen decades later, our car thrummed down Highway 281 from Grand Island toward that divide between the Little Blue and the Republican rivers, almost to the Nebraska border with Kansas. Just past the town of Millington, a flock of escaped goats milled madly at the roadside, fleeing back through a fence break, driven by a calm, long-haired woman in jean shorts and white blouse, with a black and white terrier on a lead.

Beyond, we crossed over into Catherland, proclaimed as such by the state legislature in 1965 to encompass the western half of Webster County and the county seat of Red Cloud, Nebraska, girlhood home of the author Willa Cather.

Red Cloud was founded in early 1871 by a group of settlers headed by Captain Silas Garber, a Union Army veteran who had inspected the grassy, stream-broken terrain and filed for it the previous summer. The first county election was held in Garber's dugout. Many Nebraska pioneers lived in such a fashion, scooping out a cottage-sized hole in a hillside, facing away from the prevailing winds, constructing a front wall out of blocks of sod and then roofing over the enclosed space with a web of logs and then more sod on top. These dwellings were warm in winter and cool in summer, if not always very dry. Garber later served two terms as Nebraska governor, living in slightly better accommodations in Lincoln, the capital, and over years he built up several businesses, including the bank in Red Cloud. The county seat was named for the Oglala Sioux leader born up in the North Platte Valley in 1820 or 1821, whose people had displaced the Pawnee. Red Cloud's folk, along with the Cheyenne and Arapaho, had made this part of south-central Nebraska and north-central Kansas, where immense buffalo herds roamed, an unhealthy place for white wanderers throughout the 1860s. After several years of jostling—sporadic raids and fights begun, one after another, either by the warlike tribes or the warlike whites—the hostilities had suddenly escalated and the plains had gotten particularly dangerous in 1864 and 1865 after bluecoat Colorado volunteers, under command of a former hellfire-and-brimstone Methodist preacher named Chivington, had massacred a peaceful Cheyenne-Arapaho village of 550 at Sand Creek, in November 1864. This unleashed a wave of reprisal killings and lootings that spread like a grass fire across the Kansas and Nebraska plains up well past the Platte Valley, engulfing peaceful wayfarers on the trails and surveyors, graders, and tracklayers on the Union Pacific line.

The passage of a few short years, however, transformed the old hunting and battle grounds, as the epicenter of Indian-white struggle moved northward, pushed by the railroad, settlers, and the protective bluecoats, toward the new encroachments of the Wyoming, Dakota, and Montana trails and their traffic. By 1871 Red Cloud himself was supposedly confined to a reservation up near Fort Laramie, more than 400 miles away to the northwest. The settlers were free to name their town Red Cloud, with or without the irony. Today the town newspaper is the *Red Cloud Chief*.

Willa Cather's parents moved to the area from the Shenandoah Valley of Virginia when she was ten years old, in 1883. "My grandfather and grandmother had moved to Nebraska eight years before we left Virginia; they were among the real pioneers," she told a reporter for the *Philadelphia Record* in 1913, after she had begun to be a famous author. "But it was still wild enough and bleak enough when we got there. My grandfather's homestead was about eighteen miles from Red Cloud. . . . I shall never forget my introduction to [the country]. We drove out from Red Cloud to my grandfather's homestead one day in April. I was sitting on the hay in the bottom of a Studebaker wagon, holding on to the side of the wagon box to steady myself—the roads were mostly faint trails over the bunch grass in those days. The land was open range and there was almost no fencing. As we drove further and further out into the country, I felt a good deal as if we had come to the end of everything—it was a kind of erasure of personality."

The beautiful rolling wooded terrain of Virginia was of course deeply imprinted in her mind's eye. "I would not know how much a child's life is bound up in the woods and hills and meadows around it," she recalled, "if I had not been jerked away from all these and thrown out into a country as bare as a piece of sheet iron. I had heard my father say you had to show grit in a new country, and I would have got on pretty well during that ride if it had not been for the larks. Every now and then one flew up and sang a few splendid notes and dropped down into the grass again. That reminded me of something—I don't know what, but my one purpose in life just then was not to cry, and every time they did it, I thought I should go under."

The Cathers would try farming for a year, and the little girl showed all the signs of unbearable grief. "For the first week or two on the homestead I had that kind of contraction of the stomach which comes from homesickness," she said. "I didn't like canned things anyhow, and I made an agreement with myself that I would not eat much until I got back to Virginia and could get some fresh mutton. I think the first thing that interested me after I got to the homestead was a heavy hickory cane with a steel tip which my grandmother always carried with her when she went to the garden to kill rattlesnakes. She had killed a good many snakes with it, and that seemed to argue that life might not be so flat as it looked there."

Years after these recollections in 1913, Cather elegantly rephrased them for an *Omaha Bee* reporter during one of her trips home, in 1921, a year before her Pulitzer Prize–winning novel, *One of Ours*, was published. "This country was mostly wild pasture and as naked as the back of your hand," she remembered. "I was little and homesick and lonely and my mother was homesick and nobody paid any attention to us. So the country and I had it out together and by the end of the first autumn, that shaggy grass country had gripped me with a passion I have never been able to shake. It has been the happiness and the curse of my life."

They had very few American neighbors—they were mostly Swedes and Danes, Norwegians and Bohemians. "I liked them from the first and they made up for what I missed in the country," she had said in 1913. "I particularly liked the old women, they understood my homesickness and were kind to me. I had met 'traveled' people in Virginia and in Washington, but these old women on the farms were the first people who ever gave me the real feeling of an older world across the sea. Even when they spoke very little English, the old women somehow managed to tell me a great many stories about the old country. They talk more freely to a child than to grown people, and I always felt as if every word they said to me counted for twenty." The emigrants and the pioneers of her parents' generation had started their new lives in Nebraska in dugouts or in their slightly grander counterparts, sod houses, which free-stood with solid, two-foot-thick walls and roofs of sod, symbolically making the farmers one with the earth upon which they lived. The landscape—and the stories of the people who had survived the blizzards, droughts, and cyclones, the outbreaks of cholera and influenza and the plagues of grasshoppers— made an indelible impression on Cather. After the family had moved into town and her father had opened a farm loan and land business, the young girl took every opportunity to go back to listen to more tales and drink in the countryside. The period of her youth and young adulthood marked a transformation seen all over the Great Plains. "The Divide is now thickly populated," she would write in her breakthrough novel, *O Pioneers!* (1913).

> The rich soil yields heavy harvests; the dry, bracing climate and the smoothness of the land make labor easy for men and beasts. There are few scenes more gratifying than a spring plowing in that country, where the furrows of a single field often lie a mile in length, and the brown earth, with such a strong, clean smell, and such a power of growth and fertility in it, yields itself eagerly to the plow; rolls away from the shear, not even dimming the brightness of the metal, with a soft, deep sigh of happiness. The wheat-cutting sometimes goes on all night as well as all day, and in good seasons there are scarcely men and horses enough to do the harvesting.

The grain is so heavy that it bends toward the blade and cuts like velvet.

Cather and her playmates knew where their town had obtained its name, and as they understood the folk stories, Red Cloud had, a long time before the settlers arrived, buried a daughter on the crest of a river bluff south of town. Accounts of Red Cloud's children vary, but it seems likely that he had at least one son and two daughters born in the 1850s, and five daughters and a son who survived him (he died in 1909). I could find no references in his autobiography or several biographies that located a loss there at the Republican River, but the stories repeated by Cather's contemporaries in the 1880s were widely assumed to be true in Webster County. "Her grave had been looted for her rich furs and beadwork long before my family went West," Cather once told a reporter, "but we children used to find arrowheads there and some of the bones of her pony that had been strangled above her grave." In 1923, the site of the Pawnee village assumed to be visited by Zebulon Pike was discovered not far away. So much of a farm's acreage was concerned that the Nebraska State Historical Society bought it, and later invited in archaeologists from the Smithsonian Institution. Their excavations uncovered many Spanish and English medals and coins, including an English medal bearing the image of George III, dated 1762, and a Spanish peace medal dated 1797 and bearing the head of Carlos III. Previously the site of the interchange between Zebulon Pike and the Pawnee had been presumed to be in Kansas, but these government discoveries shifted the known site into Nebraska.

All of it was buried under new farmland in Willa Cather's day. She left Webster County in 1890 for the state university in Lincoln, but carried Red Cloud and its environs with her; as she would one day write, "Some memories are realities, and are better than anything that can ever happen to one again." She worked as a magazine editor in Pittsburgh and then in New York City, began publishing her poems and stories and a first novel, *Alexander's Bridge* (1912), before she realized she was neglecting something vital about herself in her art.

"There I was on the Atlantic coast among dear and helpful friends and surrounded by the great masters and teachers with all their tradition of learning and culture, and yet I was always being pulled back into Nebraska," she explained to an *Omaha World Herald* reporter in 1921. "Whenever I crossed the Missouri river coming into Nebraska the very smell of the soil tore me to pieces. I could not decide which was the real and which was the fake me. I almost decided to settle down on a quarter section of land and let my writing go. My deepest affection was not for the other people and the other places I had been writing about. I loved the country where I had been a kid, where they still called me 'Willie' Cather.

"I knew every farm, every tree, every field in the region around my home, and they all called out to me. My deepest feelings were rooted in this country because one's strongest emotions and one's most vivid mental pictures are acquired before one is 15. I had searched for books telling about the beauty of the country I loved, its romance, and heroism and strength and courage of its people that had been plowed into the very furrows of its soil, and I did not find them. And so I wrote *O Pioneers!*"

With that novel on the front seat beside me, we pulled into town on brick-cobbled streets past a sign marking the Red Cloud population as 1,204, down about a hundred in the past ten or twelve years. We passed Catherland Auto Sales and drove into the pristine center of town—two-story, wide-awninged brickfronts from the turn of the century, most of them brightly and decoratively painted, some with arched windows and flaring, bracketed cornices, and big, clean plate glass display windows down on street level, looking out across wide sidewalks and old-fashioned street lamps. It was one of the most attractive townscapes we passed through. In Cather's works it was variously called Sweet Water, Hanover, Frankfort, Moonstone, Haverford, Black Hawk, and Skyline. We went by the Auld Public Library and the Simple Grace Cafe, and paused to look up at the sprawling Red Cloud Opera House, built in 1885, which in Cather's young years hosted many traveling theater companies. She acted in locally produced dramas there, including a benefit performance of *Beauty and the Beast* for victims of the 1888 blizzard. It's said that she and her fellow actors signed their names on a backstage wall. The Opera House was used for school programs, and it was there that she spoke many times. One newspaper account appeared when she was eleven years old: "The recitation by Miss Willa Cather was particularly noticeable on account of its delivery, which showed the little miss to be the possessor of extraordinary self-control and talent."

It would have been a short walk home, for just around the corner at Third Avenue and Cedar Street is Cather's girlhood home, which she once described in her 1915 novel, *The Song of the Lark*: "They turned into another street and saw before them lighted windows, a low story-and-a-half house, with a wing built on at the right and a kitchen addition at the back, everything a little on the slant roofs, windows, and doors." As if following these directions, we pulled around in front of it, finding the house exactly as advertised: a little beige clapboard cottage behind a low white picket fence, with red gardenias planted along the brick sidewalk. As a state historic landmark, the house is furnished much as it was around 1890—the Victorian parlor with its oriental rug, heavy upholstered furniture and lace antimacassars; mirrors, faded prints, and photographs on the wall. The plank knee-walled attic bedrooms were up under the eaves beneath exposed beams, one dominated by a central brick chimney slanting toward the roof, the other—Willa's—done in the original rose-colored wallpaper she paid for herself working downtown in the drugstore. Her bed quilt was of a matching hue, and she had a stack of books and an oil lamp on a trunk, a washstand with pale blue pitcher, and a seashell collection on low shelves. The bedroom found its way into literature. "From the time when she moved up into the wing," Cather wrote in *The Song of the Lark*, "Thea began a double life. During the day when the hours were full of tasks, she was one of the Kronborg children. But at night she was a different person. On Fridays and Saturdays she always read for a long while when she was in bed. She had no clock and there was no one to nag her."

It's this passage, I believe, among many, that spoke to my mother as she grew through her teens in the 1930s in her parents' Kansas City house. She, too, had a room of her own, and books were also her reward for the day, her retreat, her solace, her means to dream about her future. Her father would drive her over to the public library in the evenings to get books—her brother, Charles, five years older and also bookish, had done similarly while he was still at home, but he had gone off to college to study literature when she was thirteen. Serially, she pored through Willa Cather's novels, from the heroic portrayal of the resourceful Nebraska farm girl Alexandra Bergson in *O Pioneers!*; to the story of Thea Kronborg in *The Song of the Lark*, whose obsession with music seemed to mirror my mother's with music and drama; to the perilous life and survival tale of Antonia Shimerda, the Bohemian farm-girl-turned-housemaid of *My Antonia* (1918). The rural settings and immigrant characters were familiar to my mother, there in Kansas City—she *knew* them, or people like them, and Cather's heroines were the kind of women she wanted to be. And then there was the story of Marian Forrester in *A Lost Lady* (1923), whose grace, charm, and cultural sophistication were offset by the coarseness of her small-town neighbors. There were echoes of this view of Midwestern and Western communities in the boom years after the Great War in my mother's other favorite author of the time, Sinclair Lewis. Once, when she was walking between classes in her Catholic high school, her English teacher, a nun by the name of Sister Helena, saw that my mother was carrying a copy of Lewis's novel *Main Street* (1920) with its mordant portrait of smug and intolerant Gopher Prairie, Minnesota, and its smart but smothered heroine, Carol Kennicott, who leaves husband and home for adventure and stimulation elsewhere. Sister Helena was aghast when she saw that her innocent young student was reading one of those, those, those modernists, those *debunkers.* "Rosemary, I have to say that I am shocked!" she blurted out. "Sinclair Lewis is so, so controversial—hardly the thing a young girl should be reading!" Cather's women were hardly better.

By the time I was in high school, of course, Sinclair Lewis and Willa Cather were firmly in the curriculum, which gave my mother no end of amusement.

There are many other restored structures and marked sites within a few blocks of Cather's girlhood home, all with counterparts in her fiction, such as the Miner house (the Harling home in *My Antonia*), where Cather's friend Annie Pavelka worked as a maid and whose grim, trouble-clogged life was mirrored as Antonia; the Methodist church (it now serves as a Masonic Lodge) described in *My Antonia*; the Webster County Court House, a setting in *One of Ours*; and the Farmers' and Merchants' Bank building, erected in 1889 by town founder Silas Garber, who with his wife served as the models of Daniel and Marian Forrester in *A Lost Lady*. This last building, of solid brick with sills, quoins, and corner blocks of rusticated, quarried granite, was restored by the Willa Cather Pioneer Memorial and Educational Foundation in the late 1950s and now houses the Willa Cather State Historic Site and is owned by the Nebraska State Historical Society. Inside on this late Sunday morning we found a large open room with display cases and tables all around the periphery, stacked with Cather novels and stories in many editions, biographies, memorabilia, videotapes, and souvenirs. Many photographs of nineteenth- and early twentieth-century Red Cloud, paintings with local themes, and portraits, hung on the walls. That day the memorial was overseen by Katie Cardinal, a friendly and talkative transplant from Wisconsin. She had come down to the Divide as a scientist, looked around and liked what she saw, and decided to stay. "I came down to Red Cloud looking for rocks," she explained, "but the place took hold of me and it was impossible to get loose." I told her she was sounding a lot like Willa Cather. "Figures," she replied. "I hope you folks didn't wait long for me to open up. I got delayed coming in—I found an injured jackrabbit out on the road, obviously dying. I didn't know what to do about it—couldn't bring myself to put it out of its pain. I ended up taking it away from the pavement and putting it in the shade of a low-growing bush, so it could die in privacy. Do you think I did the right thing?" Mary and I shrugged. "It probably didn't live much longer," Mary said. "What else could you do?" Katie shrugged back and then was ready to change the subject. She pointed at a woman's pressed, white shirt hung in display. Behind it was a large, framed photograph of Cather, the most often published one. As usual, her eyes drew you in and put you at ease in a moment. And Cather was wearing the shirt. "She was only a size seven," Katie said, disbelief spilling from her voice. "But she had such *presence*. You look at her pictures and she seemed so much larger."

"She was." It was a new voice, and we turned around. A soft-spoken man with reverential eyes had entered, clearly moved by this place and the presence he felt in Red Cloud. He asked for the five-dollar guided tour of town. Katie explained that it took one and a half hours. "I've got the whole afternoon free," he said, quietly but fervently. "I came a long way, and I'm here to see Red Cloud." As Katie gave him street directions to where a docent was waiting to lead the next tour, I saw that my daughter had scooped up some books and added them to the pile I had already accumulated—Mimi had also picked up some kind of emotion here, whether in the light reflected off the redbrick Main Street facades, or in the little cottage two blocks away in which a girl she'd never heard of before took pains to decorate her room and repaired to it every evening as soon as she could get herself excused from family company, to read and read and read and begin to write. It mirrored her, just as it had mirrored her grandmother. For a week afterward the book she read in the middle car seat row, oblivious to the dry western Nebraska farms and plains fanning past us, was *My Antonia. O Pioneers!* was not long following it.

Later, out on the street, we stood looking up at the Red Cloud Opera House; in early 2001, the Cather Pioneer Memorial would be awarded a $275,000 challenge grant from the National Endowment for the Humanities to restore it. The foundation had raised $1.25 million to renovate the building, with funds coming from thirty-nine states and Japan. When finished, the Opera House will have an auditorium seating 200 for lectures, concerts, and dramas, as well as foundation offices, a visitors center, bookstore, art gallery, and archives.

A few minutes later, peering in the windows of the *Red Cloud Chief* and beginning to think about lunch, wondering if the absence of people on the main street meant that everyone was in church, I could at least populate Red Cloud in my mind the way Cather had in *Lucy Gayheart* (1935), talking about small-town life and how in isolated communities you saw the same faces everywhere, and sometimes that got to feeling oppressive, even dangerously so. "On the sidewalks along which everybody comes and goes," she wrote, "you must, if you walk abroad at all, at some time pass within a few inches of the man who cheated and betrayed you, or the woman you desire more than anything else in the world. Her skirt brushes against you. You say good morning and go on. It is a close shave."

"Let's eat," said my wife, showing perfect timing. Aside from the Cather memorial, the only sign of life downtown was something called Sugar & Spice, a seasonal counter service restaurant and takeout of decidely homemade character at its most whimsical. Inside, after ordering chicken tenders, fries, coleslaw, and salad from a teenaged attendant, Mary and I shepherded Mimi and David into the crowded eating area. People fresh from church were seated at sticky varnished trestle tables and benches—women in flowered dresses, men in long-sleeved shirts and tooled belts—and informal, friendly conversation drifted back and forth from table to table.

"How's Mom today?" an older man with a straw hat twisted around to ask a woman with a baseball cap sitting with another group at the next table. "Oh," she replied, her voice trailing off. "Not eating much." On the other side of us, a hearty man was booming out across the space of two tables to reach a thin fellow in brown suit pants, a white shirt and tie, all of which seemed two sizes too large for him: "Hey, Leo, still make a little muscatel there?"

"Nope." No expression.

"That's too bad, we could use some."

A woman nearby pulled out a cell phone. As everyone ate around her, she had the same kind of conversation one usually hears on Wall Street or Madison Avenue. "What time you comin' home? I'm comin' home at two. Yeah. Uh huh. Yeah. Okay, I'll call just before I leave."

Five miles south of Red Cloud is the Willa Cather Memorial Prairie, 610 acres of native grassland protected by the Nature Conservancy. We pulled off the road and got out, negotiated a wooden cattle guard in the fence, and walked out under a bright sun for nearly an hour as far away from the car as we could get: short grass; dry streambed with a cattle path along the length of it; the sound of wind and occasional songbirds—larks, of course, and bobwhites; crunch of dry grass underfoot; occasional rustle of grasshoppers. A hawk whistled far overhead. We couldn't see it, but there was a turkey buzzard drifting along the currents, following a dry wash toward a grove of cottonwoods.

The dried cow patties and paths reminded Mimi of our farm. The year before Mimi was born, Mary and I had traded the urban life of a five-room apartment with garden in Brooklyn for a more satisfying life on a 150-year-old sheep farm in the Champlain Valley of Vermont. We had saved and restored an equally old farmhouse and populated empty barns with a starter flock of sheep (ten ewes, a ram, and a surprise dividend lamb born a few days after the sale was completed, like an extra Cracker Jack prize). I learned how to deliver lambs, dock tails, give shots, and tag ears. To this little menagerie we added a Morgan horse from the Humane Society (he had been seized for neglect and we adopted and helped nurse him back to health), a cantankerous Shetland pony (that may be a redundancy), and chickens. It was an idyllic existence for about six years, and Mimi had grown up with several generations of lambs, knowing them all by name, delighting in being chased and nudged by a halfhearted ram across the barnyard, making forts where hay bales had been piled up past the rafters of our pole barn, going out with us into the pasture—twenty acres consisting of a hill, some elms, a seasonal brook, and, above the slope, a lot of thorny wild pear trees. She may have taken her first steps at ten months surrounded by doting friends at the Bread Loaf Writers' Conference, but Mimi learned how to skip and run out on the sheep paths, dodging thistle plants, kicking aside sheep pellets, and leaping over old cow flop. On the top of our pasture, called Cream Hill on the maps, we stood in a strong autumn wind and watched hundreds and hundreds of Canadian geese making for the south in succeeding waves, flying past as if in massive V-shaped commuter trains: ah, here's the 5:07! Now the 5:14! And now here comes the 5:22! Sometimes, looking up at the honkers, we'd see some catching the brilliant defining last rays of the day and they would be so bright we'd know they were snow geese, not Canadians, and that was always a breath-taking gift.

Mimi never quite forgave us when we decided to sell that farm in Shoreham. The year her brother was born, coyotes moved in on our sheep, and despite the seven-strand electric fencing and super-hot pulse charger hanging from a nail in the barn, it was like they had tacked up a sign on a fence-post: BLUE PLATE SPECIAL TODAY — LAMBCHOPS! To this trouble was added the fact that an isolated pack of neighbors had begun to make life unpleasant for everyone else up and down our dirt road, and on other roads besides. We worried for our children's safety, sold the farm for what we'd paid to buy and improve it, and moved down to the lovely Victorian-era village of Orwell, seven miles away. One was a hard place to leave, but the other was a wonderful place to leave for. The only problem was, our four-year-old girl missed her first home, the open space, the animals, and nature.

I stood there out on Willa Cather's prairie, my arm around my growing-up daughter, who'd just said, "This reminds me of Shoreham, remember that?" We warmed to some thoughts about that other home, and I drifted to thinking about the Cather girl they came to call Willie and her ties to this grassy divide. That led me to thinking about my mother when she was a girl in Kansas City, and that led me to my grandmother, Rose Donahue Haward. All of the energy she put into the Circle H Ranch, so-called, and her flock of chickens (when company came she'd just excuse herself and go out back and wring a chicken's neck for dinner), and her get-rich schemes like the goats and the turkeys, the latter of which all managed to fly up into trees and my mother and grandfather had to shinny up and bring the awful screaming birds back down again because Rose was worried they'd get chilled up there after night-fall. And somehow that memory led me back to Brooklyn, when Mary and I chopped up the concrete backyard in order to plant a vegetable garden and did so well with tomatoes and eggplant that our old Italian neighbors, all retired from the city Sanitation Department, said that we must be Italian since no one but Italians could grow eggplants that well! That led me to Mary's own family, the Duffys, and how important vegetable gardening had been to her father and grandmother, a registered nurse, who had run a little old ladies rest home on Long Island with a sprawling organic vegetable garden out back that was so big that Mary's father tilled it with a three-quarter-sized

'58 Farmall Cub tractor, which now stood in our garage back in Vermont, needing me to finish work on the carburetor.

All of us answer some sort of call, and it was the same call that Cather wrestled with, as did her characters in her novels and stories. "There are only two or three human stories," she had written in *O Pioneers!*, "and they go on repeating themselves as fiercely as if they had never happened before, like the larks in this country that have been singing the same five notes over for thousands of years." I gave Mimi a squeeze, and Mary and Davey joined in.

A herd of cows congregated a mile away below a windmill.

Shadows of clouds raced across the prairie.

Wagon ruts ran out along a ridge.

In August 1864 George Martin, a professional jockey from England who had walked the eastern part of the Oregon Trail and decided to plant stakes for a farm in Hall County, Nebraska, was out making hay in the Platte River bottom, near the present-day town of Doniphan, with his two sons, Nat and Robert. Indians appeared, wounding the father and shooting arrows at the boys.

It was a bad time to be caught out in the open, during that season of escalating tensions, just three months before the Sand Creek Massacre and the hell it unleashed across the entire Great Plains. Four months before George Martin was attacked, along with scores of other isolated civilians, Colorado territorial governor John Evans had begun a campaign to bolster Colorado's case for statehood in Washington (and Evans's chance of promotion into the U.S. Senate). Federal authorities were understandably preoccupied with the terrible toll of the Civil War and were planning to shift troops from Colorado posts to Kansas to counteract Confederate threats there. Evans needed a crisis to bolster his cause and keep the soldiers where they were. The preceding winter had been a bad one for the native tribes, with Cheyenne and Arapaho suffering greatly from malnutrition and disease. Earlier in that year of 1864 they had begun to rustle cattle and loot white settlements and freight shipments in Colorado and Kansas for food, mostly not molesting the whites. Evans, however, needed his crisis, and ludicrously inflated reports of these clashes before forwarding them to Washington. In April he unleashed the Colorado militia under Colonel Chivington. The orders were to "kill Cheyennes wherever and whenever found." At a Colorado feeder stream of the South Platte, twenty-six Cheyenne were killed, their village burned, and cattle captured and distributed among the militia. Southeast, in Kansas, even conciliators were eliminated like Chief Lean Bear, who, after approaching troops under a white flag, was blown off his saddle; on his body was a peace medal given him a year before by President Lincoln. Another conciliator, Chief Black Kettle, sought a parley with Colonel Chivington but was rebuffed. When Black Kettle's emissary, the white trader William Bent, warned the officer that many white settlers and emigrants would perish if the conflict widened, Chivington replied, "The settlers will have to protect themselves." Some 200 whites, mostly civilians, would die that summer, as Cheyenne, Arapaho, and Sioux warriors picked off stagecoaches and wagon trains and raided coach stations, settlements, and ranches in Colorado, Kansas, and Nebraska, killing, scalping, mutilating, and then vanishing before troops could respond. By now, of course, the U.S. Cavalry was heavily involved. Black Kettle and other peacemakers made further attempts to stop the killing but they went nowhere; the chief would be among those massacred at Sand Creek by Chivington's volunteer regiment in November 1864. He was shot standing below two flags he had raised above his lodge: the white flag of peace and the Stars and Stripes.

When the English farmer George Martin was wounded with a Cheyenne arrow while cutting hay near the Platte in August, he shouted for his boys to hightail it and began to drive off the Indians with his repeating rifle. Nat and Robert turned their horses loose, leaped onto a pony and took off, with the raiders in hot pursuit. The boys were shot from behind and pinned together by a single arrow, not a surprising feat since an Indian could sometimes drive an arrow entirely through a buffalo. George Martin supposed them dead, made his way back to the homestead and gathered the rest of his family. They all fled on horseback to the safety of Fort Kearney. Meanwhile, the boys' pony returned to the barn, where Nat and Robert fainted and fell off into the straw. They lay there inside the barn until they were found alive by their father the next day, and lived to tell two succeeding generations about it.

Some miles away and 136 years later, I stood in Hastings, Nebraska, about midway between Red Cloud to the south and Grand Island to the north, looking into a glass display case at a two-foot-long arrow with a filed metal point—the projectile that had pinned the Martin boys together for a day and put them into the history books for the dramatic little part they had played in a large, bloody drama. I was in the basement of an extraordinary community museum, the largest between Chicago and Denver, the Hastings Museum, which was born about 1888 in a schoolyard swap between two other little boys. One of them, Albert Brooking, had recently been dispatched from his parents' home in rural Nebraska for schooling in his grandparents' town in Illinois. One of his classmates had an old Indian spear point that Albert could not live without, so he negotiated a trade of six pennies and two unused postage stamps to acquire it. One artifact led to another, and another, many unearthed at an old Indian village site nearby. At first his collection was housed in a cigar box. When he rejoined his parents in Nebraska in 1893,

Albert Brooking turned his interest to local birds and taxidermy, which he learned through trial and error, assisted along the way by instruction books. In a short burst of time his collections of mounted birds and Native American relics filled a whole room in his parents' house.

Over succeeding years, Brooking led a peripatetic life. Through his teens various enterprises took his father back and forth from Nebraska to Colorado. The young man went to school in Chicago, after which he took traveling work—first with the Pullman Company, then as a railroad conductor. He collected relics, mammals, and fossils until his marriage in 1903, after which he and his wife finally settled for good in central Nebraska and Brooking operated a store and grain elevator. His prizes filled the couple's house. Interior photographs of their domicile suggest something between the endless joys of discovery and the mania of unbridled acquisition—shelves crowded with stuffed birds; every other square inch of wall space bristling with mounted spears, war clubs, arrows, tomahawks, beadwork, rifles, pistols, and derringers; the floors an impenetrable forest of marble busts on pedestals, giant horned steer skulls, and musical instruments. It was a nightmare that would keep a feather duster in constant use, and an impossibility to navigate.

Whether it was to restore domestic harmony or to keep his house from collapsing of its own weight, Brooking persuaded the people of Hastings, including the administration of Hastings College, to take the collection off his hands and permanently display it. Moved to the basement of the college library, it continued to grow as Brooking stepped up his collecting, and by the early 1920s it was clear that a separate building (or buildings) was required. The Hastings Museum Association was incorporated in 1926, and displays moved over subsequent years into increasingly larger vacant schools in town, substantially enhanced by other collectors' treasures. A man named A. T. Hill, for instance, contributed hundreds of firearms and thousands of Indian artifacts; another named Adam Breede donated mounted wild animals from North and South America and South Africa. By the 1930s, so many new exhibits had appeared, and so many thousands of visitors had paid to see them, that the town voted to build a new museum. Hastings contributed $38,000, and the Federal Works Progress Administration gave $28,000. Fittingly, the basement in the center of a town park was excavated by mulepower. The Hastings Museum opened in June 1939, three stories full. The basement displayed guns, Indian relics, and dozens of carriages, wagons, and vintage automobiles; the main floor and mezzanine had habitat scenes of mounted mammals and birds as well as displays portraying life and work on the plains. For decades it was widely known and publicized as "The House of Yesterday," those words emblazoned in huge neon letters on its western facade. In 1958 the museum added a planetarium, which over the next three decades kept pace with changes in technology and theatricality. New construction in the early 1970s doubled the museum's size. There was a three-story addition for permanent storage and workrooms. Then in the 1990s the building sprouted an Imax theater where large throngs of thrill seekers could scream at huge menacing images of 3-D dinosaurs.

I have no quarrel with the delicious, not-so-cheap thrills of an Imax, especially when it helps anchor such a glorious anachronism as this vast community museum. Perhaps the tens of thousands of Imax patrons drawn from all over the Plains states, funneled off the interstates from their high-speed vacation adventures, might drift from the refrigerated theater into the humbler exhibit galleries to study the unruffled feathers of a sage grouse or wild turkey, or the glassy, chilly stare of a timber wolf, bighorn sheep, or prairie dog; to plumb the prickly variations of a hundred different samples of barbed wire; to get into the vantage point of an emigrant in a prairie schooner or a speedster in a Stanley Steamer; to peer into large scale models of Pawnee earth lodges juxtaposed with sod huts; to actually read the carefully printed little caption tags and larger explanatory text placards; to meditate on the heartache, and suffering, and courage, and sheer recklessness of those who preceded us here on this old earth and took the steps necessary to convey the race to its bewildering contrasts of primitivism and modernity, of privation and plenty, of the endless acquisition of things that will not last, of ideas that will not survive, of sensations that will not bring sense.

At the Hastings Museum, comfortable padded chairs and couches are positioned throughout the three floors and they are seldom unoccupied, for the effect of an afternoon or a day there is sheer delightful exhaustion for the eye, the mind, and the lower extremities. The top floor, housing the newest and most ambitious exhibit space, may command the most time and energy. Called "People of the Plains," it traces habitation, culture, and technology from the Paleolithic to the modern ages. Moreover, beyond the one-fifth-scale cutaway earth lodge, the tipi, and the sod hut, the farmhouse interior and the walk-through general store, the storyboard-like exhibit cases and dioramas on flora and fauna, geography and geology, on hunting, gathering, farming, irrigating, and ranching, on cowboys, soldiers, and shamans, on transportation for emigrants on the Platte River Valley route and for sightseers on the Lincoln Highway, there is a sudden startling moment of recognition of a familiar cultural icon and the surprising, proudly related fact that it was born nowhere else but in Hastings, Nebraska.

Kool-Aid!

Years of Saturday afternoon sidewalk entrepreneurship fly back into consciousness, of my sister Terry and I plundering sugar bags, mixing formula, marking signs, and shouting out "Kool-Aid! Kool-Aid! Five cents a glass!" at cars and pedestrians alike in suburban New Jersey or Long Island towns, our younger siblings, Chris and Lisa, sent out to shill for customers and drinking up much of our profits, these fruity sugar highs and cascades of nickels all thanks to one Edwin E. Perkins, who was born in 1889 and grew up on a Nebraska farm. From his teens, Perkins was a tinkerer and an amateur chemist who sent away for instruction books on mixing and merchandising. His first breakthrough was a remedy for tobacco addiction, introduced in 1918 and marketed as Nix-O-Tine. A multistep program, it offered herbs to be chewed, large herbal tablets to be swallowed, a quick-acting herbal laxative, and a horrible-tasting mouthwash whose active ingredient was silver nitrate solution. Apparently no customers were killed from this patent remedy, and Perkins and his new bride, the daughter of a local doctor, settled in Hastings and opened the Perkins Chemical Company with a couple of employees.

Perkins's offerings ran the gamut, all across the 1920s. To Nix-O-Tine he added Motor-Vigor, a gasoline additive, and Ironux, a health tonic, and E-Z Wash, a laundry detergent, and Glos-Comb, a hair pomade, and Jel-Aid, a homemade jelly concentrate, and Onor-Maid, a galaxy of medicines, salves, face creams, lotions, soaps, and toilet waters. He could have been called "The Great Hyphenator." The Perkinses were soon employing most of their extended families and a good number of Hastings citizens, and Edwin Perkins created a national force of traveling salesmen and district managers, and came up with ingenious incentives to cement their loyalty and fire their enthusiasm. One of his most popular items out of more than 100 was a concentrated liquid fruit drink—just add water!—called Fruit Smack. But the little four-ounce glass bottles tended to break during shipping, or their tiny corks would pop out. So he began to tinker with ways of transforming his "six delicious flavors" to a powdered, easily soluble concentrate that could be packed into an envelope. In 1927, he perfected Kool-Aid.

He missed the summer season, though, needing to iron out details like the Kool-Aid envelope. An experiment with laminating paper with asphaltum failed when the tarry black substance migrated into the fruit powder. Plain waxed paper could not be glued shut. So he went to a laminate of paper and wax, which worked fine and took a brightly colored lithography ink. Over the winter of 1927–28 Perkins pushed toward a big launching through food wholesalers, borrowing money from acquaintances and mortgaging his factory building for $10,000. His little ten-cent, one-ounce envelopes were packed in colorful cardboard displays holding 200, which he called "Self-Selling Silent Salesmen," and were meant to be placed on front sales counters of grocery stores, where impulse buyers (or their children) would snap them up. "The product sold and repeated and they paid us for it," Perkins recalled many years later. "And we hardly slept the rest of the summer. . . . We were swamped with orders."

Even more than before, working for the Perkins family as packers and processors became a mainstay for the young people of Hastings, not to forget

housewives, and the men employed in the warehouse and shipping department. The packing process was rudimentary: seated workers would dip ice
cream scoops into small barrels of Kool-Aid powder, pour about an ounce
into funnels, extract the envelopes, seal them, and then pound them flat with
wooden mallets. The packing boxes fairly sailed out of Hastings, with no discernible dip when the Depression hit. By 1931, however, the company had
outgrown its home. Perkins relocated to Chicago, where automated assembly
lines and hundreds of round-the-clock workers awaited. Kool-Aid prospered,
even after Perkins cut prices in the deep trough of the Depression to five cents
per pack. But he never forgot where he had gotten his start. By the time he
was given the chance to sell out to General Foods, in 1953, he was ready to
retire, and his philanthropy was profound. There are libraries, hospital wings,
and elder care pavilions all over Hastings that carry the Perkins name. Edwin
Perkins's estate was estimated at his death in 1961 to be $45 million. And
both he and his wife, who followed him in 1977, chose to be buried in Hastings. Now the Hastings Museum remembers him with an annual "Kool-Aid
Days" festival, which takes over the town for a long weekend every August.
And we millions of former independent sidewalk vendors, even if we can't at-
tend, should pause in our more serious adult labors and salute that restless
and indomitable mind.

We reeled out of the Hastings Museum, awash in imagery—a dizzying
procession of whooping and sandhill cranes, mollusks and giant clams, Cam-
brian fossils and mastodon teeth, agates and fluorescent minerals, six-shooters
and Sharpe's rifles, polar bears and dung beetles, buffalo nickels and Indian
head pennies; a grasshopper plow, a Brule Sioux woman's beaded gown, a
wooden bicycle, a pie crimper made of walrus tusk, a Jubilee gasoline-powered
clothes iron; campaign buttons from John Adams to Bill Clinton; hatpins,
teddy bears, music boxes, jack-in-the-boxes, gewgaws, what-nots, and bric-a-
brac. And the metal-tipped arrow that pinned Nat and Robert Martin together for one long day. And Albert Brooking. When the Hastings Museum
founder and fanatical, lifelong acquirer, shuffled off this mortal coil, in 1946,
he was interred in the lower level of his museum—the ultimate collectible.

That evening, at a motel in Hastings, we pulled chairs outside to watch nearly
constant cloud-to-cloud lightning sheets rippling as if blown on wind-tossed
clotheslines on the eastern horizon, so far away that we could detect hardly a
rumble. In the morning we rose early and headed north to rejoin the Lincoln
Highway in Grand Island, passing through Wood River—a pretty, newly
painted, early-nineteenth-century commercial block, and neat little houses
and a towering grain silo, and then cornfields. "From a Proud Past to a
Promising Future," read the inevitable motto sign. Then there was Gibbon:

"Smile City," which, I reflected, made a certain amount of sense, as it was the birthplace of television talk show host Dick Cavett. And then there was Kearney: "Can Do Country"—Union Pacific trains passing grain elevators and a fairground, mini-malls and franchises interspersed with relic motels, bungalows, and canopied stucco gas stations, most of the latter hung with cursive old neon displays, now dark.

Kearney was named for the now vanished adobe and sodhouse army fort that stood there above the south riverbank between 1848 and 1871, both the fort and the later-established town a misspelling of their namesake, General Stephen Watts Kearny. The misspelling, says the WPA guide primly, "became statutory." "Fort Kearney was never attacked by Indians," notes the Platte route scholar Merrill Mattes, "never besieged, and no battle of size was fought within a radius of 100 miles. Its garrison seemed pitifully small in contrast to the emigrant hordes and to the savage Sioux and Pawnee bands who prowled up and down the Platte on each other's trail. Nevertheless, the Fort Kearney soldiers proved equal to their role as guardians of the trail, made innumerable patrols, kept Indians pacified, and provided emigrants with supplies, advice, and the moral support so desperately needed on their 2,000-mile pilgrimage." Fort Kearney marked the true beginning of the Great Platte River Road— there the various trail strands converged and were woven into one. "Fort Kearney was recognized as the port of call of the Nebraska Coast," says Mattes, "the end of the shakedown cruise across the prairie and the beginning of the voyage across the perilous ocean of the Great Plains, a place to pause and reflect, to recuperate, to reorganize, to get your bearings. For the fainthearted it was a good opportunity to change their minds, make a 180-degree turn, and go back where they came from before they became committed to California and later, somewhere out in the Great American Desert, reached the point of no return."

Especially in the early days of the fort, the passing wagon trains on the route were nearly constant, developing into a veritable boulevard with day-long traffic jams at the height of the California Gold Rush in spring and early summer 1849. "Yesterday 180 wagons passed here making in all 656," wrote a correspondent for the *St. Louis Republican* from Fort Kearney on May 19. "A cart load of letters started for the frontier this morning, and I presume many mothers, wives and sweethearts will soon be made happy." One hundred eighty wagons might have carried 1,000 men. On May 21 the writer counted 214 wagons, taking the total vehicle number up to 1,203, "not including the military train of 50 wagons, the advance guard of the Rifles, under Major Simonson, destined for Bear River." One forty-niner, John Milner of Georgia, wrote to his sister that week from Kearney, reporting "there are thousands of men going along the road in fact it looks like the wagons hauling cotton to Macon just after a rise in the staple. I believe there are wagons stretched in sight of one another for 500 miles." In Merrill Mattes's judgment, the greatest number of emigrants in Fort Kearney's history passed by on May 24, 1849: "a tidal wave of 500 to 600 wagons and 2,000 to 2,500 emigrants."

Some of the throngs who had followed the Platte's north bank along the Council Bluffs and Omaha route inevitably saw the fort on the other side of the river, and the dusty wagon trains rumbling by, and crossed over there for the mail, supplies, and companionship. The Platte could be two or three miles wide at Kearney, depending on conditions, and in as many as ten channels broken by scrubby little islands. Few enjoyed the crossing. "This is our experience crossing Platte River," complained Randall Hewett, "the meanest of rivers—broad, shallow, fishless, snakeful, quicksand bars and muddy water—the stage rumbles over the bottom like on a bed of rock; yet haste must be made to effect a crossing, else you disappear beneath its turbid waters, and your doom is certain."

Randall Hewett had been traveling by stagecoach, regular service of which having been inaugurated in the 1850s. In the summer of 1850 monthly mail service was begun between Independence, Missouri, and Salt Lake City, Utah, and it went to a weekly basis in 1858. The stage company built stations at intervals of ten or twelve miles, and it was at a place a mile and a half west of the fort that Kearney station was erected—a stable, barn, eatery, and office. The short-lived Pony Express of Russell, Majors & Waddell ran through Kearney in 1860–61. In that latter year the Pacific Telegraph Company line came through. In 1862 stagecoaches of Ben Holladay's company running between Atchison, Kansas, and Denver and Salt Lake City were stopping two or three times weekly, and by 1865 it became daily service. The Western Stage Company made runs westward out of Omaha, crossing the river above Fort Kearney.

West of Fort Kearney also marked the beginning of the habitat of the buffalo, at least back in the early emigration days when they were spied in numbers so vast as to defy belief. "We saw them in frightful droves, as far as the eye could reach," wrote John Wyeth in 1832, "appearing at a distance as if the ground itself was moving like the sea. Such large armies of them have no fear of man. They will travel over him and make nothing of him." Just before the Gold Rush, John Pulsipher recorded his observations. "Buffalo abounds along the Platte River in such vast numbers that it is impossible for mortal man to number them. . . . Sometimes our way seemed entirely blockaded with them but as we approached they would open to the right & left so we could pass thro. Thousands of them sometimes would run towards the river, plunge down the bank into the water, tumbel over each pile up, but all would come out right on the other side of the River & continue the race. Sometimes we would see the Plain black with them for ten miles in width."

With such number of wagon trains crossing through, though, the great herds were playing out their last act. Little forage was left along the Platte Valley by the thousands of ox teams and horses, and emigrants reveled at the blood sport of seeing how many of the wild creatures they could kill, letting wounded buffalo stumble away to a long, agonized death, leaving the animals they did manage to kill outright to rot on the plain. Buzzards and coyotes could hardly keep up with them. One traveler, Dr. C. M. Clark, condemned "the wholesale murder of these noble creatures merely for pastime." A decade later, the army of 10,000 Union Pacific graders and tracklayers passed by, bringing civilization in on iron wheels; when the railroad came through in 1866, passengers in the Pullman coaches would blaze away with pistols and rifles from the train windows whenever herds of buffalo or antelope grazed too close to the right-of-way.

Those awe-inspiring numbers thus were only a dim, nostalgic memory after the short heyday of the stagecoach was succeeded by the era of the Iron Horse, when stations, villages, and towns sprung up every ten or twelve miles at trackside. General William Tecumseh Sherman, commanding the army on the Great Plains, reported in 1866 that Fort Kearney was no longer of any military use. The buildings, he said, "are fast rotting down, and two of the largest were in such danger of tumbling that General Wessells had to pull them down. I will probably use it to shelter some horses this winter, and next year let it go to the prairie dogs." In 1871, when the Burlington & Missouri Railroad built over to a junction with the Union Pacific main line at Kearney station, on the north bank, the town was platted, and Fort Kearney was officially abandoned.

Inhabitants of the town of Kearney long entertained high hopes for their future. Unsuccessfully, they launched campaigns to have it named the capital of Nebraska because of its central location, and sometime in the boom of the 1880s or 1890s, they sponsored a convention in St. Louis to promote the shifting of the national capital from Washington to Kearney, presumably using the same rationale of its central location. That having failed, the citizens of Kearney had to content themselves with being the seat of Buffalo County, and with its historical legacy as a way station during the Great Migration, which is something considerably worthy of pride.

Twenty-five days before we rolled into the outskirts of Kearney on Route 30 and hung a left to pass through the center of town, some great excitement had occurred down at the Interstate 80 roadside. A new tourist attraction called the Great Platte River Road Archway Monument had opened its doors on June 9, 2000. We had seen announcements in Omaha and Grand Island, but the glossy red and sunburst orange giveaway cards for "America's Great

Trailblazing Adventure!" in no way prepared us for the behemoth structure that arched its way over six lanes of speeding interstate traffic. Standing eight stories tall, it was anchored on the north and south sides by large log buildings fashioned to look like palisaded forts, between which soared a breathtaking, rust-colored trusslike span, solid-walled but in three-dimensional detail seeming like a cantilevered bridge. Atop the two log buildings were bright, modernistic, heroic aluminum sculptures of winged horses crowned with sunbursts. We had to drive around for a while before we found a spot, but finally we parked and joined a large stream of vacationers heading in to the log-walled building.

Inside was an extraordinary sight. I was vaguely aware of an immense souvenir shop stretching to my right; to my left was some kind of beef and fried potato fast food franchise. But what dominated that vast two-story log interior, meant to represent Fort Kearney in its heyday in 1848, was a giant movie screen up near the ceiling, approached by a slow-moving escalator that vanished into the screen itself, which depicted a line of covered wagons rattling across the dusty plain toward some distant outcroppings. It looked like a Frederic Remington painting come alive. Climbing and beckoning life-size figures on the steep inclined planes on either side of the escalator, frozen in their action, seemed to be exhorting us to join the train. One of the figures was a Plains Indian; another looked like a mountain man in buckskin. Almost near the top of the rise, another figure held his battered hat aloft.

It was mesmerizing. I had been prepared to be skeptical but it was impossible to ignore the call. We paid admission to people stationed in pioneer costumes, were handed wireless headphones, and stepped onto the moving stairs, drawn up toward the prairie schooners and into some kind of historical dream.

The Great Platte River Road Archway had been born on May 2, 1995, when former governor Frank B. Morrison Sr. asked an Omaha marketing executive named J. Gregg Smith to develop some kind of memorial to the tens of thousands of pioneers who had followed the Platte Valley toward Utah, California, and Oregon. "The people traveling Interstate 80 have no idea they are following the route that linked America as a nation," Morrison said. "It was the greatest peaceful migration of people the world had ever witnessed." He had a point—the millions of drivers across the nation's busiest interstate could obtain no clue about their surroundings, let alone the history, beyond the median strip.

Smith proposed that the new memorial arch over the interstate. "Most developers," Smith was quoted as saying, "approach a project like this with the premise, 'Build it and they will come.' My position was and is, 'Build it because they are already here.'" The "here" decided upon was Kearney, certainly a logical notion given the fact that trails had braided together there.

The "they" was an even more compelling argument: Interstate 80 is the busiest transcontinental interstate in the country, the primary asphalt link between the East and West coasts. Nearly 13 million people use the highway through Nebraska annually, an average of 35,616 people passing per day. The Archway might attract a million visitors a year.

A sponsoring nonprofit foundation headed by Governor Morrison and a board of blue-ribbon Nebraskans was swiftly created. To plan concretely for a historical attraction over I-80, Smith assembled a team of consultants. Many were so-called Imagineers from the Disney organization. All this expertise seemed to suggest that the monument would be a grand architectural statement, but Governor Morrison had insisted that the interior exhibition be not only impressive and state-of-the-art but also historically accurate and relevant. The founders turned to an experienced historical site design firm in Cambridge, Massachusetts, Christopher Chadbourne and Associates. The consulting historian for the project was Ken Burns's neighbor in Walpole, New Hampshire, Dayton Duncan, who had worked on such series as *The West*, *The Civil War*, *Baseball*, and *The Lewis and Clark Expedition*.

Groundbreaking ceremonies took place on July 2, 1998, and the rusticated steel archway, 308 feet long and weighing 1,500 tons, was finished a year later. On the night of August 16, 1999, at 10:00 P.M., Interstate 80 was closed. All traffic was diverted in a roundabout on Route 30. The archway was jacked onto multiwheeled transporters so large that the bottom of the steel monument was twenty-two feet off the ground. It was rolled slowly across the east and the west lanes. Observers on the scene likened the process to that of an ocean liner being launched. By 6:00 A.M. on August 17, it was jacked up higher and locked into place on its platforms, and traffic was reopened, speeding under the span. It was thirty feet from pavement up to the bottom of the Archway.

Throughout the fall, winter, and spring, designers and fabricators went to work on the interior, which was some 79,000 square feet: the span itself contained two stacked horizontal decks, with the aim that visitors would walk across southward, inside interactive, "immersive" exhibits, then go upstairs and return northward through more exhibits to the finish. In the spring the two giant horse sculptures by Kent Bloomer of Yale University were erected atop the north and south towers; the height from the ground to the tip of the wings on the north tower would be 116 feet.

Meanwhile, the Smith organization mounted a large publicity campaign with billboards and other signs, television, radio, and newspaper advertisements, rack cards, and press releases. To coax some of the thousands of vehicles onto exit ramps, the Archway featured a low-power radio station in an empty frequency; passersby would see signs directing them to turn their radios there to hear a message about the Archway. It was turned on a week and a half before the opening, but so many visitors pulled in from I-80 and banged on the closed doors that they had to shut off the transmitter. After opening day on June 9, the crowds rolled back in and up that moving stairway, the longest escalator in the entire state.

We rose toward the train of wagons and passed through the screen. Our headphones had already come alive with voices reading the original pioneers' words from letters and diaries, and we passed into a darkening area where we could see a Native American encampment off on the prairie. Near us, cast figures of emigrants and their oxen were straining a covered wagon forward. Chimney Rock, that great pioneers' landmark, rose on the horizon. The room went darker, thunder rolled, and lightning flashed. The stereophonic effect of the headphones was amazing. We turned a bend and heard more thunder, but this time it was a gigantic buffalo stampede, projected on the walls as if to run right over us. Beyond there were more figures, a Mormon handcart expedition that had left the Missouri River encampment later than they should have and was caught in an early snowstorm on the Sweetwater River. Beyond was a forty-niners' campfire. The testimony in our earphones was about the mounting losses, the heartbreak, and the dreams that kept the survivors going.

Then we were somewhere else, a few years later in a Pony Express station, and outside the windows a rider galloped in, leaped off his mount, and yelled out the news that the Civil War had begun. We continued to walk, passing underneath a trestle, over which a train rumbled and clanked, throwing off a myriad of sparks. Farther, there was a Concord stagecoach, and a figure looking much like Mark Twain did in the 1860s, droopy dark mustache and tousled curly hair, leaned out the door. In the earphones I recognized a section from Twain's delightful memoir of his cross-country stagecoach journey to the Nevada silver fields, *Roughing It*. Pivoting around and ascending to the upper level, we entered the era of the first transcontinental railroad and heard snatches of the Golden Spike Ceremony at Promontory Summit, Utah.

We then leapfrogged forward into the twentieth century—Carl Fisher was proposing his idea of the Lincoln Highway, "the Main Street of America," and we were walking through roadside auto camps, souvenir shops, and attractions much like the closed, faded counterparts a few miles away up on Route 30. Further, we came upon a drive-in movie complete with carhops, and on the screen, between scenes from a Western movie and coming attractions, President Eisenhower was heard on *Fox Movietone News* announcing the launching of the interstate highway system. Just beyond the surroundings began to look more familiar. Near a simulated roadside restaurant, we came to an interactive billboard where children were pushing buttons and tracing the history of mobility from the old days of moccasins, boots, carts, and prairie schooners to the present day of big-horsepower SUVs, RVs, and tractor trailers. Underscoring this, windows overlooked the westbound lanes of the interstate. Mounted in the windows were radar guns—I took hold of one and pointed it down at the speeding trucks and passenger cars: 64 . . . 72 . . . 86 . . . 75 . . . 81 miles per hour, in air-conditioned, music-serenaded comfort. Compare that to the 8 miles a day an ox-drawn wagon train hoped to attain. Even to someone steeped in Western history who had spent years reading the letters, diaries, and journals of those people of toil nearly unimaginable today, the effect was breathtaking, sobering, exhilarating.

As an escalator took us down off the Archway into the log-walled Fort Kearney arena of fast food, interactive computer screens, and souvenir vending, I knew that there would be curled-lip "professionals" in academia who were probably already typing up jeremiads against the Disneyfication of American history, that the lessons were too complex and too important to be tossed into a theme park. But I think the creators of the Archway Monument had gone about this in the right way, relying so profoundly on the original voices. People off the interstate could breeze through this in a half hour or they could take two hours, listening carefully to all those recorded voices and examining all the collateral displays of long-captioned photographs and paintings, as we did. Neophytes as well as historical experts could not fail to be impressed by the creators' work as well as the historical lessons and legacy of the millions of humans who have followed the Great Platte River Valley route. Critics of the Archway have already derided public broadcasting's many television documentaries, but the fact remains that methods like these reach a great many people who might otherwise never be touched. What would be an alternative to the Archway, one wonders. Given the way our culture is today, I suppose one could impose roadblocks on the interstate, present lectures, administer tests. The PBS series encourage readers to go further into the subjects. At the Archway Monument, just a few weeks into what I hope and presume will be a long life, it seemed that the gift shop had far too many T-shirts, buckskin vests, cowboy hats, and Native American jewelry, and not nearly enough books. Certainly they are out there, and if a few made their way into an interstater's shopping basket along with the postcards and the T-shirts, all the better for everyone.

Heading back to our car, we had to skirt many clumps of tourists who were taking pictures of each other in front of the towering log building with its shiny, rearing horse sculpture. Costumed reenactors genially wandered about, posing for their Kodak moments. We pulled out of the Archway parking lot, not being tempted onto the interstate. The Lincoln Highway awaited just a few miles up on the north side of Kearney. There, in the summer of 1914, transcontinental driving pioneer Effie Price Gladding had urged her rattling, radiator-steaming car forward on her own transcontinental journey. Four miles west of town, she noted, "we passed the famous sign which marks the distance halfway between San Francisco and Boston," 1,733 miles each way. Naturally they stopped to take a photograph of it. "A woman living in a farmhouse across the road was much interested in our halt," Gladding wrote. "She said that almost every motor party passing stopped to photograph the sign." For many years the farm was known as the "1733 Ranch."

In Odessa, a horse grazed in a schoolyard.

A Union Pacific double-header appeared from the west hauling nearly 100 coal hopper cars from Wyoming; its lead engine's three headlights were bright in the dimming afternoon light. Clouds were massing south, west, and north of us, over the Platte's cottonwood breaks, the green cornfields, and purple alfalfa fields climbing toward an unbroken ridge of sand hills to the north. The weather looked heavy ahead. Of course we kept going.

6

Hell on Wheels

Bad late afternoon weather had dogged us since central Missouri, when big thunderheads began climbing and billowing hugely to the south and west of us and a bright blue horizon shifted ominously into the green spectrum. Arriving at a motel in Kansas City late one day, a truly furious and highly alarming wind sprung up from nowhere, driving us indoors; a man came into the lot behind us on his motorcycle, and the desk clerk, who was leaning outside and peering anxiously into the sky "for twisters," invited the man to wheel his motorcycle into the lobby. I wished there were a way to drive our car in for safety, too.

Back when I was a boy and my family would spend vacations at my grandparents' place, summer storms sometimes blew into my nightmares. There was a cyclone cellar out behind the house, a doorless opening into the side of a hill that smelled of wet earth and, my grandfather said, rattlesnakes, so I didn't venture inside the dark place but just peered in, my eyes never penetrating its depths. Once the evening news on the radio reported that a tornado had touched down just over the Kansas line, and the next day we drove out to look at the destruction. I remember clambering out of the car into a weird townscape. The tornado path had scratched a deep furrow through a neighborhood, sparing this house but obliterating that, and that, and that. Toys were scattered in the litter and debris: where, I wondered, were the children?

Now, speeding westward across central Nebraska with our weather radio switched to "alert" status so it would come alive if the National Weather Service broadcast an emergency bulletin, I saw those disquieting images in my mind's eye. For Mimi and David, the kids had enough imagery from *The Wizard of Oz* to make them nervous. Severe thunderstorms with damaging winds were predicted for late afternoon and evening and residents were cautioned about "rapidly developing conditions." We tried to focus on the sights we were passing, though always with one eye on the dark horizon. At Elm Creek, population 852, I noted a classic Fifties Chevy Bel Aire parked on a front lawn next to a farm wagon; past a little stucco and tile-roofed motel there were a few shuttered shops and one reassuringly named "Mom's Kitchen." A freight consisting of scores of black tank cars of Cargill corn sweetener rolled by. "'Elm Creek has had a history marked by misfortune,'" Mary read aloud from the WPA guide. "'Blizzard followed blizzard in the 1880s, killing many local cattle and sweeping away most of the possessions of the inhabitants; in 1906, after the town had been rebuilt, it was almost wiped out again, this time by a raging fire that destroyed all the buildings along the main street.'" She marked her place in the book and closed it. "May it get through another day without disaster, especially with us in it!"

The first flash of lightning came as we went through Overton, population 665. There was a rare original Lincoln Highway bridge there, a quarter mile east of town, built in 1914, but we only gave it a glance. All the cows were lying down, which we recognized, being from farm country in Vermont, meant something was imminent. The temperature was dropping with every mile—it had been 90 degrees in Kearney, and now the little gauge in our car's overhead panel said it was 74. We could see the outline of storms moving north toward us as we neared Lexington.

"I remember when we were camped on the Platte, the whole sky became black as ink," recalled Mary Elizabeth Munkers Estes, who was ten in 1846. "A terrific wind came up, which blew the covers off the wagons and blew down the tents. When the storm burst upon us, it frightened the cattle, so that it took all the efforts of the men to keep them from stampeding. . . . The rain came down in bucketfuls, drenching us to the skin. There wasn't a tent in the camp that held against the terrific wind. The men had to chain the wagons together to keep them from being blown into the river. . . . Finally, in spite of efforts of the men, the cattle stampeded."

The rain hit us in earnest. The weather radio came alive and warned of flash floods in southeastern Lincoln County and southwestern Dawson County, the latter just below us on the south side of the Platte. Weather spotters on the Union Pacific trains, it said, saw heavy rains to the northeast and southeast of us. Lightning was nearly constant. The temperature dropped 10 degrees in two minutes, down to 64. "Listeners in the low country, be alert," warned the radio. "It only takes a few inches of water to roll a vehicle." In my rearview mirror I caught glimpses of the anxious faces of my children. "Dad," said Mimi. "Are we in the low country or the high country?"

I reassured her that the radio was talking about the canyon lands south of the Platte, and that all we had to be worried about was staying straight on the road with the wind and rain.

But what was on both our minds was what had once happened to the two of us in a sudden bad storm that hit the Champlain Valley of Vermont when she was only two, but which had indelibly marked us. I had picked Mimi up at her play group at a farmhouse up in Bridport, the town north of us. I buckled her into her child's safety bucket in the back seat of our 1983 Subaru wagon.

Apparently a storm was at that moment flashing through the Adirondacks, just five or six miles west of us, and pouncing down onto Lake Champlain as we came out of the farm driveway and headed east on the state road. A few miles later we were heading south on State 22A, and suddenly a hard rain pelted the windshield. It was so bad I couldn't see beyond the wipers. The car shook as I tried to coast and gently pump brakes without going into a ditch. There was a terrible noise, a tree fell down onto the road right in front of us, and the car slammed into it. Mimi's little juice drink container flew past my ear. "Are you okay?" I shouted to her. She was, although scared by the noise. My chest hurt from the seatbelt (there were no air bags). I noticed through the rain-smeared window next to me that a steel cable was stretched along the length of the car. I realized dimly that the cable an inch from my face, just through the glass, was the electric power line that had been pulled down with the tree. It was tangled in the tree, just below our front wheel well. I saw that it had ignited some wood fragments just below our tire, and the glow was reflected in the wheel well. I thought about staying in the car because of the electrical hazard and I thought about the car catching on fire and exploding, and I twisted around, reclined my seat, climbed into the back and got Mimi out of her bucket seat, and kicked the back door open with my workboot. Still clutching my two-year-old, I gingerly stepped out of the car without touching metal and into a howling wind. We were soaked immediately. Staying away from the wires, I climbed with her through the tree branches and ran down to where some cars had stopped on the highway. Someone gave us a ride home. Dumbly, I'd left my keys in the ignition, so I had to break into our house to get in out of the storm. Later, the radio said that people had been calling the weather service claiming a mini-tornado had just torn across Bridport and Cornwall townships and dissipated in Middlebury, but not before knocking down trees all over the campus. Mimi and I had seen no reason to disbelieve this, and in all the years thereafter had talked about our narrow escape. I felt her fingers lightly on my shoulders now, me in the front seat and Mimi in the back, as nine years ago.

"Don't worry," I said now. "I'm on the lookout for trees."

"Dad, this is Nebraska."

"No, look, there's a tree over there."

"Well, keep clear of it."

The next town up the road was Lexington, home of the Dawson County Historical Museum. It was getting close to 5:00 P.M., but the rain didn't seem to be interested in letting up, so I pulled into the driveway of the museum. The parking lot was flanked by a restored one-room schoolhouse and a depot building, in front of which was a big old black Baldwin steam locomotive from the turn of the century; we sprinted through horizontal rain into a sprawling, one-story structure, the main building of the museum. We apologized to the director, whose name was Barb Vondras, for dripping on her floor, and for showing up a few minutes before closing time—could we quickly look around?

"Take your time," she replied. "I'm staying right here until this weather lets up anyway!"

Shivering a little as we dripped, we looked the place over. The big, open-plan museum building, built in 1967, featured many glass cases of clothing and local historical artifacts, interspersed with wall photographs and posters, depicting local history. The original settlement was called Plum Creek, and it grew up alongside the Oregon Trail from a Pony Express station and trading post founded in 1859. "The country is fast settling up," wrote Martha M. Moore in 1859 after going through and seeing the sporadic new growth. "The inhabitants build themselves sod houses and manage to live on nothing." The way station, though, did all right. Being in the heart of buffalo country, it was famous among emigrants and stagecoach travelers for its buffalo steaks.

In 1864 Plum Creek became synonymous with bloodshed. A prairie schooner carrying a family of Iowans—Thomas and Nancy Morton, man and wife, and her brother, William Fletcher, and cousin, John Fletcher, were heading west. One day earlier they had joined a train of eleven wagons for safety, but after they started on the morning of August 8, 1864, with Nancy Morton driving her team, she saw horsemen approaching. She could not make out who they were until they broke into a gallop and rode down on them.

They were Cheyenne raiders. In Colorado and Kansas that summer, Colonel Chivington had said that the settlers had to protect themselves while he made indiscriminate war on the Plains tribes, and so what might have been under other circumstances the usual shakedown for food and tobacco became just quick, cruel murder. Nancy Morton watched as all eleven men in the wagon train were killed. She was wounded by two arrows—and later removed them herself. She and the young son of one of the other families, Daniel Marble, were taken captive, beginning a long and terrifying wartime journey across four states.

The war party moved over the plains, raiding more emigrants and settlers, taking additional hostages. The boy Daniel was traded off and ultimately freed in Colorado. Nancy Morton and other prisoners were dressed "Indian fashion" and their faces painted. The Cheyenne women were unfailingly kind to the nineteen-year-old Morton and most of the captives, but the warriors made life difficult. Several proposed marriage to Morton but she replied that she would rather die. She was repeatedly tormented with the brandished scalps of her husband and brother, and taunted with war trophies after each successful raid, though she was not physically handled. When threatened with death, she retorted, go ahead, I want to join my husband. This impressed her captors. Morton had lost her two toddlers to measles a couple of years earlier, and with her husband and brothers had made several trading trips across the plains. She was tough—but not so her fellow hostages. One little girl cried and wailed unceasingly for days. When Nancy Morton tried to console and warn her that their captors were losing patience, the girl began screaming, until suddenly an arrow through the heart killed her in Morton's arms. She saw a number of captives tortured and murdered, while others like Daniel Marble were simply traded off to other bands. Over time, through her stolid endurance, she was accorded the Cheyennes' respect. It was only after an accident, when thrown from her horse, that she had any inkling of an end to her ordeal. Injured in the fall with sprains to both wrists and ankles, she had to be conveyed for days on a travois dragged behind a pony, but the Cheyenne fretted that she was slowing them down. They told her she would be sold back to the whites for provisions.

Nancy Morton had been a captive for six months when she was finally ransomed by the Indians. Her price was four good horses, three sacks of flour, forty pounds of coffee, seventy-five pounds of rice, four packages of soda, one sack each of salt and powder, thirty pounds of lead, twenty boxes of caps, one saddle, twenty yards of bed ticking, two spools of thread, ten combs, ten butcher knives, one box of tobacco, thirty bunches of beads, some paint, three papers of needles, one rifle, three revolvers, five blankets, a belt and saber, and two new coats. "I wept for joy and praised God for my freedom," she wrote later. But it was immediately clear that the Cheyenne intended to collect the ransom and then recapture her. With her small military escort she fled with the Indians at their heels until she reached the safety of Fort Laramie in present-day Wyoming.

Nancy Morton went back home to Iowa. She later remarried. Years after, she compiled an account of her capture and life as a prisoner of the Cheyenne; her manuscript has been edited and published by the Dawson County Museum. There are two side-by-side portraits of Nancy Morton at the museum—one taken back in Iowa before she set out West with her husband; the other taken after her release from the Indians. There is more than the passage of one year between those portraits, particularly around the eyes. The cemetery containing the remains of the eleven men killed is just east of the site of the Pony Express station, down on Plum Creek.

Outside, the heavy rain drumming on the roof now sounded like hailstones. There was a large Union Pacific Railroad exhibit, the tracks having been built past this point in September 1866. As had happened with many little settlements in the Platte Valley, the citizens abandoned their homes and businesses down along the wagon trails and moved up from the Platte some ten miles to trackside.

By summer 1867 the railhead had moved forward, following the South Platte into the remote northeast corner of Colorado before wandering north-westerly back into the Nebraska panhandle country along Lodgepole Creek. The Indian War had, if anything, heated up. Gold seekers and settlers intent on getting to Montana had opened up a perilous new wagon trail through the heart of Sioux hunting grounds, and the U.S. Cavalry had been ordered to protect it, constructing outposts that could barely sustain themselves, let alone the emigrants. Down in Kansas, federal cavalry forces under General W. T. Hancock and Colonel George Armstrong Custer were continuing the Chivington model by raiding Cheyenne encampments that had moved where they were told by federal Indian agents and had renounced violence against whites; this created furious backlashes as war parties struck at civilians along the Republican and Platte valleys. The army detailed Major Frank North and his companies of Pawnee Scouts to patrol the railroad line in Nebraska; they were effective at keeping interlopers on the run, but they could not be every-where at once. Little Plum Creek, Nebraska, was about to enter the history books again.

On August 4, 1867, a Cheyenne band led by Spotted Wolf—whose friendly village had been burned by Hancock in the spring and whose young warriors had been chased northward by Custer—rode up to see their first railroad train. After the puffing, smoking monster had clanked off westward, they went down to look at the tracks. Then they rode a little way north to an en-campment of the Cheyenne leader Turkey Leg. "Now the white people have taken all we had and have made us poor," they said, according to one member later interviewed by the historian George Bird Grinnell. "We ought to do something. In these big wagons that go on this metal road, there must be things that are valuable—perhaps clothing. If we could throw these wagons off the iron they run on and break them open, we should find out what was in them and could take out whatever might be useful to us." Spotted Wolf's band went back down to the railroad tracks.

Pulling down a telegraph pole, they tied it to the tracks with the telegraph wire and sat down in the darkness to wait. "Quite a long time after it got dark we heard a rumbling sound," recalled one Cheyenne, whose name was Por-cupine. "At first it was very faint, but constantly growing louder. We said to each other, 'It is coming.' Presently the sound grew loud, and through the darkness we could see a small thing coming with something on it that moved up and down." It was a handcar, with five men pumping it. They worked for the telegraph company, and were coming to investigate the break. When their car hit the pole on the tracks it flew high in the air. The men began to run away. One man, an English telegraph repairman named William Thompson, was overtaken by a Cheyenne on a horse and shot in the arm; running far-ther, he was clubbed down. "He then took out his knife," Thompson recalled, "stabbed me in the neck, and making a twirl round his fingers with my hair, he commenced sawing and hacking away at my scalp. Though the pain was awful, and I felt dizzy and sick, I knew enough to keep quiet." The process seemed to take a half hour. "He gave the last finishing cut to the scalp on my left temple, and as it still hung a little, he gave it a jerk," Thompson said. "I just thought then that I could have screamed my life out. . . . I just felt as if the whole head was taken right off. The Indian then mounted and galloped away, but as he went he dropped my scalp within a few feet of me, which I managed to hide." He could hear the warriors moving about in the dark nearby and didn't twitch a muscle.

The Cheyenne had found the workmen's tools and now decided to see what would happen if they pried up the rails. In a short time they saw lights on the horizon—two freight and work trains were approaching. The first engine hit the broken rails and jumped high in the air, and the cars behind were reduced to splinters. Forward crew members were shot and scalped, while others escaped to flag down the second train. It backed off and was not pursued.

Thompson, still lying nearby, watched as the Indians lit a bonfire and then climbed into some of the broken boxcars. As soon as he could, he crawled off, clutching his scalp, and after he had made his way to the Willow Island station, and was taken by rescuers back to Omaha, he told a visiting English journalist the story. The roving reporter was Henry Morton Stanley, later to win great fame after "finding" Dr. David Livingstone in the African jungle.

"They plundered the box-cars of everything that might prove of the least value," Stanley would write, "or what attracted their fickle fancy—bales of calicoes, cottons, boxes of tobacco, sacks of flour, sugar, coffee, boots, shoes, bonnets, hats, saddles, ribbons, and velvets. They decorated their persons by the light of the bonfire . . . their ponies were capairisoned with gaudy pieces of muslin, and the ponies' tails were adorned with ribbons of variegated colours. The scalp locks of the Indians were also set off with ribbons, while hanging over their shoulders were rich pieces of velvet." Porcupine himself recalled it similarly: they found whiskey, too, and danced a dance of triumph.

The next morning, some of the railroaders returned to Plum Creek. With a spyglass they could see that the railroad cars were now all on fire, with some of the Indians galloping in a circle. Others were carrying away the plunder. As evening began to fall, a company of soldiers appeared and went to examine the wreck and retrieve the charred bodies. They were transported back to Omaha on a train that happened also to be carrying William Thompson. At the station there was quite a push to gape at the bodies—and there was the living spectacle of the Englishman. "People flocked from all parts to view the gory baldness which had come upon him so suddenly," Stanley reported. Thompson exhibited his other wounds. "In a pail of water by his side," said Stanley, "was his scalp, about nine inches in length and four in width, somewhat resembling a drowned rat, as it floated, curled up, on the water." Although Thompson hoped that a surgeon could reattach his scalp to his head this was of course beyond medicine. William Thompson later sent the scalp, after it was tanned, to his doctor in Omaha. Ultimately it went on display at the Omaha Public Library in the children's section, and now reposes at the Union Pacific Museum, where we saw it in the hands of curator Don Snoddy.

After this derailment and plunder of a railroad train—a modus operandi soon adopted by bandits in Wyoming—Plum Creek went into a quiet period of growth as settlers came in on the railroad and got themselves situated. One amusing little struggle between locals and the U.P. occurred in 1873 when Dawson County citizens voted to construct a wooden bridge over the Platte to beef up their businesses. In an enterprising move, the county taxed its largest property owner—the Union Pacific Railroad—$50,000, most of the costs of building the bridge. The railroad refused to pay the taxes. Then one day the county clerk prevented a train from leaving at its scheduled time by chaining the locomotive to the tracks. Shotgun in hand, he ordered the engineer to wire the company president for the money. Within twenty-four hours the ransom arrived, and the railroad main line was again open for business. For years Plum Creek lived with a raffish reputation with the company and its customers alike, preying on passersby for every silver dollar and greenback they could extract, legally or not. Vigilantes finally cleaned up the town, and the good element renamed it Lexington in 1889, commemorating the Revolutionary War battle. The distrust for Plum Creek among railroaders faded with time.

We drifted around the museum, seeing age displacing age. Down the way from the U.P. exhibit with its sidelines about the Turkey Leg raid and William Thompson, and the North brothers and their Pawnee Scouts, stood some shiny, restored vehicles from the Lincoln Highway era—a royal blue 1931 Ford Model A with ecru trim and leather seating, and a bottle green 1923 Model A delivery truck with a wooden platform. Just beyond was a real rarity—the one-of-a-kind McCabe Aeroplane, built by a local inventor named Ira Emmet McCabe in 1919. McCabe had previously constructed experimental gliders and a powered aircraft. But this spindly biplane, from a design he patented, was as beautiful as it was improbable. The two wings met at their outer edges to form an ellipse. It was powered by a 22.5 horsepower, V-twin Thor motorcycle engine. One would have to be crazy to expect such a thing to fly. But fly it did, though catch on it did not. It managed extended flights, reaching altitudes of 600 to 700 feet at an average speed of 60 miles per hour—not bad for that early era—but lack of public interest and additional funds forced him to abandon the plane and the business it inspired. Emmet McCabe continued with his tinkering, acquiring some 130 patents, finding his greatest success with his invention of the mercury switch. The family donated his Aeroplane to the county museum, where it continues to inspire native tinkerers arriving in school buses and tourist cars alike.

Outside the museum, bad weather was subsiding, so we detained Barb Vondras from her dinner no more. Back on Route 30, we soon found ourselves slowing into the town limits of modest, two-story Cozad, Nebraska, a bright banner stretched over the state road to remind us that we were crossing the 100th Meridian. The town had been founded by one John J. Cozad from Ohio, who, while riding west on the Union Pacific saw the 100th Meridian sign and decided that settlers would want to live on an imaginary line drawn on maps from the North Pole to the South. He returned to Ohio, organized a company of emigrants, and moved them out to the open grassland, which settlers found to be drier than the farms they'd left behind. The 100th Meridian, in fact, marks the place where "the humid east meets the arid west." The state of Nebraska has decreed, along this line, that here the West begins.

A few years before our trip, I had some fun writing about this place, for the 100th Meridian and future Cozad played an important part in the saga of the first transcontinental railroad. The Pacific Railroad Act as amended in 1864 had stretched a finish line of sorts along the meridian; whichever railroad— Union Pacific, Kansas Pacific, or whoever—ran across the prairie and puffed out its iron chest and broke through the line would win the right to keep racing toward a meeting point with California's Central Pacific Railroad. The Omaha and Kansas roads, both dogged by financial problems, internal squabbles, and Indian threats, had been working neck and neck.

With a few hammer blows the U.P. became the undisputed main line on October 6, 1866, and the energetic, scheming vice president of the line, Dr. Thomas Clark Durant, decided to throw a blowout, and here we can see the birth of the modern press and political junket. He sent out more than 300 invitations, to President Andrew Johnson and his entire cabinet; to senators, congressmen, governors, generals, and mayors; to moguls and movers and shakers in New York, Boston, and Chicago; to ambassadors and other foreign dignitaries. Two special trains would bear them to the Missouri, where a river steamer would convey them to Omaha. A special luxury train outfitted like the grandest Eastern hotels awaited them there. Not everyone on Durant's grandiose list accepted, but the 100th Meridian Excursion, 200 strong, was filled with senators, congressmen, tycoons, newspaper magnates, and even an earl from Scotland and a marquis from France. Two official photographers would record the festivities, and two orchestras went along for entertainment.

The nine-car train pulled by two flag-decorated engines steamed out from Omaha, stopping at various points so the celebrants could admire bridges, depots, and water stations. They pulled into Columbus at nightfall to find a "brilliantly illuminated" tent encampment waiting, a sumptuous repast warming, and a coterie of Pawnee Indians preparing to entertain them with a war dance. The next morning the excursionists were awakened by shouts of "Indians! They're on the attack, boys!" but the mounted raiders who swept into camp with war whoops and shots into the air were only Pawnee play actors. Fortunately none of the Indians was shot, and no statesman or mogul perished from fright. Later in the morning, after everyone had reassembled in their Pullman cars to be conveyed out to the Loup River bridge, the railroaders organized a mock Indian battle between the Pawnee and some of their number dressed up as Sioux. Many of them would later go to work for the Buffalo Bill Wild West Show. That night the dignitaries camped opposite Fort McPherson, 279 miles from Omaha, enjoying the ample food and copious champagne to orchestral accompaniment. And on the last day they were taken out to the end-of-track, where a triumphal 100th Meridian arch had been erected over the dry sod, and everyone posed for the photographers. On their return trip, the train steamed through the night past an immense prairie fire set by their hosts to entertain them.

Of course, the Union Pacific reaped many political and public relations benefits from that blowout in the autumn of 1866. And though there were two and a half years of extraordinary struggle ahead of the railroaders, the 100th Meridian excursionists could feel a sense of destiny abroad on the plains. "The laws of civilization are such that it must press forward," wrote one of their number, seeing some poignancy in the presumed fate of the Plains tribes but of course no genuine empathy, "and it is in vain that these poor ignorant creatures attempt to stay its progress by resisting inch by inch, and foot by foot, its onward march over these lovely plains, where but a few years since, they were 'monarchs of all they surveyed.'

"The locomotive must go onward until it reaches the Rocky Mountains, the Laramie Plains, the Great Salt Lake, the Sierra Nevada, and the Pacific Ocean. Lateral roads must also be built, extending in all directions from the main line, as veins from an artery, and penetrating the hunting grounds of these worse than useless Indian tribes, until they are either driven from the face of the earth or forced to look for safety in the adoption of that very civilization and humanity, which they now so savagely ignore and despise."

In following the Platte Valley in July 1843, John C. Frémont and Kit Carson had encountered a grand herd of buffalo here, and thanks to the writing talent of Frémont's wife, Jessie, he published a vivid account of the hunt. "A thick cloud of dust hung upon their rear," he recalled, "which filled my mouth and eyes, and nearly smothered me. In the midst of this I could see nothing, and the buffalo were not distinguishable until within thirty feet. They crowded together more densely still as I came upon them, and rushed along in such a compact body, that I could not obtain an entrance—the horse almost leaping upon them. In a few moments the mass divided to the right and left, the horns clattering with a noise heard above every thing else, and my horse darted into the opening. Five or six bulls charged on us as we dashed along the line, but were left far behind, and singling out a cow, I gave her my fire, but struck too high. She gave a tremendous leap, and scoured on swifter than before. I reined up my horse, and the band swept on like a torrent, and left the place quiet and clear."

In mid-June 1846, the Donner-Reed Party of emigrants passed by. Because James Frazier Reed could afford to pay drivers for his wagons, he was able to wander from the rolling emigrants to hunt grouse, antelope, and wild goats. On June 15 he went on his first buffalo hunt. "The plains appeared to be one living, moving mass of bulls, cows and calves," he wrote. "With me it was an exciting time, being in the midst of a herd of upwards of a hundred head of buffalo alone, entirely out of sight of my companions." He killed two calves and mortally wounded a protective bull. Then Reed rested—and counted 597 buffalo within sight. Helping some of his friends kill another bull, and then chasing and killing another calf, the travelers secured as much of the meat of the calves as they could carry and headed back to the main camp, "leaving the balance to the wolves, which are very numerous."

An even more vivid account was left by a Mormon pioneer, Appleton Harmon, who traveled with leader Brigham Young's vanguard just four years later, in 1847. Young had called the men together and issued instructions that no more game was to be killed until needed for food, "for it was a Sin to waste life & flesh." In such a decree Young was showing a humanity seldom seen in those early times, in which every white emigrant seemed to find it necessary to shoot as many wild beasts as possible, leaving most of the carcasses for scavengers. Several days later, Harmon wrote, they "had to drive the buffalo out of the way whare we halted the buffalo seemed to form a complete line from the river their watering place to the bluffs as far as I could see which was at least 4 m[iles]. [T]hey stood their ground appurently amased at us until within 30 rods of the wagons when their line was broken down by some taking fright & running off others to satisfy that curiosity came closer within gun shot of the camp snuffing and shaking their Shaggy heads, but being pursued by the dogs ran off, at this time I could stand on my waggon & see more than 10,000 Buffalo from the fact that the Plain was purfectly black with them on both sides of the river & on the bluff on our right which slopes off gradualy."

During covered wagon days, we were informed by the WPA guide, emigrants often wrote on buffalo skulls and shoulder blades along the route. They recorded and dated their names and wagon companies, and indicated the whereabouts of springs and grass for the benefit of those who would follow their tracks. It was not far from here, I think, that Frank Young wrote a beautiful account (in his journal, not on a buffalo skull) of a night spent out on the wagon trail. In reprinting it, the Platte Valley road scholar Merrill Mattes assembled it from Young's fragmented notes; I reproduce it here with its multitude of ellipses removed: "In the splendor of the night soon a distant rumbling breaks in and approaches rapidly. [I]n a very few minutes the over-land coach, with a dozen horses or mules on a steady gait, glides by on the road with three or four shadowy figures on top, all in absolute silence except for the dull roll of wheels. It disappears down in the western sky among the stars."

For us, we followed the tracks with the echoes of their words in our ears, and with markers mostly of the historic highway variety. West of Gothenburg Route 30 plunges into the sand hill country; on either side of the road the sharply eroded preglacial forms washed down from the mountains and deposited here. It was so odd to see these hills after many days of Nebraska's rolling, essentially featureless terrain, almost a comfort to the eye and something over which everyone in the car exclaimed. Then we were nearing Brady, population 331, named for Brady Island in the Platte, where Frémont camped in 1843. About 1859 a man named Jack Morrow built a station here that was popularly called Junction House and known as the halfway point between Omaha and Denver. Frank Young wrote about it in 1865. "Jack Morrow's ranche, said to be the finest on the whole route. . . . He is one of the characters of the Plains, famed as a scout, and happy in thee possession of a squaw, or some other kind of influential Indian connection said to insure him against attack. He has a handsome and extensive log building, filled to the roof with a general stock of Plains staples . . . tier on tier of sardines, tomatoes, and peaches . . . and stacks of log cabin bottles of Drake's Plantation Bitters. . . . These empty bottles blaze the trail."

If, in forging west along the Oregon Trail almost to the fork of the Platte River, Frank Young noticed empty liquor bottles littering the route, he should have seen it just a few years later, when the Union Pacific work crews came through, and wintered over. Now that was a time.

Two hundred and seventy-two miles west of the Missouri River is the confluence where the Platte River is created by two headwaters. The South Platte River hews to a mostly east–west plane to that joining, its snowmelt Rocky Mountain waters flowing into Nebraska from the northeastern plains of Colorado. The North Platte River enters Nebraska from Wyoming from the northwest through eroded, increasingly rougher terrain, having negotiated many improbable twists and turns through hundreds of miles of Wyoming ranges from its source in Colorado's Medicine Bow Mountains. At the place where the North and the South Platte waters mingled, emigrants had a choice. The southern branch would lead them across the Colorado plains toward the growing city of Denver. Travelers with farther sights on their minds—Oregon and California, or Utah—would follow the northern branch, heading for the safety and supplies at Fort Laramie, Wyoming, and the continuing months of toil westward.

Many thousands in wagon trains or pushing Mormon handcarts rolled past that confluence during the 1840s and 1850s. In 1863 the Overland Stage Company established a station called Cottonwood Springs, but later that year the army arrived, reacting to many Indian raids on emigrants, and planted a large log-palisaded stockade called Fort McPherson. In the frosty late autumn of 1866, the army of Union Pacific tracklayers worked their way past the fort to where chief engineer Grenville Dodge was laying out a new town on the delta bottomland between river bluffs, to be called North Platte. The mostly Irish laborers settled into dormitories for the winter, watching as the storekeepers, saloon owners, gamblers, and brothel operators swarmed in, erected flimsy buildings of boards, canvas, tin, and what-have-you, pegging it all together with nails, bolts, spit, ambition, and greed.

It proved to be a wild winter by all accounts. It became easier to find a drink of whiskey than a drink of water. Saloons and more saloons, dice and roulette parlors, and houses and tents of prostitution proliferated. Day and night the streets were jammed with drunken railroaders, prostitutes, pimps, pickpockets, and cardsharps—but no lawmen. Winter snows arrived in December and if anything North Platte heated up.

A young army surgeon, Dr. Henry C. Parry, visited North Platte in early 1867. "Indians had driven all the traders and miners in from the mountains," he wrote, there joining the throngs of idle tracklayers, "and at North Platte they were having a good time, gambling, drinking, and shooting each other. There are fifteen houses in North Platte: one hotel, nine eating or drinking saloons, one billiard room, three groceries, and one engine house, belonging to the Pacific Railroad Company. The last named building is the finest structure in the station. I observed that in every establishment the persons behind the counters attended to their customers with loaded and half-cocked revolvers in their hands. Law is unknown here, and the people are about to get up a vigilance committee."

Some weeks later the English journalist Henry Morton Stanley took a fifteen-hour stagecoach ride out from Omaha to look over the place. He paused at North Platte's outskirts, deeply impressed. "Piled up, and covered over with sailcloth, and lying in every direction, are large quantities of Government freight," he reported. "Some officials state the amount awaiting transportation at 15,000 tons. Encamped in the immediate vicinity of this town were 1236 waggons, divided into trains of 27 waggons, each train officered by 29 men and a superintendent. There were Mormon emigrants bound to Utah, settlers for far Idaho, and pilgrims to mountainous Montana, who were emigrating with their wives, children, and worldy substance. The prairie around seemed turned into a canvas city."

But Stanley, who was a man of the world, was truly impressed once he got into town. An immense amount of freighting was being done to Idaho, Montana, Utah, Dakota, and Colorado, so to the masses of emigrants, traders, miners, and tracklayers he added "hundreds of bull-whackers [who] walk about, and turn the one street into a perfect Babel." North Platte may have been the place to hibernate before the spring thaws opened opportunities for work and travel again, but nonetheless North Platte never slept. "Every gambler in the Union seems to have steered his course for North Platte, and every known game under the sun is played here. The days of Pike's Peak and California are revived. Every house is a saloon, and every saloon is a gambling den. Revolvers are in great requisition. Beardless youths imitate the peculiar swagger of the devil-may-care bull-whacker and blackleg, and here, for the first time, they try their hands at the 'Mexican monte,' 'high-low-jack,' 'strap,' 'rouge-et-noir,' 'three-card monte,' and that satanic game, 'chuck-a-luck,' and lose their all."

North Platte all but emptied out in the spring of 1867—the emigrants moved on, the freighters and traders moved on, and the army of tracklayers went off along the South Platte following the railroad grade. Saloons, gambling establishments, hotels, and stores were disassembled and transported out to the next end-of-track at the tiny stagecoach station of Julesburg, Colorado. From a population in the thousands North Platte was reduced to about 300, with perhaps 20 houses remaining. What prevented the devastated settlement from withering further and blowing away with the first summer storm was its being a Union Pacific division point, with machine shops, a twenty-stall roundhouse, and a railroad hotel. As the railroad prospered, so did North Platte, and during the years that the Indian Wars continued, the town's safety was guaranteed by the presence of nearby Fort McPherson.

It was at Fort McPherson on May 20, 1869, just ten days after the driving of the Golden Spike in Utah, that the 5th U.S. Cavalry rode in to its new regimental post from Fort Lyon, Kansas, having skirmished along the way with a Cheyenne war party led by Tall Bull. The soldiers were still resting from their journey on May 29 when a large traveling circus rolled into North Platte on the Union Pacific for a one-night stand. It was called Dan Castello's Great Show, Circus, Menagerie, and Abyssinian Caravan, and it was on a much heralded coast-to-coast tour. Major Frank North, the fearsome leader of the Pawnee Scouts, had seen the show while on an official trip in Omaha three nights earlier. On May 29, another army scout named William Frederick Cody, attached like Frank North to the 5th Cavalry, was (it has been presumed) in the big circus tent at North Platte, eyeing the pretty showgirls, marveling at the trained lions and tigers, and taking in the fancy riding tricks. Cody, the former Pony Express rider whose exploits as a buffalo hunter for the Kansas Pacific Railroad and as an Indian fighter and scout for the army had earned him the nickname Buffalo Bill and made him famous in the West, was an avid horse racer and trick rider himself. For weeks afterward at North Platte and at the stockade, Cody practiced some of the tricks he'd seen, including leaping on and off a galloping bareback horse. And the seeds of spectacle had been planted in him, to bloom forth as the Wild West Show more than a decade after he had seen his first-ever circus.

A few months later in July 1869 Cody met a traveling temperance lecturer and dime novelist from the East, E. Z. C. Judson, better known by his nom de plume as Ned Buntline. Their time together was brief, but when Buntline went home he had a new hero for a dime novel series, Buffalo Bill, whose tales came more out of Buntline's imagination than actual fact. The stories would bring Cody a national reputation, one he had no trouble living up to out in the Platte Valley where the Indian War continued to crackle.

Supervising the military situation since early 1869 was Major General Philip H. Sheridan, commanding the Division of the Missouri and overseeing a vast territory stretching from its headquarters in Chicago to the western borders of present-day Montana, Wyoming, Utah, and New Mexico, and from the Rio Grande northward to the Canadian border. Sheridan, the Civil War hero of Stone's River, Missionary Ridge, and Cedar Creek, a favorite of Ulysses S. Grant's, was quite a hero to me when I was between the ages of eight and fourteen and a Civil War buff, my room decorated with battle maps and portraits. And there were personal reasons for the hero worship. Until her death in 1958, my grandmother, Rose Donahue Haward, never lost an opportunity to remind me that we were related to Sheridan. Her maternal grandmother was Mary Sheridan Coughlin, a first cousin to the general, which forever in Rose's eyes simplified him to "Uncle Phil." Sheridan died in 1888, the year before Rose was born, but one could almost get the impression that "Uncle Phil" had dandled her on his knee. One time we were all down visiting Arlington National Cemetery and Rose got the idea to find Sheridan's grave. She marched us all off the marked paths, getting lost almost immediately among the crowded hillside stones. We searched everywhere, the younger of us quickly exhausted and complaining and my father's usually obscured Southern sympathies and lack of patience with his mother-in-law nearing him then to rebellion, but abruptly Rose veered downhill between markers, crying, "Uncle Phil! Uncle Phil!" and we came upon it, stolid and austere, and eminently satisfying to my grandmother.

As I grew toward my late teens and read further into the complexities of war and politics there were fewer reasons to be proud of my family connection to Philip Sheridan. A modern-type commander whose philosophy was total war, in the Grant and Sherman mold, which included punishing the civilian population by destroying their supplies, crops, and livelihood as a way of denying support to enemy forces, Sheridan ruthlessly employed these tactics after the Civil War out on the Great Plains. He is remembered today as the author of the sentiment "the only good Indian is a dead Indian." Though he always denied he ever said those words, he may as well have. Apparently once some bleeding-heart civilian tried to lecture him on the difference between "good" Indians and "bad" Indians, and Sheridan retorted that "the only good Indians I ever saw were dead." The general saw Native Americans as an inferior race with a primitive culture. As biographer Paul Andrew Hutton has written, "he felt them to be inordinately barbarous in war, which he attributed to a natural, ingrained savageness of the race. They formed, in Sheridan's mind, a stone-age barrier to the inevitable advance of white, Christian civilization. Sheridan not only favored this advance but also proudly saw himself as its instrument."

Total war on the plains included, in Sheridan's view, destruction of the great buffalo herds. The vast numbers afforded Plains tribes a good life. Treaties permitted them to stray far from reservations in the buffaloes' pursuit. Extermination of the buffalo would result in the Indians staying where they were told, impoverished and deprived of their central cultural quality, until they were assimilated into the dominant society or died out. As biographer Hutton and the Nebraska writer Mari Sandoz have noted, it was very conservatively estimated that the herds contained some 15 million animals in 1867. Thanks to army policy and economics—buffalo hides became all the rage across America and Europe and professional hunters swarmed onto the plains in the 1870s, amassing great, boxcar-sized stacks of treated hides at rail depots from Texas to the Dakotas—the buffalo herds were reduced by the mid-1880s to "a pitiful remnant numbering in the hundreds," as Hutton observed.

Sheridan had an exquisite understanding of the value of public relations and political power, and that is how, as I drove my family into North Platte, both "Uncle Phil" and Buffalo Bill together came to mind. Sheridan treasured Cody, the brave and resourceful army scout, both for his value in the Plains War and because Cody reminded him of himself: ponder the younger Sheridan's legendary Civil War ride to rejoin his troops against Jubal Early's raiders at Cedar Creek, or how he stood in a rifle pit at the base of Missionary Ridge and saluted the enemy guns bristling overhead on the summit with his pewter flask of brandy. Fearlessness and flair—that's what it took, and both had it in excess. The buffalo hunts that Sheridan sponsored and Cody led out of Fort McPherson and North Platte in 1871 and 1872 required a little fearlessness—riding into a large herd at a gallop, rifle at the ready, was, after all, dangerous—but also needed a large amount of flair. The great publicity afforded these hunts helped create the legend of Buffalo Bill among Americans everywhere.

General Sheridan sent many civilian hunting parties from the East— always friends and dignitaries who could do him some political good—to the point that some post commanders groaned under the weight of the letters of introduction and the calls for military escorts and supplies. Whenever he could excuse himself from administrative duties in Chicago, Sheridan went along. He was there for the 1871 expedition, which took many high army officers and civilian dignitaries and included James Gordon Bennett, renowned editor of the *New York Herald*; Charles L. Wilson, editor of the *Chicago Evening Journal*; and Henry Eugene Davies, who wrote a small book about the adventure, *Ten Days on the Plains*. Davies, a major general of volunteers during the Civil War, vividly described William Frederick Cody as he first glimpsed him, and it went into the collective mind's eye of the public. "The most striking feature of the whole was the figure of our friend Buffalo Bill riding down from the Fort to our camp," he wrote, "mounted upon a snowy white horse. Dressed in a suit of light buckskin, trimmed along the seams with fringes of the same leather, his costume lighted by the crimson shirt worn under his open coat, a broad sombrero on his head, and carrying his rifle lightly in his hand as his horse came toward us on an easy gallop, he realized to perfection the bold hunter and gallant sportsman of the plains."

The party required sixteen wagons for baggage and supplies, including one for ice, and three four-horse ambulances for comfortable riding when the tenderfeet got saddlesore. Some had brought greyhounds along for the side hunts of antelope and rabbit. As deliriously described by General Davies and by Cody's delightful biographer, Don Russell, at dinnertime the dignitaries sat at sumptuously outfitted tables with linen, china, silver, and crystal; waiters in evening dress served multicourse dinners prepared by French chefs. Champagne flowed like water. Cody himself recorded in his memoirs that "for years afterward travellers and settlers recognized the sites upon which these camps had been constructed by the quantities of empty bottles which remained behind to mark them."

Champagne fueled carnage as much as gunpowder: in ten days, following a trail nearly 200 miles long from the Platte Valley to the Republican, Solomon, and Saline rivers, the hunters killed more than 600 buffalo and 200 elk, and hundreds of antelope and wild turkeys. This, of course, was far beyond sport. Some months later, even this record for indulgence and promiscuous destruction was in several ways surpassed, when in early 1872 Sheridan sponsored a royal hunt for the Grand Duke Alexis of Russia, twenty-one-year-old son of the czar, whose extensive grand tour of the United States had nearly caused a national shortage of newspaper ink. With Sheridan riding next to Buffalo Bill at the head, the hunters trailed out across a great, clattering, dusty mile, including the royal party, American dignitaries both civilian and military (not only Sheridan but Lieutenant Colonel George A. Custer), escorted by two companies of cavalry, two of infantry, and the 2nd Cavalry's regimental band. Sheridan personally detailed Cody away from the 5th Cavalry, which had just been transferred to Arizona, to have him as the centerpiece, and Cody's first duty had been to ride northward to the Brule Sioux reservation and persuade Chief Spotted Tail to join the hunt with 100 of his warriors. As had been the case with the railroaders and their Pawnee entertainers, Spotted Tail's men entertained Grand Duke Alexis with war dances, mock raids, daredevil riding, and shooting matches.

The party was so successful that Sheridan granted Cody a paid leave of absence from the army and encouraged him to go East so that one of his important friends, James Gordon Bennett, could entertain Cody in New York. Once there, feted by publisher Bennett and his illustrious friends, Cody became the toast of New York, finding his way into the society as well as the news columns. In New York he was reunited with Ned Buntline, who had tirelessly inflated the Buffalo Bill legend with serialized newspaper tales like *Buffalo Bill, the King of Border Men* and the soon-to-be-published *Buffalo Bill's Best Shot; or, The Heart of Spotted Tail*. (It would have been a surprise and disappointment to the real Spotted Tail, who had just contributed so much drama to Grand Duke Alexis's extravaganza, to learn that he had been killed in print by his real-life friend Buffalo Bill.) Buntline's first epic had been dramatized for the stage, and Cody went with him on February 20, 1872, to the Bowery Theater on opening night. The audience thrilled to the wild and woolly plot as smoke from blank cartridges filled the auditorium, but when it was announced that Buffalo Bill himself was in attendance, the people roared for him to stand and bow from his box. Afterward, the theater manager offered Cody $500 a week to play himself onstage. Pay for an army scout was $500 a month.

Though Buffalo Bill demurred—he was due back at Fort McPherson to join the 3rd Cavalry—public acclaim and the jingle of coins had filled his ears as the high life had dazed his eyes. He would never be the same. By November 1872, Buntline had wheedled him off the plains in favor of the footlights of a Chicago theater. "On the whole it is not probable that Chicago will ever look upon the like again," said the *Chicago Tribune*. "Such a combination of incongruous drama, execrable acting, renowned performers, mixed audience, intolerable stench, scalping, blood and thunder, is not likely to be vouchsafed to a city a second time—even Chicago." The critic erred only in thinking such a spectacle would never be repeated. Cody would star in a series of fantastically popular frontier melodramas all around the nation for a decade. The run was only broken in 1876 with a brief summer return to real-life scouting with the 5th Cavalry and a retaliation against the Cheyenne for the Sioux's killing of Custer and his five companies of cavalrymen up at Little Big Horn. During his years in the limelight, as Don Russell has calculated, no fewer than 1,700 dime novels (including reprints) were published by an assortment of writers with titles like *Buffalo Bill's Bonanza; or, The Knights of the Silver Circle* and *Buffalo Bill's Brand; or, The Brimstone Brotherhood*. As his biographers have taken pains to show, on stage and off, on the page as well as in public and private, the line between life and legend continued to blur.

For thirty years William Frederick Cody was North Platte's most famous citizen. In 1878 the traveling thespian built a home he called Welcome Wigwam for his mostly stay-at-home wife, Lulu, and their two daughters in North Platte, and invested in the cattle ranching business nearby with Frank North. In 1882 during a home stay from the theatrical circuit, he learned that there would be no organized Fourth of July celebration, and Cody could not let that stand. He got the owner of the town's racetrack to reserve the space, persuaded a friend to donate his small herd of buffalo and local businessmen to put up prizes for riding, shooting, roping, and other stunts, and put out a call for 100 cowboys to compete. A thousand signed up. Cody sent out 5,000 handbills for the "Old Glory Blow-Out." As one contemporary described the public's reaction, "the attendance was unprecedented for that section—the whole country for a radius of over one hundred and fifty miles being temporarily depopulated." Cody had been mulling over the idea of exchanging the floorboards of one-night-only stages and a small theatrical company for the sawdust of big tents and the dust, noise, and stupendous excitement of an outdoor spectacle with a large, rough-and-ready cast, much like the circus he'd seen at North Platte thirteen years before (and much like other traveling Western shows). Buffalo Bill's Old Glory Blow-Out has been credited with being the first organized rodeo, and although that is not accurate, it certainly became the best known. From that was born Buffalo Bill's Wild West and Congress of Rough Riders of the World (as it would come to be named), the bright, celebrated troupe that toured America, England, and Europe from its opening at Omaha on May 17, 1883, until it sputtered and winked out three extraordinary decades later.

In 1886 Cody built a larger and grander house, named Scout's Rest, on a sprawling, 4,000-acre ranch north of North Platte, not far from the racetrack that had given him a new direction for his theatricality. By then his

marriage had grown complicated—he was seldom home, and when away enjoyed the company of other women, which infuriated his wife, who berated
him and sent private detectives on his heels. For some years, therefore, they
maintained separate residences in the same town. Scout's Rest was a fifteen-
room house with porches ten feet wide, designed in the popular French Second Empire style with mansard roof, dormer and bay windows, bracketed
eaves, and a splendid central tower. It looked like a large white wedding cake
trimmed with green icing and set down by accident on the Nebraska sod.
Cody left the designing, building, and furnishing details to his sister and
brother-in-law, who were to be resident managers, with only the stipulation
that the bibulous Buffalo Bill's room be equipped with a sideboard stocked
with whiskey decanters and glasses and a large bed. That room had a secret
second door, which aided Cody in entertaining female guests. "Sprees at
North Platte between tours," noted Cody's biographer Don Russell, "became
proverbial." Other visitors to Scout's Rest included the army scout North
brothers, Kit Carson, Philip Sheridan, assorted members of the European nobility, capitalists, and show business luminaries of the era, as well as rougher
remnants of Cody's exciting frontier life. He lived there until 1913 (he and his
wife were finally reconciled after years of storms and a messy, uncompleted
divorce case), when the Codys moved to Cody, Wyoming, a longtime investment property of his. Forty-eight years after Buffalo Bill's death in Denver in
1917, Scout's Rest became a state historical park in 1965, preserved by the
Nebraska Game and Parks Commission.

The house and rambling stock barn sit beneath statuesque cottonwoods
on sixteen manicured and shrub-tended acres, North Platte having overspread the rest of the original acreage. Driving there through residential
neighborhoods that varied from hundreds of tiny, crowded trailer lots hard by
the railroad yards, to neat bungalows, ranch houses, and turn-of-the-century
houses with well-watered lawns, we passed many places trading on Cody's
name, like Buffalo Bill's RV Park and the Scout Motel. In the center of wide
lawns the two-story house was well restored on the outside, its interior furnished in overstuffed period furniture. But we found the exhibits, though
certainly resonant with Buffalo Bill's legend, to be amateurishly curated—
original period photographs, many in dimestore frames with glass pressing
against the old emulsion, others matted with acid-rich oaktag; lithographs
and posters "protected" with do-it-yourself plastic adhesive laminates; exhibition cases with interiors lit by standard Westinghouse Econowatt fluorescent bulbs, which over time were ruining Cody's stetson hat, marriage license,
stock certificates for the Wild West Show, dime magazines and novels, holo-
graph letters in Buffalo Bill's handwriting, felt pennants, and many other treasures. Outside in a paddock near the barn was a "Wild Buffalo Display" of
exactly three—a bull, a calf, and a cow, indolent in the high summer heat. In

the barn, numbers of original saddles ridden in the great show sat slowly deteriorating astride horse stall partitions in the sunlight and open air. Overlooking battered and peeling, though attractive, old carriages and wagons, were a great many brittle original Wild West Show posters tacked directly to the wall, some with the yellowing plastic adhesive laminate, many without, one with a fire sprinkler pipe cut through it.

Despite the dreary and careless presentation, the brightly colored and exciting lithographs with scenes of high drama lent a strongly cheerful air. On one wall was a heroically sized painted portrait of Cody, white shoulder-length hair flowing down to his buckskin suit, having doffed his Stetson while his horse performed a practiced bow, the standard farewell that closed out every show and brought the thousands to their feet. Though empty of life other than tourists and tour guides, the place seemed full of echoes of wintering race horses, show ponies, even dromidaries, and not even the scratchy, noisy, unevenly running and jumping biographical film, no better and no worse than any of the 16-millimeter school documentary movies I sat through back in the 1950s, could diminish the brawny Buffalo Bill legend that lit the house, barn, and grounds from within.

On the other side of town, between the South Platte and the interstate, is another tribute to Buffalo Bill at a large souvenir emporium called Fort Cody Trading Post and disguised as a palisaded log fort with tall corner watchtowers. Fort Cody is advertised as Nebraska's largest souvenir and Western gift store, and in our travels across the state we saw nothing to rival it. Spilling off shelves and display cases are moccasins, T-shirts, toys, candles, Native American crafts and sand paintings, figurines, key chains, stuffed animals, and tray upon tray of turquoise jewelry, bone feather jewelry, silver, belt buckles, and leather goods. Commendably, there's also a large book section, strong on Western history, Native Americana, Western art, and fiction. There is also an Old West Museum, displaying an array of weaponry, vintage cowboy clothing and gear, cavalry and Indian Wars memorabilia, and Native American beadwork, as well as mounted buffalo, moose, coyote, and elk, not to forget a little stuffed two-headed calf said to have been born in the Nebraska sand hills.

But the showpiece is the Buffalo Bill Miniature Wild West Show, a display of tiny figurines hand carved and painted by local citizens Ernie and Virginia Palmquist—more than 20,000 pieces. Below small outdoor reviewing stands filled with individually carved viewers is a center ring display of Buffalo Bill, cowboys and Indians, and trick shooter Annie Oakley. Nearby, the grand parade of cowboys, Rough Riders, hussars, cossacks, Indians, Abyssinians, musicians, circus wagons, covered wagons, and stagecoaches march or ride beneath waving flags and banners on a little motorized moving belt, while tinny marching music emerges from hidden speakers. Alongside the animated show are vignettes of circus acts and sideshow displays of snake charmers, a knife thrower, and a fat lady. Behind the scenes are the traveling troupe's kitchen and dining areas, sleeping tents, a barber shop, and even a crooked card game. It is all witty in the whimsical Forties fashion made famous by Norman Rockwell and Reginald Marsh, perhaps with a little Rockwell Kent thrown in, but it's also inimitably Western. Not even after bending and peering for an hour did one get a fill of it. The Palmquists took twelve years to carve and paint the display. After they retired to Florida they apparently lost touch with North Platte and with their exquisite handiwork. One day, a clerk told me, the Fort Cody phone rang. It was a long-distance call, and the high, ancient voice on the other end, inquiring if they still had that old Buffalo Bill display at work, was Ernie Palmquist himself, pleased to be reassured that his tiny, beloved Wild West Show was still amazing tourists.

North Platte was born as a railroad town and there's no reason to think it will ever stop being a railroad town—2,500 Union Pacific employees still call it home because of the Bailey Yard, the largest railroad classification yard in the world. Every twenty-four hours, coupled in 130 trains, some 10,000 railroad cars pass through the Bailey, a third of them requiring transferal to a different train. There are no fewer than 114 tracks in the classification yard. It sprawls for what seems hundreds of acres, and there is a constant thunder of steel wheels and a pall of diesel smoke in the air—music and ambrosia to train buffs with their elaborate amateur film gear, who can be found in avid numbers at overlooks day and night, delirious at the cacophony and the bright palette of boxcar colors. In the spring and summer of 2001, the Union Pacific would open a modernistic glass visitors center at the Bailey Yard, with a three-screen surround film called *On Track for America*, a Hall of Fame, and interactive and static displays. All the railroadiana would be overlooked by a shining, gold, fifteen-story observation tower fashioned to look like a golden spike, with dual elevators ready to speed enthusiasts up to a panoramic view of the busy yards and its thousands of railroad cars and scores of engines.

That not being open in the summer of 2000, we made our way to humbler Cody Park and its historic railroad museum. There one can climb aboard a massive black 3900 class Challenger steam locomotive, built by the American Schenectady Works and looking like it had rolled right out of a 1940s movie, or hang off the side of a bright yellow 6900 class Centennial diesel locomotive, built by General Motors in 1969 and retired from service in 1985. Other rolling stock on display include a caboose, baggage car, and mail car, beside an old-fashioned depot brought there from Hershey, Nebraska, and now filled with railroad equipment and memorabilia. The park is particularly striking during summer evenings, when displays are open until 9:00 P.M. and the locomotives' headlights, running lights, marker, and class lights are on, as are the block signals and depot platform lights.

Two days—July 3 and 4, no less—were hardly sufficient to do North Platte justice, but the pull westward was palpable. Following an old railbed out of town, we had the sight of the South Platte River, shrouded by cottonwoods but much as it had appeared to Henry Morton Stanley in 1867. "On our left was truly a beautiful stream, notwithstanding the bad repute given to it by emigrants," he wrote.

> Its shifting shoals and treacherous quicksands may render fording in many places dangerous and impracticable, but they do not detract from the view. The surface of the river appeared nearly on a level with the meadow lands, and glistened like a tin roof in the noonday sun. The stream is choked with myriads of small islets, tufted with wild grass and tall weeds, that lie basking in the sunshine like huge bouquets tossed by some wandering Peri into the stream. Over one hundred of these little islets were counted within a short distance.

North Platte is on the boundary between Central Standard and Mountain Standard time, giving us, when we crossed the line on the western city limits, an extra hour in the day as we drove past auto graveyards and a defunct drive-in—white paint peeled in large strips from the screen, and some aquamarine paint still hung on the weed-surrounded Art Deco refreshment stand, worthy of its place on Route 30. "I would like to see *My Man Godfrey* with Carole Lombard in that drive-in," said Mimi. "On a double feature," David replied, "with *It Happened One Night* with Clark Gable." Their parents agreed.

Beyond, there were cattle ranches and feed yards, corn and soybean fields, and a green and yellow John Deere thresher moving through alfalfa. At one driveway, two mailboxes on poles: one the regulation thirty inches, the other ten feet higher with a little silver model plane affixed, and a sign—AIRMAIL. At O'Fallon's and Sutherland the sand hills had crept down to a half mile north of us, covered by short, buff grass and clumps of sage. At Paxton there were bungalows and ranches and a one-block main street with an Art Deco neon sign, "Ole's Big Game Bar," and next, "Ole's Big Game Restaurant and Lounge," said to contain 200 mounted trophies from five continents, personally shot by the former owner, Ole Herstedt. Then, sandstone blocks and three big elliptical windows looking like a big radio. After that a yellow brick crenellated and corniced edifice called "Swede's Lounge." Paxton, population 536, was evidently a jumping place, the reason for which we may have found on a big yellow sign near the ramshackle depot: "Enjoy Beef: Real Food for Real People. —Nebraska Cattlewomen and Cattlemen." We bought fresh popcorn at a store called "Paxton's Pit Stop." "Do you want to carry it away in a sack?" a pretty fortyish blond woman in a yellow T-shirt asked. "You can use it for trash later." Shelf upon shelf of pint bottles rose behind her. The store was so tightly packed with merchandise she could have boasted of having one of everything.

Westward, the sand hills came right up against the road, considerably rockier. There were yellow hay ricks in the fields outside of Ogallala (pronounced with the third syllable accented), named after the Ogallala branch of the Sioux tribe. There was a Pony Express and stage station there in the early 1860s; the eloquent diarist Frank Young found it occupied by soldiers in 1865. It looked, he said, "as comfortless as a Siberian picture in a storybook." In 1867, though, the town went into a long period of prosperity with the arrival of cowboys driving Texas cattle herds up to the Union Pacific line. Ogallala became a fabled cow town, as wild as any wild place in the West. Outside town rose Boot Hill cemetery, where many a desperado and many good men as well were buried with their boots on. "There has not been a burial since the 1880s," Mary read from the 1939 WPA guide. "Boot Hill today, except for a faded sign bearing its name, is like any other hill. No mounds are visible and there are no grave stones." We stood at its summit in dry weeds near a sign warning us of rattlesnakes, eyeing a number of ersatz board grave markers with whimsical epitaphs put up by local civic boosters. Downhill, beyond our car, was a tidy, greenery-cooled suburban home development with tricycles abandoned in asphalt driveways under the 100 degree heat.

The Ogallala of today thrives as much as the cow town, with many streets and active businesses. We could not bypass a block of stores fashioned for tourists like a Wild West street, and we threaded through a busy parking lot to the free Western Museum attached to a gift shop called A&D Collectibles. Inside the former there were lifesize dioramas of a sheriff's office, barbershop, and undertaker, and well-curated display cases with photographs, artifacts, and explanatory texts; in the latter, presiding over the cash register, was owner Alice Hackbarth, who with her husband, Dale, had bought the shop and museum in 1994. "Dale died last year," Alice said. She was auburn-haired, arthritic, and quick to smile from where she looked up, leaning over a cane. "We owned a restaurant down at the dam on the lake," she continued, "ran it forty-five years. We started with a lunch counter and six stools, and built it to where it seated eighty-five people. When you're young, you can do things like that. You have the energy. When we retired and bought this store, and then Dale died, I didn't want to make any sudden changes, but I think I'm just about ready now." The shop looked good, and, judging by the number of patrons drawn in by the free museum, she would get a good price when she retired the second time.

A few doors down was the newly opened Kenfield Petrified Wood Gallery, owned by the identical twins Howard and Harvey Kenfield. Born in Albion, Nebraska, in 1928 and raised in Brule, they served together with the 24th Infantry Division in Korea. After their hitch they worked for an electric company for years, during vacations going on expeditions for rocks, gems, fossils, petrified wood, and Indian artifacts. Their forty years of collecting had earlier been on display on a road south of Ogallala and off Highway 61 where it had been flanked by the brothers' houses and set behind an elaborate, manicured garden, infrequently found by wanderers. Now moved to the downtown Wild West block, sold to the Western Nebraska Community Foundation, the seventy-two-year-old twins continued to run it, hoping the gallery would attract more people. In their exhibit were palm and bamboo from Wyoming, sequoia from North Dakota, fossilized fish and leaves from Colorado, agates from Nebraska to Brazil and South Africa, black walnut and oak from Oregon, sycamore from Oregon, and cypress from Washington State, all attesting to different climates in ancient times.

"Old rockhounds never die," read a little sign. "They just slowly petrify." The gallery was filled also with little folk art pictures, wall sculptures, and free-standing models rendered by the brothers out of tiny slices and pieces of petrified wood, depicting desolate ghost town streets, ranches, barns, and windmills. Perhaps from years of patient prospecting, or from the methodical steps of rock polishing, or the incredibly detailed little architectural renderings of Great Plains vignettes, the brothers were soft-spoken and serene so that visitors leaned forward to hear what they said about the process of fossilization, and listeners felt almost blessed by their presence. "They're bodhisattvas," murmured my wife at one point, as we leaned over an amythyst geode from Brazil as big as an elephant's tusk, while Harvey went to find a leftover brochure from the old gallery, wanting to illustrate a point. And from that we later read their credo. "We thank our Lord for the creation of beautiful rocks and providing us with the ability and talent to create so many pieces of art," their brochure pronounces. "Many people ask the age of petrified wood. It is controversial, since many Christians date the earth between 5,000 and 50,000 years. . . . While those who study paleontology date petrified wood in the millions of years, to us it's still a mystery. What really is important is the beauty of each piece and the mystery it holds."

With both sides of the ancient science versus faith debate seemingly sent back to their corners by the gentle Kenfield brothers, we bid them farewell to meander west and southwest toward the site of the wildest Hell on Wheels town, Julesburg, just down Route 138 and barely a point-blank shot over the Colorado line.

In 1861 the twenty-five-year-old tramp printer and steamboat pilot Samuel Clemens left St. Joseph with his brother, Orion, on a stagecoach trip to Nevada Territory. There Orion had secured a patronage job as secretary to the territorial governor, a reward for political work toward Abraham Lincoln's presidential election some months before. On their fifth day and 470th mile out from St. Joseph, they found themselves alighting at the Julesburg stagecoach station. "It did seem strange enough to see a town again after what appeared to us such a long acquaintance with deep, still, almost lifeless and houseless solitude," Clemens would note in the whimsical travel journal he'd publish under the name of Mark Twain. "We tumbled out into the busy street feeling like meteoric people crumbled off the corner of some other world, and wakened up suddenly in this." An hour later, they were breathlessly off, leaving no details on Julesburg beyond the observation that it was "the strangest, quaintest, funniest frontier town that our untraveled eyes had ever stared at and been astonished with."

Four years later, the whole town was destroyed during an Indian raid. It was rebuilt three miles away, but with the arrival of Union Pacific work crews in 1867, Julesburg picked itself up and moved over to trackside, where for a brief moment in Western history it burned very bright.

Henry Morton Stanley arrived there in August 1867 on a stage from Omaha, stepping out into "a mixed crowd, composed of gamblers, teamsters, and soldiers," finding that much of the frenetic excitement and air of dangerousness had traveled down the railroad line with the portable buildings and tents from North Platte. After dinner he walked around, "astonished at the extraordinary growth of the town, and the energy of its people. It was unmistakable go-ahead-it-ative-ness, illustrated by substantial warehouses, stores, saloons, piled with goods of all sorts, and of the newest fashion." Gambling and drinking establishments were full.

> I walked on till I came to a dance-house, bearing the euphonious title of "King of the Hills," gorgeously decorated and brilliantly lighted. Coming suddenly from the dimly lighted street to the kerosene-lighted restaurant, I was almost blinded by the glare and stunned by the clatter. The ground floor was as crowded as it could well be, and all were talking loud and fast, and mostly every one seemed bent on debauchery, and dissipation. The women appeared to be the most reckless, and the men seemed nothing loath to enter a whirlpool of sin. Several of the women had what they called "husbands," and these occasional wives bore their husbands' names with as much ease as if both mayor and priest had

given them a legal title. The managers of the saloons rake in greenbacks by hundreds every night; there appears to be plenty of money here, and plenty of fools to squander it.

These women are expensive articles, and come in for a large share of the money wasted. In broad daylight they may be seen gliding through the sandy streets in Black Crook dresses, carrying fancy derringers slung to their waists, with which tools they are dangerously expert. Should they get into a fuss, western chivalry will not allow them to be abused by any man whom they may have robbed.

At night new aspects are presented in this city of premature growth. Watch-fires gleam over the sea-like expanse of ground outside of the city, while inside soldiers, herdsmen, teamsters, women, railroad men, are dancing, singing, or gambling. I verily believe that there are men here who would murder a fellow-creature for five dollars. Nay, there are men who have already done it, and who stalk abroad in daylight unwhipped of justice. Not a day passes but a dead body is found somewhere in the vicinity with pockets riffled of their contents. But the people generally are strangely indifferent to what is going on.

A month later he returned and found the place being deserted. "Julesburg is now an overdone town, a played-out place. It was built in a single month. . . . It is now about to be abandoned by the transient sojourners, and many of them are shifting their portable shanties to some prospective city west—Cheyenne, or some 'prairie-dog town,' where cash can be made without work, and by any means that will not subject the operator to an indictment before a Grand Jury for obtaining money under false pretences."

We found the fourth Julesburg where the maps and guidebooks said it would be, three miles away from the featureless Colorado soil that covered all traces of Gomorrah, lying among broken hills in a curve of the South Platte River: grain elevators and railroad tracks, a one-story town with a little museum in the depot and more exhibits down the street marking the history of Fort Sedgwick, operated between 1864 and 1871. The Colorado WPA guide found almost nothing to say about modern Julesburg other than that it had been founded in 1881 and lay on the U.P. main line to Denver.

A Union Pacific double-header freight thundered by as our children clambered aboard a miniature engine standing on the depot grounds next to a tipi, which had been donated, we learned later, by a movie company after filming the Kevin Costner epic *Dances with Wolves*. Inside the museum was an entertaining jumble of artifacts from a vanished lifetime on the frontier, dominated by its simulacrum, a vintage movie poster from Cecil B. DeMille's sprawling

Union Pacific, featuring Joel McCrea nuzzling Barbara Stanwyck. We were shown around by the society's director, Lenora Troelstrup. ("That's a Danish name," she explained. "My husband's name is Nels.") They lived on the ranch her husband's family had owned for 100 years. Nels's namesake and grandfather was a cattle baron whose work and play took him all over the world, and who was a personal friend of Buffalo Bill's—"Grandfather had a check for $300 from him," Lenora told us, "but he never cashed it." The ranch in those olden days was 36,000 acres, in recent years, sadly, considerably diminished. "We have four sons," said Lenora, "and wanted to keep the ranch in the family. *National Geographic* did a story on it a few years ago. But we've had to sell it off, one piece after another, to cover taxes and expenses. These haven't been friendly times for anyone, landowners or not, sad as it is to see Grandfather's spread crumbling away."

Outside and across the street at Julesburg Auto Sales, there was one Ford van out on the lot, among the weeds. Another freight roared by. It was only my imagination, but the big train's wake seemed to make the few buildings of Julesburg shudder a little. I thought about how, in October and November 1867, the tracklayers spiked their way off the Nebraska panhandle, across the Wyoming territorial line, and stopped at the marked-off depot lot in a new, growing town named Cheyenne by the Union Pacific's chief engineer and land agent, Grenville Dodge. On November 13 the first train whistled into town. Behind the locomotive stretched a long series of flatcars, upon which were stacked the broken-down shacks and dismantled frame buildings, tents, whiskey crates, brass beds, mattresses, looking glasses, barstools, and gaming tables of what was now a dusty, desolate ghost town in the northeastern corner of Colorado.

"Gentlemen," called a guard as he swung down off the train into an expectant platform crowd, "here's Julesburg!"

Hell on Wheels—and the Course of Empire—were continuing to edge westward. As were we.

7

The View from the Bluffs

A s we left the South Platte there was an image in my mind's eye of a watercolor made of the stream many years ago. It showed the popular fording place where, from the 1840s on, the Oregon- and California-bound wagon trains that had followed the south bank of the Platte past the river fork at Fort McPherson, and past Julesburg by a few miles, turned to strike northward toward the North Platte and the continuing trail to Fort Laramie. In ruddy tones, the painting shows the wide South Platte with lines of ox-drawn covered wagons threading down to the rutted bank and venturing out into the brown, muddy current. There are knots of emigrants standing and watching the progress, worrying about upsets and pools of quicksand in the streambed. A Sioux tipi stands nearby—stolid witness to the procession signaling the beginning of the end of Native American culture. The painting was drawn from memory and a hurried sketch done by a twenty-three-year-old artist an incredible seventy years after the events he depicted. The long-lived artist was William Henry Jackson, and one of our goals over the next several days was to see some of the master's original sketches and watercolors and the subjects he saw along a stretch of the Oregon Trail.

Jackson, from upstate New York, had worked as a colorist in photographers' studios in Rutland and Burlington, Vermont, but a soured romance and restlessness sent him West, where, in 1866, penniless and desperate, he signed on as a driver, or bullwhacker, in a freight wagon train. It was heading from the Missouri River landing at Nebraska City to Fort Laramie, Wyoming Territory, and then all the way to Salt Lake City; Jackson had no experience but the trail boss was desperate. Wages were a measly $20 per month, and young Jackson would be in hock for supplies from his boss for two months before he would begin to clear anything.

It was a brutal journey at an agonizing pace; they averaged fourteen miles a day. And there was no end to a bullwhacker's duties—tending, yoking, and driving oxen, keeping them in line with a sharp, cruel snap of the driver's leather bullwhip, wrestling and winching the wagons across defiles and up and over slopes, and fording streams. The South Platte at Julesburg was the most dramatic crossing. Just a few years before Sam Clemens had gone through, finding it "a melancholy stream straggling through the center of the enormous flat plain," and hardly remarkable given his pilot's familiarity with the mighty Mississippi. He couldn't refrain from sneering. "The Platte was 'up,' they said—which made me wish I could see it when it was down, if it could look any sicker and sorrier." Clemens's stagecoach did encounter some anxious moments with quicksand at midstream, but he was not impressed.

By contrast, William Henry Jackson saw it at a higher season, crowded with puny human activity and seething with drama, and his diary entry brings it all to life, as does the painting he rendered so many decades later. "The river was filled from bank to bank with wagon trains passing over and oxen recrossing for another wagon," he wrote. "The drivers as a general thing were minus pants and attired in the light and airy costume of the shirt and hat only.

"The bulls when they first enter the stream are timid and it is extremely hard work to get them started, and the yelling, the hooting, the hurrahing and the cracking of whips of the drivers make [for] pandemonium. The cattle plunged and tugged and geed and hawed & after a deal of trouble pulled out of the plunge from the bank of the stream. Clear across the river it was the same yelling at the oxen—'yaw-ho, whoa haw,' all the time laying on your very best and might with cudgel or whip."

All the talk on the trail was of the Union Pacific Railroad, now building up the Platte Valley. Jackson did not know it, but in a few years, after adventures on wagon roads all the way to the Pacific Ocean and back, the artist would become the railroad's official photographer, capturing exquisite scenes along the right-of-way and out into the framing and defining mountains and plains. We would see more of him and his artistry soon.

As Jackson's wagon train rattled off in the direction of the North Platte River, which they would trace northwesterly as had the thousands of Oregon/California Trail emigrants and freight haulers before them, location engineers for the Union Pacific were staking the railroad right-of-way from the South Platte crossing in a more directly western route along Lodgepole Creek toward Wyoming. Fifty years later the Lincoln Highway would adopt a parallel route. Heading up near the old rail line to Route 30 from our brief sojourn in Colorado, we shuddered as we ducked under Interstate 80 with its screaming semis and rushing RVs, and turned westward onto the stately vintage highway a few miles later at Chappell. High limestone bluffs rose to the north. Occasional brush-choked canyons with mostly dry creekbeds opened beyond well-tended ranches. At Lodgepole: concrete Thirties-era motels and gas stations, with cottages beyond. For a while we paced a long U.P. freight drawn by seven engines. Outside Sunol, a herd of 100 buffalo were grazing behind barbed wire fences. A dark column of smoke from a wildfire climbed from the far side of a hill to stain the clouds. It was the first of many wildfires we would pass in a summer that would be known for its terrible dryness and disastrous fires, all the way to California.

We drove through Sidney, ONE OF AMERICA'S FAVORITE STOPPING PLACES SINCE 1867, proclaimed a sign. It had been named after Sidney Dillon, a U.P. executive, who had cause to rue the honor. In 1868 on a nice summer's day, two conductors, Tom Cahoon and Wilkes Edmundson, went fishing on their day off down at Lodgepole Creek. They were attacked by a Sioux war party, shot, and scalped, but drove off the raiders with a desperate burst of energy and some lucky shots. Both men survived. Cahoon worked for the Union Pacific for many years, wearing his conductor's hat well to the back of his head to cover his bald spot. Depredations like this dogged the railroaders all across 1867 and 1868, but Sidney Dillon really abhorred his namesake town in the Lodgepole Valley because Sidney, Nebraska, was a notorious hell town for many years. The railroad warned its through passengers that if they disembarked from the train, they did so at their own peril. Sidney never closed—at one time, the WPA guide told us, there were twenty-three saloons in one block. "Shootings were a daily event that drew little attention," I read to Mary (who was driving) and the kids. "Someone was shot at a dance one night and instead of stopping the dance the incident only served to heighten the entertainment. The corpse was propped up in a corner and the dancing continued. During a later blast of gunfire, another man was killed. His body was set up beside that of the first victim. It was not until a third corpse was added to the group that the party came to an end."

Just west of Sidney was a marker for the trailhead of the Sidney–Deadwood Trail, a supply wagon road between the U.P. tracks and the Black Hills gold rush town of Deadwood, South Dakota. Farther down the highway, passing through blink-of-an-eye hamlets and intervening ranches, we finally came to the relatively bustling Kimball, population 2,574, with its busy brickfront Main Street and quiet bungalow lanes shaded by big cottonwoods. We went north there, following four-lane Route 71 arrow-straight into the hills between hayfields decorated with round bales, climbing a long, long divide for miles until we crested to behold dramatic cream-colored bluffs ahead marching east and west. All of us were excited to see landforms again, especially the urgent buttes, synonymous in our minds with the West and offering rock-solid proof that we had most certainly crossed that boundary.

Twenty miles away, a curtain of rain obscured the east. We were in the Wildcat Hills now, winding and threading through and down toward the North Platte Valley where we would rejoin the wagon ruts of the California/Oregon pioneers. It was growing late in the day and gravity was pulling us down toward the stream, but there was time for a stop at the Wildcat Hills

Nature Center, emptied of visitors and employees and shut tight, but affording us an overlook facing east to canyons, cliffs, and glimpses of irrigated green fields beyond.

It towered so high over the wagon trail that emigrants saw it marking the horizon like a needle or a pencil line for three or four days—thirty, forty, or even fifty miles—as they urged their teams forward, tracing either the north or the south bank of the North Platte. It was the most remarkable natural feature any of them had ever seen, and detailed, awe-inspired descriptions appear in more pioneer diaries than any other landmark on that long and toilsome California/Oregon Trail. "To emigrants," said trails scholar Merrill Mattes, "Chimney Rock was indeed the eighth wonder of the world." Elsewhere, he wrote, "Today there may or may not be more people who have heard of South Pass or Independence Rock or Scotts Bluff than there are people who have heard of Chimney Rock. The point is, in historic times, in Oregon Trail days, Chimney Rock seems to have been the number one attraction."

It is thought to have been named in 1827 by Joshua Pilcher or perhaps by one of the forty-five fur trappers he guided toward the rendezvous at Bear Lake. The first printed description was noted by Captain B. L. E. Bonneville in 1832, and published by Washington Irving four years later. Five hundred fifty miles west of St. Joseph it stood, "a singular phenomenon," Irving-Bonneville reported, "which is among the curiosities of the country. It is called the chimney. The lower part is a conical mound, rising out of the naked plain; from the summit shoots up a shaft or column, about 120 feet in height, from which it derives its name. The height of the whole, according to Captain Bonneville, is 175 yards." Estimations of height varied wildly over those early years. Another of the earliest descriptions was left by Warren A. Ferris, who saw it in 1830. "[It] appears at the distance of fifty miles shooting up from the prairie in solitary grandeur, like the limbless trunk of a gigantic tree."

Travelers were immensely grateful for something to look at after Nebraska's numbing terrain, and outdid each other in subsequent similes. To Father Pierre DeSmet in 1841 it was an inverted funnel; to Rufus Sage in 1841 "like the spire of some church. . . . A grand and imposing spectacle, truly;—a wonderful display of the eccentricity of Nature"; to John C. Frémont's surveyor Charles Preuss in 1842, a "long chimney of a steam-factory establishment, or a shot-tower in Baltimore"; to Elijah White in 1842 "like the contemplated Washington Monument" in the nation's capital; to Philip St. George Cooke in 1845 "like the pharos of a prairie sea"; to William Clayton in 1847 like "the large factory chimneys in England"; to William G. Johnston in 1849 like "the great Egyptian obelisks, towering high above the vast deserts which surround them"; to William Kelly in 1849 like London's

Wellington Monument set atop a Danish castle. Observers finally got so fanciful it turned almost comical—"like a spire in the heavens, its top gilded by the setting sun, looking like its crest was burnished with gold" (Kirk Anderson, 1858), or "like the neck of an ostrich" (Daniel Gelwicks, 1849). Merrill Mattes enterprisingly examined hundreds of unpublished trail diaries and published travel accounts that mentioned Chimney Rock, and one of his favorites was that of a forty-niner on his way to stake his claim in the California Gold Rush—Elisha Douglas Perkins—whose diary finally found its way to the Henry E. Huntington Library in California. "No conception can be formed of the magnitude of this grand work of nature," scrawled Perkins on June 27, 1849, "till you stand at its base and look up. If a man does not feel like an insect then I don't know when he should."

Usually in those days Chimney Rock was surrounded by hundreds of camping covered wagons—there was an excellent spring nearby, and travelers left their names and dates carved or painted on the soft sandstone. "I saw hundreds of names out in the rock some at a dizzy height while others less ambitious had been content to subscribe their names lower down," recorded Charles M. Tuttle in 1859. "I wrote mine above all except two and theirs were about 8 feet higher than mine but I should have written mine as high if not higher than theirs if I had not left my knife back on an island in the river." "We all fell to deciphering the names and dates," wrote Benjamin M. Connor in 1863, "cut on almost every portion of its surface that could be reached by even the most agile climbers."

One of the most well-traveled to regard the rock was the flamboyant English explorer and linguist Richard Francis Burton, who had been born in 1821, dropped out of Oxford to join the Indian army, left the army seven years later to travel to the forbidden city of Mecca (which he penetrated disguised in a burnoose), and who had explored darkest Africa. In 1860 he crossed North America mainly to study the Mormon colony at Salt Lake City, publishing a wildly popular account of his journey. One day on the Nebraska trail in 1860 his party "nooned" for an hour at midday "at a little hovel called a ranch, with the normal corral." Burton sketched Chimney Rock. "One might almost expect to see smoke or steam jetting from the summit," he commented, adding that "the weather served us; nothing could be more picturesque than this lone pillar of pale rock lying against a huge black cloud, with the forked lightning playing over its devoted head."

Many observers predicted that the rock could not tower much longer with such frequent electrical storms hitting the needlelike landform; with strong winds, eroding sand, and splitting winter frosts, with constant damage from toes and boot heels of climbers and knife-carved names and dates, with scars from potshots from pistols and rifles. It was said that an army column once used the tip for target practice with a field cannon. It seems from the more trustworthy accounts and from period sketches and paintings as if the height of Chimney Rock was severely reduced between the 1830s and 1860s. Today, with the top standing some 470 feet above the level of the North Platte—the buff-colored chimney itself climbing 325 feet from its talus-widened base—it may be 50 to 100 feet shorter than in the days of the fur traders and Captain Bonneville, but there is no way to tell. It is still a remarkable thing to see, as one drives up the two-laned Route 92 westward through green irrigated fields and occasional cottonwoods, not far from the Burlington railroad tracks.

We had stayed overnight in a motel in the city of Scottsbluff, a sprawling, busy, and good-natured place with no vacant storefronts, and the next morning drove back eastward on 92 to see Chimney Rock from the side that the wagon pioneers would have seen, craning their necks and squinting through the choking dust and glare as they bumped and jostled up the riverbank. When young William Henry Jackson cracked his whip over his ox team in 1866 and turned onto the North Platte road, he took in the rough, eroded old bluffs and buttes and the new palette of colors with a joyful artist's eye—"yellows were turning into reds and saffrons, while the blues were becoming deep purples." He sketched from the driver's seat and at every campsite, and a lifetime later the images of what he took in, such as the sight of Chimney Rock rising on the horizon while a wagon procession rolled its cautious way down the steep incline of Ash Hollow to the river road, were recaptured in warm-toned watercolors.

There was little traffic on the highway, so we could slow to a few miles an hour to do some neck-craning ourselves, and I poked a video camera out the passenger-side window to study the pillar as we went east, overshooting by some miles before turning around and heading back toward the visitors center, only recently opened in 1994 and run by the Nebraska State Historical Society. A half mile away from the visitors center, Chimney Rock stands solitary and magnificent on eighty-three acres donated to the state of Nebraska by the family of a local rancher. Designated a National Historic Site in 1956, it is administered jointly by the Nebraska State Historical Society, the National Park Service, the U.S. Department of Interior, and the city of Bayard. The visitors center looks like a trim ranch building and sits low on the ground, respectful of the flat terrain and the dramatic contrast of the Chimney. Inside the attractive space are exhibits of pioneer clothing and belongings, Native American artifacts including a marvelous long-feathered headdress, and a large mounted buffalo head. Children are invited to play the "fill your wagon box with supplies without tipping over the weight limit" game, which never failed to attract Mimi and Davey; parents like us, who want to read all the wall texts and minutely examine the exhibits and pictures, bless museum planners who buy us an extra half hour or forty-five minutes of browsing time. In Mattes's day one could get up to the base of the rock only by clambering down through ravines, stumbling over geological debris, and worming under a barbed wire fence. One can get there a little easier nowadays, but with the number of rattlesnake warnings on signs and in brochures, combined with my overprotective worry about our two excited children, we elected to save that hike for another time.

Many pioneers didn't make it past Chimney Rock. Rebecca Burdick Winters, the daughter of a Revolutionary War veteran, was originally from New York, and she and her husband, Hiram, were early converts to the Mormon church. Persecutions of the Saints in Ohio, Illinois, and Iowa drove them ever westward, and they were part of the great exodus that pushed off from the Missouri River in 1852. Along the Platte River a number of the party fell victim to cholera, Rebecca Winters among them, though she fought the disease for more than a month while on a pile of quilts in the back of her husband's wagon. She finally succumbed perhaps a day after the emigrants passed Chimney Rock. Many of the great numbers of emigrants who perished were buried right in the middle of the trail with no marker so wagon wheels would tamp down the earth and discourage animals from disturbing the graves. But Hiram Winters buried his wife very deep, to the side of the trail. A friend bent a metal wagon wheel rim into an oval shape, like a tombstone, after chiseling "Rebecca Winters, Aged 50 Years," into the metal. The sorrowful party moved on to Wyoming and then Utah.

Forty-seven years later, in 1899, surveyors for the Burlington Northern Railroad found the lonely grave. Out of respect, they supposedly moved the route so it would pass five or six feet from the grave. Over subsequent decades, the whereabouts of that Mormon pioneer's resting place became known to historians, historical tourists, and Latter-day Saints congregants. Enough people found their way to the site—across the unfriendly, snake-ridden terrain and barbed wire fences and inches from an increasingly busy railroad track—that Burlington officials finally traced the Winters's descendants and got permission to move the grave. Archaeologists from the Nebraska State Historical Society exhumed the remains in 1995. In a large ceremony—125 members of her family attended, including her sixteen-year-old great-great-great-granddaughter, whose name was also Rebecca Winters—the pioneer was again laid to rest. Now prominently marked and easily accessible by car, the site is on the north bank of the river between Chimney Rock and Scotts Bluff. That wagon rim engraved with Rebecca Winters's name in 1852 is still there. In sunlight it's warm to the touch, and something more than mere heat passes into one's fingertips.

For narrative reasons I wanted to focus on Chimney Rock and then Rebecca Winters's gravesite in the order that westering pioneers would have seen them, before moving on to Scotts Bluff. In reality we had come down out of the Wildcats the night before to see the bluff—that massive pale iconic triangle dominating the valley in the late afternoon light and visible from everywhere in the nearby cities of Gering and Scottsbluff, even after dark. The subject of so many nineteenth-century artists' sketches and paintings, most durably from the watercolor brush of William Henry Jackson, the sudden sight of the prominence as we sped down ramplike Route 71 took my breath away and might have impelled us off the road had I been driving. Checking into our motel, going to dinner and then breakfast, heading down onto Route 92 to make the twenty-mile trip to Chimney Rock, the somber and striated 800-foot bluff stood there, seeming to invite us just a little farther westward, just as it must have done to the people riding and trudging beside the lines of white-topped wagons. Who could have described it better than Alfred J. Miller in 1837? "At a distance as we approached it," Miller noted, "the appearance was that of an immense fortification with bastions, towers, battlements, embrazures, scarps and counterscarps." His sketch, the earliest we know about, romantically made it seem so, from its monumental triangular white stone southward to the abrupt split of Mitchell Pass and then the sudden-rising wall of South Bluff with the standout Sentinel Rock and Crown Rock, attended by the eerily graceful Dome Rock. Many pioneers would see it two mornings after they had camped at the base of Chimney Rock; Scotts Bluff would sometimes be rosy in the early light, beckoning them on.

Scotts Bluff was named in memory of a fur trapper, Hiram Scott (c. 1805–1828), who went West at age seventeen or eighteen. Accounts of his death vary. It's known that he was with the Ashley-Chouteau caravan returning from the fur trappers' rendezvous of 1828 at Bear Lake (at the present-day Utah-Idaho border). He sickened at the rendezvous or on the return trip, suggests Merrill Mattes. Scott may have been abandoned by his two companions when they capsized in a bull boat on the North Platte River junction with the Laramie River. Having actually survived the river spill, Scott may have then crawled sixty miles to the foot of the bluff now bearing his name where he gave up the ghost. Alternatively, he may have simply died at the bluff. His bleached bones were found and buried by William L. Sublette in the spring of 1829. Whatever the actual circumstances of Hiram Scott's death, the landform inspired many emigrants to eloquence and imagination. "They encamped admidst high and beetling cliffs of indurated clay and sandstone," wrote Irving-Bonneville upon seeing them in 1832, "bearing the semblance of towers, castles, churches and fortified cities. At a distance it was scarcely possible to persuade one's self that the works of art were not mingled with those fantastic freaks of nature." In 1841 Rufus Sage tried to describe it. "The spectacle was grand and imposing beyond description," he essayed.

> It seemed as if Nature, in mere sportiveness, had thought to excel the noblest works of art, and rear up a mimic city as the grand metropolis of her empire. No higher encomium could be passed upon it than by employing the homely phrase of one of our voyageurs. In speaking of the varied enchantments of its scenery at the season, he said: "I could die here . . . certain of being not far from heaven.

J. Henry Carleton said in 1845 that "it constitutes a Mausoleum which the mightiest of earth might covet." Philip St. George Cooke saw it the same year and called it America's Gibraltar. And Richard Burton, who had actually seen Gibraltar, along with the cities and temples of India, the Near East, and Africa, was just as impressed in 1860. "As you approach within four or five miles," he wrote,

> a massive medieval city gradually defines itself, clustering, with a wonderful fullness of detail, round a colossal fortress, and crowned with a royal castle. Buttress and barbican, bastion, demilune and guardhouse, tower, turrent and donjon-keep, all are there; on one place parapets and battlements still stand upon the crumbling wall of a fortalise like the giant ruins of Chateau Gaillard. . . . Quaint watch and ward upon the slopes, the lion of Bastia crouches unmistakably overlooking the road. . . . Travellers have compared the glory of the mauvaises terres to Gibraltar, to the Capitol at Washington, to Stirling Castle. I could not think of anything in its presence but the Arab's "City of Brass," that mysterious abode of bewitching infidels, which often appears at a distance to the wayfarer toiling under the burning sun, but ever eludes his nearer approach.

In the shadow of the great bluff, not far from the low-slung Oregon Trail Museum, built of adobe in the 1930s to anchor the 3,000-acre monument for the National Park Service, stands a replica covered wagon. Nearby, wagon ruts are scored across the ground, leading west toward Mitchell Pass. In emigrants' minds this place marked the boundary line between the Great Plains and the Rocky Mountains, although appreciable mountain terrain was still days ahead of them. It was a hot day, mid-90s an hour before noon, and I stood for a minute in what little shade there was, reflecting on the lives that had been contained in those small wagon boxes of the people heading west into the unknown, whether it was Rebecca Winter ending her life in western Nebraska or Rose Donahue beginning hers in eastern Kansas.

We headed inside out of the sun, to find an attractively laid out exhibition space of artifacts of the wagon pioneers and the tribal peoples, but for us the greatest interest was the number of journal entries, notebook sketches, and finished watercolors of William Henry Jackson, superbly and respectfully presented. In one place here were his views along the length of the Oregon Trail—the sand hills along the Platte; the South Platte ford; the approach to Chimney Rock; the snakelike way through narrow Mitchell Pass with Scotts Bluff looming overhead; Wyoming's gracefully shaped Independence Rock with its hundreds of inscribed names; the ascent of South Pass with the backdrop of Wind River Mountains snow-covered even in midsummer; the buttes near Fort Bridger silhouetted in late afternoon light; the descent of pilgrims' wagons down Wasatch canyons into Salt Lake Valley. All were here, with realistic vignettes of emigrants, mountain men, Native Americans, buffalo hunters, Pony Express riders, and the Mormon handcart brigades. In a compact gallery in a modest-sized building, the effect was as stupendous as all the Great West.

Jackson, born in 1843, had spent many decades photographing the West for the Union Pacific, for a succession of government exploring surveys, for a large commercial publisher, and for himself. He published two extraordinary memoirs in 1929 and 1940. In 1935, at the age of ninety-two, he was commissioned to paint four heroic-sized murals—each thirty feet by sixty feet—for the new Interior Department headquarters in Washington, a project taking him eighteen months. As if this were not enough, as research secretary for the Oregon Trail Memorial Association he not only maintained a winter research schedule, with many breaks for lectures, but went back out on the trail every summer. During this period he also painted scores of watercolors like the ones at the Scotts Bluff museum, which he personally presented during its dedication in 1936.

Later, after a well-done slide show about the trail, a seasonal park ranger named Patty Wagner stood to deliver an appropriate benediction in the spirit of the emigrant thousands and with an echo of the first site director, Merrill Mattes: "Hope you make at least fifteen miles today," she said, "and please, make sure you have plenty of water and grass for your animals."

As with Chimney Rock, a number of the emigrant thousands were not satisfied with having reached such a place after weeks of hard work—they had to climb Scotts Bluff, partly to be able to say they had, partly to see from where they'd come and to where they were going. The Donner-Reed Party had passed by Scotts Bluff on June 24, 1846, leaving a scanty diary entry and no record of a climbing exploration, but one who did was Francis Parkman in that same year. Having wearied of the commonplace company of the covered wagon emigrants, he had pulled ahead with three others and left them behind, following the North Platte toward Fort Laramie. On the climb he found Hiram Scott Spring and was somehow moved to think of home. "All these bluffs are singular and fantastic formations—abrupt, scored with wooded ravines, and wrought by storms into the semblance of lines of buildings," he marveled. "Midway on one of them gushes the spring, in the midst of wild roses, currants, cherries, and a hundred trees; and cuts for itself a devious and wooded ravine across the smooth plain below. Stood among the fresh wild roses and recalled old and delightful associations."

Today there is a 1.6-mile trail that gracefully ascends to the top of the bluff, and my eight-year-old son was avid to try it. Despite the midday heat it briefly looked inviting. But there were admonitions everywhere: "For your safety, stay on the paved path!" the brochure warned. "The rock along the Summit Trail is soft and crumbly; leaving the pavement can be extremely dangerous. Rattlesnakes in the area are shy but will strike if threatened." My imagination for disaster needed to go no further: enough was an image of him pinwheeling off the cliff, snakes fanged and swinging from feet and hands. To Davey's great disappointment, we took the car and the summit roadway, climbing a long horseshoe curve from the visitors center and making for a smooth tunnel opening on the bluff's south corner. The summit road had been laid out in the 1930s on a 7 percent grade, with three tunnels, at a cost of $500,000. For two-way traffic it barely accommodated two vehicle widths, and our passage through the tunnels was unnerving because we were being followed by some yahoo who blew his horn to hear the echoes.

But at the top there was ample room to walk the summit trail through clumps of fragrant juniper, twisted ponderosa pine, and stunted cacti, some with wilting coral-colored flowers. Still picking our way gingerly past BEWARE OF SNAKES signs—me barking one warning after another at a heedless little boy—even with the glare and the haze, the prospects in nearly every direction were wonderful. To the south there was the castlelike South Butte and Dome Rock, looking like something Richard Francis Burton had spied in Arabia Petrarca. Easterly by squinting one could see Chimney Rock, twenty-three miles away. To the northeast and north spread the cities of Gering and Scottsbluff, with their commercial blocks and towering beet sugar processors, and many irrigated fields. High above the alfalfa, at eye level, tens of thousands of yellow butterflies were riding the thermals in billowy clouds. To the west, we had been told, one could see the dim mass of Laramie Peak, 120 miles away, but we did not know what to look for and west of us the air was beginning to look solid, anyway, from the haze. We passed through a large squadron of dragonflies at one sharp overlook. Far below, badlands were deeply and sharply cut by ravines, and it was nightmarish to think of straying off the wagon trail as it emerged from claustrophobic Mitchell Pass and set off toward the Rockies.

Back in 1936 when William Henry Jackson had helped dedicate the little adobe museum, he had returned to the site months later for the annual meeting of the Oregon Trail Memorial Association. Still unbent and sure on his feet at ninety-three, Jackson went out to Mitchell Pass to show where he had camped as a bullwhacker seven decades before. "We had one of the steepest and worse gulches to drive through that we have yet had," he recorded in his journal of 1866. As if that miserable transit through a pass barely accommodating a wagon-width were not enough, while camping in one of the narrowest places of the pass, "where the walls rise up perpendicularly on either hand," they discovered their water keg had leaked. Some "swearing and damning" boys had to foot it three miles with buckets to the nearest water, the North Platte—they had entirely missed the plentiful Scott Spring. The site—indicated by Jackson in 1936 and then again two years later for site director Merrill Mattes—was later marked with an interpretive sign. Jackson also showed Mattes the exact place he had paused to do his sketch of Mitchell Pass.

Today, a handsome marble bench sits some three or four hundred yards down the trail that leads to Jackson's campsite marker. The bench is dedicated to the memory of Merrill Mattes, who died in 1996. As first director of the Scotts Bluff National Monument, national parks historian, and founder of the Oregon-California Trails Association, he did much to capture the experiences and mark the traces of the pioneers' progress out West. We'd spent part of a day back in Independence, Missouri, at the splendid museum and library that houses his unmatched collection, and now here we were in Scotts Bluff. And a trailside bench, on which to rest and reflect on the thousands who had passed on, was the best kind of memorial I could imagine—that and the book of Mattes's I had carried since Independence, *The Great Platte River Road*, a most authoritative companion crammed with voices from the trails.

There is a superb illustrated biography of William Henry Jackson (he died in 1942 and was buried at Arlington National Cemetery). The author, Douglas Waitley, tells of the time in the late 1930s when Ansel Adams selected some of Jackson's early prints for a show at the Museum of Modern Art. Jackson appeared at the preview as a guest of honor. "Making his way among photos of the Civil War and the frontier," says Waitley,

Jackson paused in front of one and remarked, "That's a pretty good picture, Mr. Adams. Who took it?" Adams answered with a smile, "You did." Jackson mused for a moment, then nodded,

"Why so I did. But I can do better now and in color with this . . ."
he said as he pulled a small camera from his hip pocket, laughing.
"And no need for a string of mules!"

A couple of hours later, we were back heading west on the Lincoln Highway through Kimball, Nebraska, where the major Main Street industry is its taverns. West of town there was a defunct drive-in movie with a big old harvester parked on the weedy concrete lot.

Now, finally, Wyoming was exerting its gravitational pull and we were moving lickety-split across panhandle Nebraska, primed for new terrain. The mountains, after all, were promised ahead. For a number of miles we drove parallel to Interstate 80, keeping pace with tractor trailers at 65 miles per hour without having to contend with them, but just over the state line, at Pine Bluffs, population 1,034, altitude 5,047, bucolic old Route 30 was overrun by the loud and oily sprawl of the long-distance asphalt, and we had no choice but to join the jostling crowd. Bluffs ran toward the southwest, then unbroken plains. Next to us on the Union Pacific tracks, an eastbound coal train passed a westbound container train in a thunderclap of wind and noise.

PART III

8

Magic City

Nothing gleams so enticingly as polished gold in bright sunlight, exciting forgotten dreams and stifling old disappointments. As we shot across the broad plain in maddened westward traffic, before the Cheyenne city limits, the twenty-four-carat gold leaf dome of the Wyoming State Capitol stood as a beacon, homing us in. A breath or two past the green glimpse of an exit sign, and we were down on familiar ground again on a brick and stone streetscape out of 1900, commercial blocks with their figured cornice signs, their brackets and balconies, their professional offices, and their plate glass storefronts—a scale and a pace we could handle.

In Cheyenne the state capitol building, a graceful neoclassical structure that echoes the national Capitol building in Washington, looks down eight city blocks to the magnificent old Union Pacific depot with its rusticated sandstone, turreted high clocktower, and the big, bright, red, white, and blue shield emblem of the Union Pacific. They regard each other, government and railroad, from their respective anchors at either end of Capitol Avenue, symbolic of their tangled and sometimes tense relationship stretching back to the city's founding in 1867. If there was any question as to which held more power for a century—their edifices rising about equally high above the plain— the terminal had its 425-acre railyard sprawling behind, backing it up with its noisy freights carrying the commerce and the energy of the nation right through, and right past, downtown Cheyenne, heading everywhere else.

Cheyenne, "The Magic City," as its finger-crossing, wood-knocking boosters dubbed it, was created by the railroad, named, surveyed, staked, platted, and sold by chief engineer Grenville Dodge. We stood and stretched in the July sun next to a parking space between depot and capitol, taking in the bustle of downtown. I tried to imagine the scene as Dodge and his large retinue of railroad surveyors and army escorts would have seen it from their campsite on an empty, grassy plain on July 4, 1867, after their long 100-mile ride from Julesburg. Dodge had camped there before, where the waters of Clear Creek trickled into Crow Creek. A few miles away, in running a survey line from Julesburg up Lodgepole Creek just three weeks before, one of his best advance teams had been overrun by an Arapaho war party. Their head surveyor had been killed, his body riddled by five bullets and bristling with nineteen arrows. But on July 4 Dodge and his railroaders were bound to the task of continuing; they spent the morning measuring and planting survey stakes for the new division town he would call Cheyenne, just as their large army escort was there to establish a big military post northwest of town. Given the significance of the day, though, work ended around noon with the rest of the day off for celebration. The combination of beautiful scenery and patriotism inspired one of their number, the army surgeon Henry C. Parry, to eloquent imagery. "Here on the rushing, clear waters of Crow Creek, flowing through a prairie adorned with beautiful flowers and rich, tall grass, with the towering heights of the Rocky Mountains and the long range of Black Hills before us in the west and north, our national anniversary was not forgotten." After speeches, cheers, and congratulations, the Fourth of July toasts went on through the afternoon and into the night.

One must assume under such festive conditions hangovers were the rule the next morning. When Sioux raiders swept up from nowhere then, daring to strike against a nearby wagon train of Mormon graders camped at a slight distance from the big body of soldiers, the cavalrymen could not even saddle their horses before the Sioux galloped off, leaving two Mormons dead on the ground. After the victims were buried, their headstones would read, "Killed by Indians." They became, Dodge wrote later, "the first inhabitants of that city."

Those two lonely mounds amidst the prairie flowers, with the creek burbling past and the foothills rising toward mountains a few miles westward, would enjoy only a short space of peace and quiet. Word that Cheyenne would be a permanent settlement spread to all the dying Hell on Wheels towns like Julesburg, and people began to drift in, lured by the jobs or the easy money of a railroad division point with shops, roundhouse, and yards, and attending suppliers and stores, and ultimately with a branch line heading ninety miles south to Denver. Claim jumpers who tried to squat on land titled by law to the railroad found themselves staring down the rifle barrels of the bluecoats from nearby Fort D. A. Russell; they either bought or moved meekly on. The Magic City grew magically if rambunctiously over that summer and autumn of 1867. As the WPA guide to Wyoming reported, by November 1867 "Cheyenne's population had reached 4,000, and lots that sold originally for $150 were bringing speculators $1,500. Citizens lived in anything that would shed rain and sun: covered wagon boxes, dugouts, tents, shacks. More than 3,000 such dwellings had been erected within six months. A wag described the houses of Cheyenne as 'standing insults to every wind that blows.'"

The first Union Pacific Railroad train arrived in November 1867, bringing all the dismantled emporia of the last railroad rush town of Julesburg—"piled high with frame shacks, boards, furniture, poles, tents, and all the rubbish of a mushroom city." Every succeeding train brought in more furnishings and more denizens eager to fatten themselves on opportunity. Many others drove up the deepening ruts of the trail from Denver. Soon there were more than 300 businesses in Cheyenne and its roughness and easy money made it legendary in the West as the greatest Hell on Wheels town yet.

Standing out of car and truck traffic at a concrete curb on Pioneer Avenue, I thought about all the old stereographic cards I had held up to my eyes in viewers at the Library of Congress and in Omaha at the Union Pacific archives. I had lost myself on those muddy dung-splattered 1868 streets of Cheyenne—the false fronts and garishly painted signs, the raw-looking hotels, saloon after saloon, and crowds of elbowing workers, drifters, grifters, and emigrants.

Not surprisingly images from that wild era are everywhere in the Cheyenne of today—in fact, they are wrung for every ounce of promotion one can extract—from the figure of a cowboy astride a furiously bucking bronco that appears on every Wyoming license plate; from the murals painted high up in the statehouse and Senate chambers in the gold-domed capitol; from most store window displays, not only souvenir shops and Western clothiers but real estate, restaurants, and RV dealers; from street names in residential developments north of town (Custer and Cody streets, Maverick and Wrangler avenues, Buckskin and Red Cloud trails); down to Gunslinger Square in old downtown Cheyenne, within a whistle of the railroad tracks.

Interestingly, the statehouse murals and Wyoming's buckaroo license plate image were rendered by the same artist, a Colorado native named Allen True (1881–1955), who had studied at the Corcoran School of Art in Washington, D.C., before taking classes with the renowned illustrator Howard Pyle at the artist's home and studio at Chadd's Ford, Pennsylvania; one of True's classmates was N. C. Wyeth. True began as a magazine illustrator but then apprenticed with the British muralist Frank Brangwyn doing murals at the Panama-Pacific Exposition in San Francisco. After 1915, True went out on his own, specializing in Western scenes. His murals are in the state capitol buildings of Missouri, Nebraska, and Colorado, as well as Wyoming. We stood in the West Wing and the echoing, empty Senate Chamber, beneath a brilliantly sunlit Tiffany glass ceiling, gazing up at True's four murals crowning four corner pillars: a Cheyenne Indian chief; a group of railroad surveyors and builders; a frontier cavalry officer (resembling George Armstrong Custer); and a Pony Express rider (looking considerably more mature and filled-out than the young skinny fellows sixteen or seventeen years of age who actually rode the trails). True's murals in the East Wing's House Chamber were unfortunately off limits, obscured by scaffolding and plastic tarps during that between-sessions month. But the ubiquitous bronco rider silhouette, a wonderful Thirties-era rendering, crept by us by the thousands out in the Cheyenne streets on yellow Wyoming license plates. Allen True was paid the princely sum of $75 for the logo in 1935, and it has been hanging on vehicles and adorning state literature ever since.

Summertime tourists strolling past old downtown Western-theme curio shops or trying on tall hats of various gallon capacities in a Western clothing store would, around 6:00 P.M. on weekdays and at midday on Saturdays and Sundays, hear cascades of sharp pistol shots echoing off city walls. Even in these terrorized times of high school and post office shootings, out-of-towners would likely be drawn by curiosity toward the source of the sound at Cheyenne's Gunslinger Square at the bottom of Carey Avenue. We heard, we investigated, and we found ourselves places on the grass while the Cheyenne Gunslingers Association began its daily show. The group is a nonprofit organization founded in 1983 by local businessmen headed by Ron LeVene, a real estate agent, to promote tourism in downtown Cheyenne. LeVene, born in Chicago and raised in the small central Illinois farming town of Sullivan, enlisted in the air force at seventeen and retired with a captain's rank in 1981, having picked up along the way bachelor's and master's degrees; the air force transferred him to Cheyenne in 1975. "At the time I founded the Gunslingers," he told me, "I owned an art gallery in downtown Cheyenne and thought that we needed more Western attractions here in the summer as tourists came to Cheyenne. Also, I had seen many staged gunfights and always thought that it looked like fun. Being an Old West history nut and raised with Gene Autry and Roy Rogers and being a ham, I knew it would be a great time."

The members, mostly middle-aged men of ample proportions and with grizzled beards, come from all walks of life—air force technicians, retired military, computer engineers, lock specialists, business owners, printers. They attire themselves in Wild West clothes to perform what is billed as "scripted, family-oriented gunfights," which means in traditional American cultural terms that there is a lot of violence but no blood, no profanity, and no nudity. "We've been doing this stuff for eighteen years," Ron LeVene said. "Back when we started I had black hair and a black beard, and I could fall down and jump right up just like that. Now they give me parts so when I fall down I can lie there for a little while. Another thing—all those years of gunfights have also cost me a little hearing, because I'm too dumb to wear earplugs." The members meet twice a month during the off-season, and once a week in the summer, to write scripts and rehearse their shows. "We are not a reenactment group nor are we a stunt group," their advertising explains. "Some of our scripts are loosely based on historic events, but in our research we have found that most goings-on were pretty boring back in the 'good old days,' and figured most folks wouldn't stick around to watch, so we livened things up a bit."

Gunslinger Square is a small grassy park bordered by busily trafficked streets and faced by antique shops. The railyard is in sight, and there are usually freight trains in motion behind the players. Stage-set flats behind the performance area depict a Wild West block not unlike those in the A. J. Russell stereographs of the 1860s. Aided by wireless microphones, the actors begin every performance with a safety lecture. "One of the things that we were concerned with when we started," LeVene told us as his fellow actors ambled and scuffed around the grassy area, "was that we didn't want people to think we were just playing around with handguns. We're very concerned that people understand that we respect the guns we carry and we want everyone to know how dangerous a handgun can be. We use .38 Special handguns in our gunfights and the question we're probably asked more than any other question is, 'Do you use real guns in a gunfight?' The answer is, yes—they are real guns and they're fully capable of firing real ammunition. However, the first year we tried that it cost us too many gunslingers—we couldn't get any volunteers to come out here. So we have gone to blanks. But we don't want anyone to think that blanks are not dangerous."

He warned the audience that it would commence getting noisy around Gunslinger Square for a good while, and asked if anyone in the crowd had brought a dog. "Yesterday we had a lady out there with a dog on a lead rope. First shot the dog went out heading for Laramie at 93 miles an hour, dragging that lady. We ain't seen her since, but we hope she's all right."

With great solemnity, a cowboy set an empty soda can on the ground, and another unholstered his gun, took aim from some inches away, and shot at the can with a blank cartridge. The can flew ten feet into the air. When it was held aloft, it had been crushed and pierced just by the concussion of the shot. "A gun is fully capable of killing, even loaded with blanks," Ron LeVene told us. The onlookers were impressed; some of us recalled a news report about a prank on a Hollywood film set, in which an actor shot himself in the head with a blank-cartridge-loaded pistol, dying instantly.

Gunslingers performed six different scripts during the week, giving a different show every day. The scripts were simple and to-the-point, dialogue and movement serving as a series of excuses to fire the weapons—variations on "You're a dirty yellow dog, Slim, and I'm callin' you out!" or "Okay, Huss, just reach for the ceilin' and no one'll get hurt!" It was done studiously and energetically. By the end of the show most of the gunslingers were sprawled motionless out on the grass in the unyielding sun, waiting through the moments until the audience began applauding and they could struggle to their feet and take a bow. Afterward, on the day we were there, it came time for their "audience interaction"—and master of ceremonies Ron LeVene began to talk about "proper Western attire for a summer day in Cheyenne." From where I sat on the edge of things I saw he was heading my way. "Okay, folks," he called, "we're going to vol-un-teer a volunteer for our demonstration." He nodded to me. "Let's give him a hand, folks!" I had no interest in letting these fellows in Western getup try to make me a part of their tableaux so they could humiliate a tourist—*all in good fun, folks!*—so, in my comfortable baggy cargo shorts and T-shirt, I looked stonily up at the sweating and puffy master of ceremonies in his heavy boots, tight denim trousers, long-sleeved shirt, vest, and big black bowler hat standing above me in 95 degree noontime heat. *"Get lost, pal!"* I rasped. *"Get somebody else!"*

Taken aback, he looked over at the other gunslingers who were just as surprised and didn't know what to do. No doubt in nineteenth-century Cheyenne I would have been riddled with bullets or dragged down the street behind a galloping horse, and perhaps some of these players were wishing we were all back there long enough for that activity to be accomplished. "Ladies and gentlemen," the MC stammered, "I've been doing this for eighteen years and no one ever refused before!" He laughed, winked, shrugged, and gave up on me. (*"Daaaaa-d,"* my deeply embarrassed daughter was hissing from behind.) Confidently striding across the performance space to another knot of onlookers, Ron found another patsy in short pants. They stood him up and had a lot of "proper Western attire" fun at his expense, hanging a huge cowboy hat down over his ears, draping a vest and chaps on him, and making aesthetic remarks about his knees—during which he managed a sickly grin. The reward for his ordeal was a modest gift certificate at the big Western clothing shop down the avenue—advertising for one of the show's sponsors.

Ron LeVene came over after hawking gunslinger belt buckles, pins, and "Old West ball caps." "No hard feelings, bub," he reassured me, and invited us over to the Cheyenne Gunslingers Soda Saloon, a block away, "where the Gunslingers can be found restin' their bones and sippin' a cold sarsaparilla." We went by for cold drinks, turning down the opportunity for microwaved hamburgers and chips. The place looked approximately like a Wild West saloon squeezed into a storefront, with period photographs, wanted posters, and mounted animal heads on the wall. But all the assembled Gunslingers were a little dazed from the heat, and seemingly were having reentry problems back into the present time. All seemed tired and dejected, as actors and musicians often are after the curtain is down. I had performed in public as both, and know a few reenactors: a high school friend of mine, now in his early fifties and selling insurance on Long Island, spends his summer vacations in a gray uniform eating beans and thrashing through underbrush down South on Civil War battlefields; acquaintances in Vermont towns near our home don itchy, stained, and badly patched Revolutionary-era garb at Mount Independence encampments and Hubbardton and Bennington battlefields. For some reason all of the reenactors I know are twice the age of the historical characters they depict—but nevertheless part of the preparation is a kind of self-hypnosis. I knew what the Gunslingers felt like as they just sat around in stunned silence. The jukebox music was up very high—they were all partly deaf from all the shooting—and finally it drove us out into the street.

Several days later, when I was working and Mary gave the children respite from history for some window shopping and escape from Cheyenne heat in a refrigerated multiplex, they saw that the Cheyenne Gunslingers were hard at work right there, at the Frontier Mall.

In late 1867 a French scientist by the name of Louis L. Simonin visited Cheyenne, and he was impressed by the citizens' certitude that their new burg had a mighty future.

> The little city, the youngest, if not the least populated of all cities in the world, which no geography yet mentions, proud of its hotels, its newspapers, its marvelous growth, and its topographic situation, already dreams of the title of capital. It does not wish to be annexed to Colorado. It wishes to annex Colorado. As it is the only city in Dakota, and as this territory is still entirely deserted, it does not wish to be a part of Dakota. It dreams of detaching a fragment from this territory and from Colorado and Utah, of which it will be the center. So local patriotism is born, and so local questions arise, even in the midst of a great desert.

Not for long. The Texas cattle drives up to stockyards south of the U.P. tracks at Cheyenne, which began in 1869, helped solidify the town economically. Cattlemen spread northward across the Wyoming Territory. Their great herds and vast ranching claims became the source of immense power and wealth on the plains. North of Fort Laramie and the North Platte River in the early 1870s was designated Indian Country, permanent reservations given in the peace talks ending the First Sioux War in 1868. But the cattle interests pressed to be permitted to expand into those pristine tribal hunting hills—as did prospectors convinced the mineral-rich lands would yield gold. In 1875 hundreds of gold seekers slipped in, determined to strike it rich before the end of that rain-sodden summer, if they could retain their scalps. Then, in August 1875, General George Armstrong Custer publicly reported that his military expedition had discovered gold on French Creek in Dakota. America was at the time wracked by a depression brought on by the Panic of 1873, with many thousands out of work. Hearing that gold could be found wherever one bent down in the Dakota hills was all desperate people needed to hear. Despite firm attempts by General Philip Sheridan and his commanders to keep the hordes of gold rushers out of military departments where resentful tribes waited to prey on them, unheeding prospectors poured into the Black Hills—some 15,000 evaded the authorities within two months of the first leaked reports, concentrated outside of the biggest camp, called Deadwood. Such trespass into the reservation (and the army response to depredations) brought on the Second Sioux War. It lasted a year—and claimed foolhardy Custer and his 264 men, hardly ten months after the words "gold in them thar hills" were first uttered.

The boom built Cheyenne even further; it was the nearest and most logical railhead for supplying the Black Hills camps, and the biggest and most secure settlement for assaying and depositing mineral wealth. The wagon trail from Cheyenne to Deadwood supported, as state historian Agnes Wright Spring would write in 1949, "a ceaseless rumble of wagon wheels [that] reverberated across the plains, through the valleys, and into the foothills, as thousands upon thousands of gold hunters, freighters, carpenters, merchants, soldiers, and adventurers surged toward a new Eldorado." In early 1876, with the army establishing permanent camps along the line and sending escorts, the trail became safe enough to support regular Concord stagecoaches among the increased wagon traffic. The average travel time in good, dry weather was fifty hours; rain and mud increased the transit by ten to twenty hours.

Business, though, was good no matter what the weather. It seemed then as if every issue of the *Cheyenne Daily Leader* contained reports of riches transported to Cheyenne—two men with "gold dust in buckskin bags and belts strapped about their waists containing $4,000 each" . . . "Fourteen thousand dollars worth of gold was brought in from the Hills by the stage last evening; Colonel C. V. Gardner and other gentlemen who came by the same stage brought $10,000 in dust with them" . . . "A seedy individual arrived on the coach, having in his possession a dilapidated carpet-bag in which we have since learned was stored $103,000 in gold." Tales into early the next year, 1877, only inflated as the Black Hills were officially opened to exploitation. One miner named Billie Gay climbed off the Cheyenne stage with a "watch chain about two feet long composed entirely of gold nuggets from the size of a pea up to the size of a hickory nut, with two or three larger ones for pendants." The stagecoach "treasure box" contained $30,000 in gold dust on January 21, and $25,000 on January 26. With warmer weather came even richer loads, most excitingly the "spring cleanup" from the Black Hills, some $200,000 in gold, escorted by special shotgun messengers riding on and mounted alongside the stage.

As Agnes Wright Spring would comment, "it was little wonder . . . that palms itched with greed" at such reports. Rustling and mail robbery had already increased in Wyoming and Dakota, and on March 26, 1877, telegrams reached Cheyenne reporting that "road agents" had stopped the coach out-side Deadwood, murdered the driver, and shot up the other escorts. It seemed to unleash a tide of lawlessness along the trail as not one but many bands of outlaws preyed on the wagon traffic, blowing the locks off reinforced iron treasure boxes and fleecing the customers of dust, currency, watches, and other valuables. Sometimes the crew and passengers were treated to the in-dignity of being stopped by several gangs in one trip; the first got the money and valuables, while the second had to be content with blankets and clothing. Sometimes extra shotgun messengers or troop escorts discouraged the ban-dits, but the infestation grew so heavy that often robbers tricked or got the drop on the guards, even when they bristled with extra weaponry. The more egregious of the holdups prompted the Cheyenne and Black Hills Stage Line to offer rewards: $1,000 for the capture and conviction of the road agents, or $200 for each of their dead bodies. But the lure of easy money kept the out-laws busy, and it was with a sinking feeling that travelers on that bright red and yellow stagecoach continued to peer outside as the horse teams were sud-denly reined in to a stop far from a station, and mounted, masked men or-dered them out to stand in a line with their hands up while bags and boxes were torn apart, pockets emptied, and women were forced to let down their pinned-up hair where, more often than not, a roll of currency was hidden.

It is with these experiences in mind that we stood alongside a Cheyenne–Deadwood stagecoach, with its bright red body and vivid yellow wheels, not far from where the original line would have taken such a coach in the 1870s down into Cheyenne, whether bereft of valuables or no. Lovingly restored wagons stretched out before us, with the magnificent stagecoach the unques-tioned star. We were in Cheyenne's brick-pavilioned Old West Museum, in a big indoor gallery holding what was said to be the third largest horse-drawn vehicle collection in the country. In other galleries we found an excellent exhibit, "Past Ties, Present Lives," on Native American culture and history, and another on rodeo lore and artifacts. The Old West Museum sits on the grounds of the Cheyenne Frontier Days fairgrounds where, annually during the last week of July, the "Daddy of All Rodeos" attracts tens of thousands of visitors and hundreds of contestants, performers, and exhibitors.

Frontier Days was first held in 1897, and seems to get bigger every year. Before the huge red, white, and blue grandstand, contestants vie for awards in saddle and bareback bronco riding, steer roping and wrestling, calf rop-ing, bull riding, and women's barrel racing. There are parades, processions, demonstrations, Western art shows, a midway, and an air show put on by the air force's Thunderbirds. One of the nicer touches of Frontier Days is the free pancake breakfasts doled out to thousands. But we were three weeks shy of the opening ceremonies—standing what felt like a hundred years away from timed events, grand parades, and aerial displays, and clean, well-groomed

Frontier Nights music stars like Randy Travis, Reba McEntire, and John Michael Montgomery—fingering the wheel dents and old gouges in the body of the Cheyenne & Black Hills stagecoach, from back when what plowed out those gouges wasn't ersatz and wasn't even history, it was the latest news, splayed across the big-sheet *Leader* in handset display type, with no extra charge for the exclamation points.

As for the wave of terror up on the trail between Cheyenne and Deadwood, the stagecoaches continued to be halted and robbed by road agents well into 1878. Then, at Canyon Springs, Wyoming, a gang ambushed a heavily laden stage as it slowed for a change of horses, engaging in a running gun battle with many wounded on both sides and one of the guards killed. With the surviving coach men tied up or run off, the bandits went at the coach safe with a sledge hammer and cold chisel, finally breaking through after two hours. They removed cash, bullion, and jewelry worth some $27,000.

The boldness of their crime and the high prize they won let loose a furor of reaction: lawmen, private detectives, and vigilantes swarmed up the trail, stimulated by a big reward. Over subsequent weeks, the outlaws were tracked as far as Nevada and Iowa (some managed daring escapes later), while many others of different gangs were found strung up wherever they were accosted. The Cheyenne–Black Hills line became too dangerous for outlaws, and they—or the next generation of bad men—moved on to the Powder River basin in Montana or westward across Wyoming, following the Union Pacific toward the Laramie Plains, Rock Springs, and Green River.

There were some colorful characters associated with the Cheyenne–Deadwood trail, many of them given capsule biographies by Agnes Wright Spring in the appendix of her book on the trail—people like Henry Chase, stage agent and postmaster at Raw Hide Buttes, a Dartmouth graduate at whose Virginia home Harriet Beecher Stowe wrote *Uncle Tom's Cabin*, and who went West to ranch after his wife died; "Owl-Eyed Tom," born Thomas Cooper in New York City, a Civil War cavalryman and Union Pacific construction crewman before becoming the night driver on the Cheyenne stage route, earning his nickname and becoming known as "the best six-horse stage driver in the United States"; "Friday," a Chinese stage station cook who wore a long queue, loved to gamble, and bought himself a big ranch near Custer, South Dakota, with his winnings; John Higby, a transplanted New Yorker known as one of the fastest six-horse stage drivers in the nation: "he always hit the bridges and grade roads in such a way," it was said, "that he kept his coach ride side up"; H. E. "Stuttering" Brown, a stage driver "killed in April 1876 by Persimmon Bill or Indians"; and Tom Duffy, driver, who later drove the Concord stagecoach for Buffalo Bill's Wild West show during its European tour. But one of the most memorable characters was known as Calamity Jane.

She was born Martha Jane Cannary in Princeton, Missouri, in 1852, learned to ride as a young girl, and migrated West with her family at twelve, living in Montana, Salt Lake City, and southwestern Wyoming as the Union Pacific was forging toward the Wasatch Mountains and its meeting point in northern Utah with the Central Pacific Railroad. She lived in cow towns and hell towns and mining camps, took on risky and hard physical work, and for practical reasons adopted men's clothing. She may have served as an army scout in various Wyoming Indian campaigns, ranched, and drove ox teams, but her main occupation seems to have been simply that of a colorful character. She called her long succession of lovers "husbands," as was common in the camps and hell towns. One "husband," she always claimed, was Wild Bill Hickock, the scout and lawman shot in 1876. She talked rough like a man, relished improbable yarns, and bought drinks for the house whenever she had money in her pocket. A number of dime novels by Easterners who never met her turned Calamity Jane, as she was universally known, into a female Buffalo Bill, and she no doubt added to her own mythology with almost every utterance. "As she sits astride her horse there is nothing in her attire to distinguish her sex save her small, neat-fitting gaiters and sweeping raven locks," marveled Denver's *Rocky Mountain News* in 1877. "She wears buckskin clothes, gaily beaded and fringed and a broad-brimmed Spanish hat. . . . Donning male attire in the mining regions, where no restraints were imposed for such 'freaks,' she 'took the road' and has ever since been nomadic in her habits."

Mainly, though, she seems to have been a barfly. She once said that she was resolved never to go to bed "with a nickel in her pocket or sober," and judging from the many accounts of her good-timing exploits in Western newspapers, she stuck to her intentions. One of her most enduring anecdotes comes from the *Cheyenne Leader* in 1876, soon after she was released from the hoosegow for some petty civil disturbance, and concerned a trip she intended to make between Cheyenne and Fort D. A. Russell, three miles distant.

On Sunday, June 10, that notorious female, Calamity Jane, greatly rejoiced over her release from durance vile, procured a horse and buggy from Jas. Abney's stable, ostensibly to drive to Ft. Russell and back. By the time she had reached the Fort, however, indulgence in frequent and liberal potations completely befogged her not very clear brain, and she drove right by the place, never drawing rein until she reached Chug[water], 90 miles distant. Continuing to imbibe bug-juice at close intervals and in large quantities throughout the night, she woke up the next morning with a vague idea that Fort Russell had been removed, but still being bent on finding it, she drove on, finally sighting Fort Laramie, 90 miles distant. Reaching there she discovered her mistake, but didn't show much disappointment. She turned her horse out to grass, ran the buggy into a corral, and began enjoying life in camp after her usual fashion. When Joe Rankin reached the Fort, several days later she begged of him not to arrest her, and as he had no authority to do so, he merely took charge of the Abney's outfit, which was brought back to this city Sunday.

Calamity Jane published her autobiography, a tissue of tall tales, in 1896, and died in 1903 in a Dakota mining camp. Her last request was to be laid to rest in Deadwood next to her "husband," Wild Bill Hickock, and, there being no objections heard from him or anyone else, such was done. The citizens of Deadwood buried her in a white dress.

The territory of Wyoming was created in 1868 with the Wyoming Organic Act, which carved territory out of Dakota, Utah, and Idaho. The territorial government commenced, with Cheyenne designated capital, on April 15, 1869, when the governor and secretary took their oaths of office. The legislature convened on October 12 of that year, and representatives immediately began working on a law that granted equal rights for women. This landmark legislation, the first in the world, was swiftly approved and signed on December 10, 1869. With the new year, women began serving on juries in Wyoming—it has since been known as "The Equality State"—and the first appointment of a woman to a civil post, that of justice of the peace of South Pass City, on the Oregon Trail, was also made in 1870. The appointee was Esther Hobart Morris, an active suffragist who had lobbied strenuously for the rights legislation. Her first case as justice of the peace was a lawsuit against her predecessor; he refused to turn over the court docket to a woman. When prosecution and recovery seemed in doubt, Morris simply ruled in favor of "a new, clean docket" and blithely launched into her work. Her bronze statue now stands outside the state capitol in Cheyenne, a replica of the larger version in Statuary Hall of the United States Capitol. When we paused there to pay homage, a pink sash had been recently stretched across the statue's torso with the words "Race for the Cure," referring to a local runners' marathon to raise funds for breast cancer research. Esther Hobart Morris lived out her declining days in Cheyenne. Celebrated for her activism, she was visited in 1895 by Susan B. Anthony, who said her pilgrimage to Wyoming was one of the proudest moments of her life. Morris died at eighty-seven in 1902.

Ella Watson was another Wyoming woman whose name is connected to equality, though no doubt she would have preferred to forgo the distinction. She was originally from Kansas, and settled in the Sweetwater Valley just off the Oregon Trail near historic Independence Rock, where thousands of westering pioneers had scratched their names. Her husband, Jim Averill, raised a building for a combination store and saloon in a newly established town called Bothwell, while Watson tended their ranch a mile out of town. Cowpunchers from other nearby spreads left a goodly proportion of their wages in the till of Averill's saloon, while the couple's ranch expanded correspondingly. Neighbors grew to resent their small but growing herd of cattle and envy can do terrible things. The fact that Averill held the monopoly on firewater in town and could charge what he wanted did not help their case. False rumors of rustling began to be heard; a knot of vigilantes threatened them to vacate the county, but they ignored all the warnings. On July 20, 1889, six ranchers abducted the pair, took them to Spring Creek Gulch, about five miles away, and summarily lynched them from a pine tree. A deputy sheriff from Casper discovered the bodies several days later and returned them to Bothwell, where they were buried near Averill's saloon. Prosecution of the six murderers went nowhere, it being said that two witnesses had disappeared and another mysteriously died. Ella Watson was nicknamed "Cattle Kate" in the frontier press, and the victim was not surprisingly blamed for her own end by the false charges of rustling and other insults to her character, such as the baseless accusation of being a prostitute. Ella Watson was the only woman ever hanged in the Equality State.

Her beaded moccasins—the ones she was wearing on her last day, which the deputy found at the base of her hanging tree—are now exhibited, with an account of her tragic circumstances, in the Wyoming State Museum, just a few blocks away from the capitol building on Central Avenue.

The State Museum interprets Wyoming history, culture, and natural science in a modern fashion, cleverly and attractively, in ten large exhibition galleries. My family went around the good-sized museum accompanied by the personable Kansas-born curator of interpretation, Ellen Stump, who was working on a new gallery called "Living in Wyoming," which would cover subjects such as conflict, religion, recreation, entertainment, and fashion.

The galleries house a lively history, crammed with human stories. There were those beaded moccasins of Cattle Kate, eloquent in their simplicity. There were Cheyenne stagecoach artifacts like a gold strongbox, handcuffs, and a mean-looking .44 caliber Winchester Henry rifle from 1866. There was a washboard carved from a single plank of wood, inscribed "1849," and probably abandoned until it was picked up out of the weeds seven decades later. There was a large deerhide painting by the son of the famed nineteenth-century Shoshone tribal leader Washakie, which depicted buffalo hunts and Shoshone engagements with Blackfoot, Lakota, Cheyenne, and Crow warriors.

And there were many artifacts from the "good roads boosters" who promoted the Lincoln Highway, underscoring how crucial the issue was to settlements scattered across the rough and unfriendly terrain that makes up so much of the state. Standing before a still vivid red, white, and blue "L" sign pointing the way to Cheyenne and Laramie I thought of Alice Huyler Ramsey, the first woman to drive across the country in 1909, and what she encountered in Wyoming in the years before a continuous, graded, and graveled route was scratched out. "Roads in Wyoming were scarcely what we would designate as such," she recalled many years later. "They were wagon trails, pure and simple; at times, mere horse trails. Where the conveyances had usually been drawn by a team, there would be just the two definite tracks—or maybe ruts—often grass-grown in between. On the other hand, where many one-horse rigs had passed, a third track would be visible in the middle through the grass or weeds. With no signboards and not too many telegraph poles, it was an easy matter to pick up a side trail and find oneself arrived at the wrong destination." Often Ramsey and her companions could only traverse a path across privately owned ranches for sheep and cattle. "To keep the cattle in, there were gates we had to open and close behind us as we passed through. No matter how inconvenient it was, no one would think of neglecting this little chore in return for the right to pass. This part of the road was a mere trail from here into Cheyenne as it crossed the ranches and hills. But it was a beautiful ride, with ever-expanding views, as one looked across the rolling land to the distant horizon."

Four years later, on opening day, December 13, 1913, joyful residents lit a string of bonfires 450 miles long following the new Lincoln Highway route across Wyoming. Resulting road traffic into the state in the Teens, Twenties, and Thirties brought many cultural changes and opened economic opportunities for people living along the route. We could see some of the surviving remnants along Cheyenne's major east–west boulevard running parallel to Interstate 80. It is called the Lincolnway, commemorating the old highway, and in the summer of 2000 it was torn up for miles in a repaving project, traffic bumping around cones and detour signs. Cultural archaeologists could have had a field day in that overturned earth, judging from the string of little 1930s-era neon-lit, adobe-walled motels, tourist courts, filling stations, and beaneries standing on the southern curbs of the Lincolnway. It had been built back in the days when there were twenty-one hotels and fourteen automobile tourist camps, two railroad depots and three railway lines, two bus stations and the two airlines out at the little municipal airfield. At night people went to the movies or vaudeville or listened to the radio, like everywhere else in the nation.

At the Wyoming State Museum one can trace the route discovered and mapped out by Grenville Dodge for the Union Pacific, later traced by the Lincoln Highway and then shadowed—with many of the curves knocked out—by the Eisenhower interstate. One simply steps into the museum's story gallery and is presented with the most delightful sight of a huge relief map of the state, some twenty-five feet to a side. I stood with my kids peering down at the Laramie Hills, Cheyenne and Laramie, the Laramie Plains, the Medicine Bow and the Wind River ranges, the Continental Divide, the Red Desert, all the way to the mountain men rendezvous out on Green River, with the Wasatch Range rising beyond, all along the lower third of Wyoming—through which we would be passing over the next several weeks. With the aid of colored buttons and their corrresponding lights I could show the children the trails, rail routes, and elderly highways we would be following, past old hunting plains, forts, battlegrounds, station towns, mining camps, get-rich-quick schemes, pipe dreams, and pies in the sky. The upper two thirds of Wyoming stood for a future year's sojourn—Fort Laramie and 250 miles of the Oregon Trail, Cody, the Bighorn Mountains, and of course, Yellowstone—but we had their impressive expanse of terrain and history stretching out to be beheld in relief on the museum's big map.

"This is probably our most popular feature," curator Ellen Stump told me, as children and adults leaned out over the railing into the Wyoming sky to look down at the view. "It's kind of ironic. When we put this museum together, and as we continue to refine, expand, and interpret, we're under a lot of pressure from local people over the exhibitions and interpretations, some of which don't appeal to old-fashioned attitudes about what a museum's job should be, or what history tells us. Some people would be happier if we just kept it to cowboys, cavalry, coal, and dinosaurs. While all of that is important, there's so much more to the story and so much more that can be understood and investigated. We're proud to be told from Wyoming citizens who've traveled, and from out-of-state visitors, that we compare favorably to large establishments like the Smithsonian Institution." She looked across the map as first the fur trappers' domain, and then the emigrant trails, lit up in little colored lights across the mountain ridges, defiles, and plains. "Simple pushbutton history," she laughed. "As old as the hills, or nearly so. It's great to see how people will linger through the rest of the galleries, taking time to read the wall texts, to stop and think, but in the end they'll always gravitate back here for a little button pushing."

It was time to move on. I pointed down to the terrain map to show Davey and Mimi how the Union Pacific Railroad had crossed the territorial boundary in 1867 and then moved beyond rough and raw little Cheyenne to be confronted with the spine of the Laramie Hills—the easiest way to get across the Rockies—and how on the western side it would curve north and west at a new settlement to be called Laramie.

The next morning we crammed our belongings into the Dodge and rooftop carrier, and after a hearty breakfast left Cheyenne, driving warily past the line of intercontinental ballistic missiles standing as haughty monuments at Fort Francis E. Warren, the successor to the nineteenth-century cavalry stockade, Fort D. A. Russell. We were to cross Sherman Hill to Laramie, but I postponed joining Interstate 80 in favor of the once well-traveled Happy Jack Road, named for a rancher who settled there in 1884.

Happy Jack Hollingworth lived in an adobe house he built himself, and made his money hauling wood down to Cheyenne from the Medicine Bows. He always sang in his work, thus earning his nickname, but Happy Jack was unhappily killed in a brawl down near the Mexican border. Under a bright, partly cloudy sky, we drove past numerous trailer lots, a large ranch raising big Black Angus beef cattle, and a white stone mausoleum on a hillside. We climbed steadily between buff short-grass hills, finally breaking into open ranchland and then began ascending the foothills.

The day was growing warm, even at our elevation, and I hoped we would be able to swim. Ahead was Curt Gowdy State Park, named for the prominent television sportscaster who had been born in Green River, Wyoming, in 1919. Back in the 1940s, Gowdy had dated my mother sporadically; he was working as a sportscaster for a radio station in Oklahoma, and my mother was writing continuity scripts and hosting her own radio interview show on WHB ("the Voice of the *Kansas City Star*") in Kansas City. "Whenever Oklahoma basketball teams came to Kansas City," she told me, "we'd go out to dinner. He was a nice man, very good manners. I never thought he'd become so famous as to have a state park named after him!" About twenty-five miles west of Cheyenne we drove into the park and threaded downhill in a big natural bowl toward Crystal Lake. We were all primed for a good swim in a cold mountain lake, but a ranger on duty told us swimming was forbidden. "This is drinking water for the city of Cheyenne," she told us severely. Down at lakeside campers and RVs were parked at sites all the way around, and people had taken outboards into the middle of the lake to go trout fishing. "You can't swim in the lake because it's drinking water and some kid might pee," I said sourly to myself, "but the cancer-causing emissions from gasoline engines are permissable." We skipped stones for a while, though that may have been illegal, and then we followed a path up a steep, gravelly hillside and out through dramatic outcroppings until we were halted by a cliff. Above us on the hillside were stacked and tumbled boulders big as trucks. The high mountain winds made our perches too tenuous to remain long.

Turning away from Curt Gowdy's picturesque but dissatisfying lake, and then a few miles later from Happy Jack Road, we bumped off onto a smaller dirt road leading into the Medicine Bow National Forest. We passed dreamlike reddish granite pedestal formations scoured by glaciers and smoothed and shaped by high winds—here looking like a giant child's building blocks, here like a seal, there like a buffalo. We wound ever upward through junipers and cedars, stopping several times so the children could climb some of the strange rocks. This area was called Vedauwoo (pronounced Vee-da-voo) Glen, and it had spiritual associations with the native tribes. The name Vedauwoo was from the Arapaho, meaning "earthborn." Farther was something called Devil's Playground—a procession of steep cliffs and giant balanced rocks rising from the forest. Why was it that in naming so many magnificent natural formations in the nineteenth century, people connected them with the Prince of Darkness? This "playground" at Vedauwoo seemed less devilish than mischievous—an abrupt escarpment several hundred feet tall, decorated with boulders bigger than houses, the whole thing stacked neatly, almost studiedly. That morning the cliff was aswarm with rock climbers dangling from ropes in the fierce wind. The climbing opportunities are well known to University of Wyoming students who drive up from Laramie, and indeed climbers come from all over, putting the semiprimitive forest area into some danger from overuse, though small-engined vehicles are doing more to scar the thin soil and erode formations. As we lounged there, watching climbers made tiny by the massive cliffs, seeing the plays of light and cloud shadows and hearing nothing but the sound of wind, the peace was ruined by the mad hornet swarming buzz of a squadron of dirt bikes and four-wheeled ATVs that zoomed up the road, raising horrific dust, before darting down a side trail. But others enter Vedauwoo gingerly, almost on tiptoe. The night before, Mary had driven up here with the children for a program held under the light of the full moon and sponsored by the National Forest. It had been the "Fiddler's Moon," and so the weirdly beautiful rock formations echoed with music made there at the base of the cliffs, while the participants were urged to "smell the flowers, feel the wind, hear the water, and sense other wildlife."

Beyond the cliffs we came out in the clear and we were suddenly confronted again by the interstate, but before consigning ourselves to it I had a pilgrimage to make. I steered the car east and south beyond an exit and down a badly rutted road toward a grand anomaly there at the summit, a massive granite pyramid. Standing some sixty-five feet high and sixty feet on a side, it was built between 1880 and 1882 by the Union Pacific to commemorate two brothers, Oakes and Oliver Ames of North Easton, Massachusetts, who on several occasions had saved the railroad from going under during some of its most dramatic struggles for life. Oakes Ames had also been at the center of one of the most tumultuous political scandals of the nineteenth century.

The brothers had inherited their father's Ames Iron Works, maker of the famous Old Colony shovel that had become indispensable to so many thousands of fortune seekers in the California Gold Rush. Both had been early investors in the Union Pacific as well as other railroads. Oliver Ames (1807–1877) had occupied a quieter niche than his brother, running the family factories and taking the presidency of the Union Pacific, but Oakes Ames, born in 1804, had been in the public spotlight for years. Elected to the U.S. Congress as a Republican in 1862, Ames was assigned to the Pacific Railroad Committee and played a strong part in legislation creating the Union Pacific corporation and designating California's Central Pacific Railroad as its east-building competitor in the first transcontinental railroad sweepstakes. As an incorporator and investor, Representative Ames was tireless in his support of the road.

As the Civil War wound down in January 1865, just months before President Lincoln's second inauguration and assassination, Oakes Ames was summoned to the White House for a private meeting. "Ames," Lincoln said, "I want you to take hold of this." The Union Pacific was in the hands of shrewd speculators and was in danger of failing. If Ames and the moneyed, influential men he knew would step in, Lincoln said, Ames would "be the remembered man of your generation." It was of national importance—not just for the war effort but for helping to bind the nation together once peace was finally attained.

Representative Ames took this as an anointment, especially after Lincoln (the Pacific railroad's greatest friend) was dead. In time his brother, Oliver, ascended to the U.P. presidency. The brothers were forced many times to oppose the boldly corrupt efforts of the Union Pacific's vice president and general manager, Dr. Thomas C. Durant, saving the company from disaster. Meanwhile, Oakes took advantage of his political position to safeguard and advance the railroad's interests.

This is where he got himself into trouble. In 1867 and 1868, Ames decided that the railroad would fare better politically if his colleagues in the House and Senate were personally interested, and so he began to discreetly distribute stock in the Union Pacific's construction contracting arm, a company called the Crédit Mobilier Corporation of America; The Crédit Mobilier was managed by the U.P. directorate so it could inflate building charges and make huge profits whether the railroad was finished or not. Some congressmen were allowed to buy the Crédit Mobilier stock at half of par; others were "loaned" the purchase price but permitted to collect all dividends; others were simply given stock as gifts. When the dividends came in at 200 percent, Oakes Ames found himself the most popular man in Congress, besieged under the Capitol dome by politicians in a veritable feeding frenzy. He could not accommodate everyone, and a number went away bitter and unsatisfied. That the arrangement was not only questionable but crooked—these public servants would be voting on legislation affecting the railroad and their own pocketbooks—was shown by the fact that many had the certificates issued in the names of spouses, sons-in-law, or hometown business associates. Obviously, one could see that this was not the first time this had happened in Washington, nor (as we know) the last. The Crédit Mobilier erupted into a dramatic political scandal three years after the Pacific railroad was completed.

During the hotly contested Grant reelection campaign in 1872, President Grant was challenged by the Democratic populist and newspaper publisher Horace Greeley. Grant's administration had been plagued by corruption and mismanagement, but he was, after all, the savior of the Union. In September 1872, however, the muckraking *New York Sun* revealed that many key Republican figures—including a sitting vice president, a new vice presidential nominee, and a number of congressional leaders and committee chairs—had corruptly accepted the Crédit Mobilier stock and sought to hide it.

The resulting scandal filled the front pages of virtually every newspaper in the nation for months to come. For half a year barely any governmental arm functioned beyond the most perfunctory efforts.

President Grant was too revered to lose the election, though the scandal was too big to blow over. Beginning in December 1872, congressional inquiries convened to get to the bottom of it. For nearly two months, inside committee rooms and out on the Capitol steps and corridors, politicians did their best to put daylight between them and the corrupting betrayer of democracy, Oakes Ames. Friends of many years vied to denounce him. Finally, in February 1873, Representative Ames had had enough. He saw he was being made a scapegoat, and in a hearing room he produced the little red memorandum book in which he had recorded the names of twelve public servants and the amounts they had received; a thirteenth had been identified by others' testimony. With this damning evidence entered into the record, the congressional committee retired to deliberate.

The result handed down a week later should not have surprised anyone. Ames was seated as if at the gallows, between the chaplains of the House and Senate, and it was noted that throughout the reading of the report he smiled ironically, derisively, while most of his named colleagues seemed smug and sanctimonious. Only the thirteenth named—Senator James Brooks of New York—was mortally shaken, for he had reason to feel as if he were on the block as will be seen. In the end, Ames was found guilty of bribery. His twelve Republican colleagues were found innocent of both accepting bribes and perjury in denying it before Congress. The lone Democrat—Senator Brooks—was pronounced guilty. Brooks and Ames avoided being expelled but were roundly denounced and were censured, four days before the end of the congressional session. One of the last acts of the 42nd Congress was to raise members' salaries by 50 percent (from $5,000 to $7,500) and to award each a bonus of $5,000 for the session just ending. While they were at it, they doubled the president's salary (to $50,000) and gave raises to other high Republican officials.

The opposition press denounced the whole affair as a whitewash: the vice president, the vice president elect, the chairmen of the Ways and Means Committee, the Appropriation Committee, the Judiciary Committee, the Naval

Committee, the Banking and Currency Committee, and several stalwart party leaders, had slipped out of the noose. (One, Representative James Garfield, would later ascend to the White House, albeit briefly.) Furthermore, other important party members had since escaped across the jurisdictional borderline into private life. The shaken Senator Brooks died within weeks. Back in Massachusetts, Oakes Ames lasted two months then died after a stroke at age sixty-nine.

Two years later, in 1875, stockholders of the Union Pacific voted to raise an appropriate monument to Oakes Ames somewhere on the line of the Pacific railroad. Construction did not begin until 1880, three years after Oliver Ames had also died, so the monument would be dedicated to both. At a cost of nearly $65,000, the Union Pacific sent an army of stonemasons and other artisans from Massachusetts to Sherman Summit to construct a granite pyramid designed by the famed architect H. H. Richardson, whose other works included halls at Harvard, Trinity Church in Boston, and many church, educational, and public buildings around the country. Stone carvers incised large letters into the north face, facing the railroad: "In Memory of Oakes Ames and Oliver Ames." The famous American sculptor, Augustus Saint-Gaudens, fashioned eight-foot-square granite plaques set on the east and west sides—relief portraits of the brothers with their initials and life dates. The monument was completed and dedicated in 1882. Perhaps on some level the railroad hoped to crush the memory of the scandals that had stopped the nation in its tracks and that had come to symbolize the dark underlay of the Gilded Age.

In a fierce wind, I was dwarfed by the great pyramid on Sherman Summit. The nearby railroad was long gone. The town of Sherman was long gone, leaving foundation stones for a roundhouse, a bay-fronted depot, and a few scoured markers in the graveyard (most residents were exhumed and moved elsewhere). Circumnavigating the base, I squinted up at the relief portraits. The noses of Oakes and Oliver were vandalized—shot off by rifle fire decades ago. Perhaps there was some moral—or at least an ironic comment—to be gleaned from this, something about justice or history or the politics of money, but the only phrase that came to mind was, "the more things change, the more they stay the same." Every generation has its Crédit Mobilier scandal, and power corrupted in absolutely every direction one could see from this bare summit, then as now. Even the Ames Monument itself was drawn in. A few years after it was completed, an official down in Laramie—his name was Murphy, the WPA guide had reported, and he was a justice of the peace—was poking about in the public records and discovered that the Ames Monument had been built on public land. Murphy hastily filed a homestead claim on the property and wired the Union Pacific to take the pile of stone off his "farm."

He intended to hold out for a huge cash settlement. The U.P. attorneys finally quashed his suit by threatening to impeach him for conspiracy.

Justice, in that wry little footnote, may have been served. But the Ames brothers' noses are still gone, the monument is still out in the middle of nowhere, the railroad is long gone, and conclusions are in about as short supply—and unless one had a private excursion train to get from Cheyenne over Sherman Hill to Laramie, the only way to do it was by automobile. The wind was picking up. My children had tired of the pyramid's novelty after a few minutes and run off, but there was so much broken glass around—the place was densely littered with shattered beer bottles, .22 caliber shells, and spent firecrackers—that Mary and I had to get them into the car. We bumped off across the summit toward the interstate, that eight-laned successor to Mr. Lincoln's railroad and Mr. Lincoln's highway, then descended in roaring high-speed traffic through evergreens, aspen, and sage into a steep, hazardously curving declivity called Telephone Canyon toward the Laramie Plains, abode of outlaws and dinosaurs.

9

Road Tested on the Red Plains

Pity poor French Canadian Jacque LaRamée, the legendary free fur trapper. He hunted the wildlife-choked feeder streams of the North Platte River through the southeastern mountains of present-day Wyoming, and disappeared around 1820.

In 1821 his concerned fellow trappers organized a search party that included mountain man Jim Bridger, but when they reached his cabin at the confluence of the North Platte and an unnamed tributary, they found it only half completed, an ominous sign. There were no other traces of LaRamée beyond a broken beaver trap. It remained a mystery until two years later, when Bridger encountered some Arapaho tribesmen. They admitted they had killed LaRamée and placed his body behind a beaver dam, under the ice. The trapper's name was given to the North Platte tributary beside which he had intended to live, to the mountains through which it tumbled, and to a peak some 10,272 feet in altitude, one of whose headwaters was called Lost Creek. Twelve or thirteen years after his death when traders Robert Campbell and William Sublette built the first permanent fortified trading post in Wyoming, the place came to be known as Fort Laramie, from 1849 the famous army garrison on the Oregon Trail. Meanwhile, LaRamée, whose isolated, unknown grave had no marker, continued to be memorialized with the naming of the high, grassy plain through which the ancient Laramie River meandered between the Snowy Range of the Medicine Bow Mountains and the Laramie Hills. Then, in the spring of 1868, Union Pacific graders and tracklayers edged down the slopes of Sherman Hill to the head of the plains and past the palisaded Fort Sanders and the flag-fluttering survey stakes of the new railroad division town, which chief engineer Grenville Dodge had named Laramie. As the WPA historians would indicate, the river, valley, range, peak, fort, plains, town, and county would be marker enough for such a man as Jacque LaRamée.

The twin bands of iron of the Union Pacific joined at Laramie with the wagon ruts laid down by emigrants and by the coaches of Ben Holladay's Overland Stage Company, which had abandoned the central Wyoming route of the Oregon Trail because of Indian troubles in 1862. The Overland Trail followed a path along the South Platte through Colorado to Fort Collins, and then due north to the Laramie Plains, where it struck westward toward Bridger Pass. To Holladay's simple log stage station house were added hundreds of tents of speculators, entrepreneurs, and fast-buck artists, awaiting the Union Pacific's town lot auction in April 1868. Some 400 plots were snapped up within a couple of days, and Laramie shortly grew into a metropolis of shacks. The first train chugged into town on May 9, its coaches raucous with saloon keepers, gamblers, peddlers, tradesmen, brothel owners and their "prairie flowers," the flatcars spilling over with all of their various paraphernalia and with towering stacks of dismantled building sections. Hell on Wheels had advanced a little farther into the West.

Legitimate merchants swiftly formed a provisional government and elected a mayor and town trustees—but they were beaten to the punch by the lawless element, which had formed a union of its own. Within weeks the elected officials all resigned in fear of their lives. Disorder ruled across the summer and autumn; brawls, shootings, and robberies were everyday occurrences. Not even the commander at Fort Sanders would intervene—he simply issued strict orders that his bluecoats were forbidden from venturing into town. The worst miscreant was a sadistic bully who called himself "The Kid." When a little band of shopkeepers and U.P. laborers turned vigilante in August and lynched him, the other outlaws of Laramie terrorized the town. By late October, however, enough solid citizens had settled that some 500 armed vigilantes rose up in a coordinated campaign to raid a number of outlaw dens on October 29. Plans went awry, but in desperation the vigilantes swarmed to the Belle of the West dancehall, where a pitched gunfight erupted with most of the town citizens on one side or the other. Five were killed and fifteen wounded, and when four bad men surrendered to the vigilantes they were hanged from telegraph poles. Many others moved on down the line, or else joined the vigilantes.

Crime thereafter subsided, and in time civil government returned. Churches and schools began to make their appearance in early 1869. Small industry grew, there was a short-lived gold rush, and Laramie began to come into its own in the 1870s when small ranchers imported Texas longhorns to fatten and breed on the Laramie Plains, thus becoming large ranchers in the resulting boom. Easterners and Europeans looking to enrich themselves on cattle herded in to Laramie, some eventually building mansions and founding clubs. Spreading out from the U.P. tracks across a brickfront business district and through residential neighborhoods with their immature rows of poplars and cottonwoods, Laramie became a permanency, seat, from 1887, of the University of Wyoming.

When we drove into Laramie we naturally gravitated to the Union Pacific yards, climbed the long pedestrian metal truss bridge over the tracks, and watched a few freights pass below. We weren't surprised to find a couple of men with home video equipment set up on the bridge. "What are you filming today?" I asked them. "Whatever comes by," one answered. We were hard by Old Laramie, the first permanent streetscape of Laramie with its late-nineteenth-century brick business blocks, listed on the National Register of Historic Places and home to many restaurants and shops. We had just missed the street closings and strolling crowds of Laramie Jubilee Days, and now it was something close to normal for a sleepy July Sunday; traffic was less bustling and potentially dangerous than it was in Cheyenne, and with the weekend crowds gone we could walk into any restaurant at noon and find a table.

We picked the Overland Restaurant for its trail associations as well as for the building in which it was housed, the old Johnson Hotel building. A bright, two-story brickfront with white cornices and green sidewalk awnings, it was erected in 1900 on Grand Street across from the now long-gone U.P. passenger depot, and considered the premier hotel of the era. We had buffalo burgers and Mexican dips and struck up a conversation with a friendly couple at the next table, Kathy and Bruce, who were traveling on an extended, six-week family driving trip with their towheaded twin boys, aged eight, the same age as Davey.

"Wyoming has a state dinosaur!" one of their sons (Keith, I think, or maybe it was Russell) announced at lunch. This we had known since the Wyoming State Museum in Cheyenne when Mimi and Davey found that the state dinosaur—triceratops—had been determined by vote of Wyoming elementary schoolchildren in 1994 and thus legislated. We had several stops planned for the next day but the first destination was prehistorically pertinent: we were headed to the University of Wyoming Geological Museum, which dated back to the university's founding.

It is housed in the sandstone Geology Building in the northwest corner of the campus; a life-size, eighteen-foot-tall copper-plated statue of *Tyrannosaurus rex* stands guard before the entrance. Inside, towering fossil skeletons dominate a high-ceilinged room overseen by balcony displays. The state dinosaur is represented by a seven-foot-long skull; nearby is the skull of a cousin, Anchiceratops, the only one on display in the United States. These horned, armored herbivores occupied Wyoming and the region some 65 to 70 million years ago in the late Cretaceous Age, long before the geological uplift that created the Rocky Mountains. Their habitat was warm, low, and moist savannas and plains. Not far away is a skull cast of their contemporary predator, *Tyrannosaurus rex*, found near Rock Springs. The carnivore's head is some fifty-five inches high and fifty-nine inches long; its fifty-eight teeth and jaw look strong enough to contend even with the official state dinosaur's armor plate—indeed, to crack it like a lobster.

But these impressive skulls are dwarfed by the full, seventy-five-foot-long skeleton of the Apatosaurus (it was known as Brontosaurus when I was young), which at twenty-five tons in life lumbered through warm lowland floodplains in the Jurassic Period, munching vegetation some 130 to 190 million years ago, about eighty miles northwest of where we stood. It was discovered near Sheep Creek in 1901 and snapped up by agents of wealthy industrialist Andrew Carnegie for the Pittsburgh museum he sponsored—funding paleontology projects around the world was a good way to launder the Carnegie name after the brutal suppression of striking steelworkers in 1892. The stockpile of bones was acquired for a return to Wyoming some decades ago, and mounted here between 1959 and 1961. Apatosaurus now stands tail to tail with a newcomer—one it would not have liked to see up close—the full cast of Allosaurus, a twenty-five-foot-long, meat-eating young adult dinosaur, a Jurassic Period contemporary who has been named "Big Al."

Big Al may stand for Big Albert or Big Allison—the gender is unknown but popularly identified as male—and he is now a worldwide celebrity, being the most complete carnivorous dinosaur unearthed in fossil-rich Wyoming. The skeleton was found fully articulated (attached in life position) in 1991 in the Big Horn Basin near the town of Shell in north-central Wyoming. A commercial Swiss fossil-hunting company had been working a fossil bed on private property under license, but apparently by mistake the hunters worked their way about 100 yards onto public land administered by the federal Bureau of Land Management; there they discovered the skeleton. A bureau employee happened to be flying his plane overhead and recognized that they had strayed over the unmarked public boundary. He tipped off authorities. Artifacts on public land belong to the people of the United States—they cannot be removed by a commercial company and sold, out of the reach of scientists, to private collectors. Moreover, there is great sensitivity in Wyoming that so many valuable artifacts from its prehistoric past were exported to museums all over the world, with relatively few to be exhibited in the state.

The Swiss were sent back where they belonged, apologetic over their error but no doubt disappointed not to have bagged a prize worth at least a quarter million dollars in the private market. A team of experts from the University of Wyoming and Montana State University's Museum of the Rockies took over; they were beset by bad weather, their precious discovery endangered by quick erosion. But the fossilized bones were safely and chemically stabilized and removed with small hand tools and brushes from the surrounding rock outcropping in only eight days. The discovery of Big Al made international news; some 4,000 students were also present to view the unearthing. The skeleton was transported to the Montana facility with its better resources—because the fossil was found on public land it was required to go to a recognized federal repository. Big Al's bones remain unmounted there and ready for study, while Laramie's permanent mounted cast shows him stalking on his hind legs with a height of ten feet, his short arms with three fingers outstretched. Had he lived to full adult size, he would have been forty feet long and weighed two tons. With his sharp talons and teeth and prodigious leg muscles for running and leaping on his prey, Big Al would have been a formidable predator, even had he encountered the much larger Apatosaurus somewhere on a warm and humid Wyoming floodplain. As Brent H. Breithaupt, the museum director, commented at the time of Big Al's installation, "It's not as big as a *Tyrannosaurus rex*—but you wouldn't want to have met one."

We saw excellent displays on these recent discoveries in Wyoming, along with older exhibits. There was the head of a Columbian mammoth, a young adult found in a bog near Chicken Spring, thirty-five miles southwest of Rawlins; it is a true skull, not fossilized, and its five-ton body would have stood eleven feet tall. There was the full fossilized skeleton of a Mosasaur, an aquatic lizard that had lived in shallow seas and was found fifty miles northeast of the town of Lusk, Wyoming. There was a six-foot-long gar, a 50-million-year-old freshwater fish. There were full camel skeletons from the Miocene and Oligocene (5 to 37 million years ago).

Wyoming today is known for its habitat of grouse and partridge, pheasant and turkey, its bighorn sheep, mountain goat, black bear, mountain lion, mule deer, moose, elk, white-tailed deer, and of course pronghorn antelope. But at that compact and crammed little Geological Museum at Laramie, one sees how that is just the latest of a fascinating cycle of stories; the ground underfoot seems to pulse with those stories, with echoes of vanished lives. In a few days we would briefly visit one of the most famous of the dinosaur graveyards, at Como Bluff, an unescorted and unscientific traverse—alas without the good instructive company of Brent Breithaupt, who knows Como Bluff like the back of his metacarpals. To get there we would have to cross terrain that had, thereafter, crunched under the weight of a particularly noteworthy band of predators—those of the genus *Cowboyiia*, sub *outlawus.* Preparing us for this transition from the prehistoric era to that of rustlers, highwaymen, and train robbers, I intrepidly led my family out of the museum under the baleful gaze of the copper Tyrannosaurus and over to the parking lot, for a law-abiding ride over to the Wyoming Territorial Penitentiary.

The prison picturesquely presides over acreage on the west bank of the Laramie River: a large, two-story building with mansard-roofed wings extending north and south, all of it of brown-trimmed, buff limestone and sandstone. Constructed in 1872, it burned down soon thereafter, was rebuilt, and operated until 1903 when it was superseded by a new facility at Rawlins. The cell blocks were dismantled and dispersed to needy towns around the state, and for seven decades after, the main prison building served as a cattle and sheep barn for the University of Wyoming's experimental stock farm. Apparently during this era no inmates went over the wall. But by the 1970s the structure was so deteriorated that it was unsuitable even for livestock, and the animals were humanely moved elsewhere. In 1986 the building and grounds began to be restored as a museum complex and Old West theme park, giving birth to the Wyoming Territorial Park and Penitentiary, with re-created cell blocks and exhibits in the main building. The park also includes the National U.S. Marshals Museum, a Frontier Town, a Union Pacific Surveyors' Camp, and a Cheyenne-style tipi campground.

I stood in the original 1872 north wing of the prison, strong afternoon light streaming through the windows onto the restored three levels of cages, with their steel latticework grates and doors facing out to the corridor. The brick cells, each six feet by eight feet, with eight-foot-tall arched ceilings, contain bunks, chamberpots, and little else; two prisoners often inhabited them. Forty-two cells stood in this wing, overlooked by a barred watchtower high in the southeast corner of the interior space. One cell on each level was equipped with a bathtub, washbasin, and toilet. The self-guided tour proceeded through the eating room and guards quarters, and into the south wing, constructed in 1889, housing the women's quarters (women were incarcerated for the same types of crime as men, from theft to manslaughter), laundry, and the "dark cell"—used for bread-and-water solitary confinement. One particular inmate noted here was a tough 1890s rustler, Kinch McKinney. He did "hard time" for several successful escapes (he was always recaptured), threatening the lives of his guards, and attacking the warden. When McKinney finally got out of solitary for this last offense he plotted his final escape with a cell mate, William Doughterty, in May 1896. First, they secretly sawed their cell bars until they could be easily removed. At night, McKinney pulled out the bars, wormed out the cell door, and slid into the shop area where he cut a hole in the roof. Outside, he shinnied down a rain gutter and disappeared. Cell mate Doughterty, however, was unable to follow: he became wedged in the cell door hole, which is where he was found by guards.

Two doors down, in a regular cell only five feet by seven and seven feet high, was the abode of a much more notorious inmate who, while a guest of the state of Wyoming for horse stealing, followed all the rules and hoped to get time off for good behavior against his two-year sentence. His name was Robert LeRoy Parker—better known by his alias, Butch Cassidy.

He was born in 1866 in Utah to a devout Mormon farming family, but left home early, falling in with drifters and rustlers. Parker and accomplices pulled off their first bank robbery at Telluride, Colorado, in June 1889, in which they obtained nearly $21,000 in greenbacks and gold. Soon after, living off the proceeds and operating as a rustler in the central mountains of

Wyoming, Parker shucked his born name for "George Cassidy." There are differing accounts as to how he became known as "Butch"—some writers say he worked briefly as a butcher in Rock Springs, Wyoming, and in those days rustlers often became familiar with the back entrances of butcher shops—but he was primarily rustling when he was arrested and convicted of grand larceny for buying a stolen horse for $5. His time in the Laramie prison (July 1894–January 1896) was spent unremarkably, and Cassidy obtained a release six months before his two-year sentence was over. It has been said that he was paroled after promising the governor that he would not rustle again in the state of Wyoming, which, at least, was a promise he appeared to have kept.

Butch Cassidy returned to armed robbery, as it was much more lucrative than rustling. By August 1896, he had drifted to western Wyoming, and with two accomplices pulled off a bank robbery just over the state line in Montpelier, Idaho, which netted them $7,165. In April 1897 a payroll heist of $8,800 was tied to him in Castle Gate, Utah, south of Salt Lake City. During this time he was using a hideout in Utah called Robbers' Roost. By then Cassidy was forming the nucleus of a permanent gang, eight or nine men who would come to be known as the Wild Bunch. Their base of operations shifted to central Wyoming, to an inaccessible canyon called Hole in the Wall. One of the members was Harvey Logan, known as "Kid Curry." Another was "Flat Nose" George Currie. And another was a criminal all the way from just outside Philadelphia whose name was Harry Longabaugh. His alias was "the Sundance Kid." Longabaugh had a little experience robbing trains; it was known that such a pursuit could bring in much more money than robbing banks, and it was more dependable than banks and payroll jobs, with fewer witnesses out on a lonely right-of-way. Apparently the Wild Bunch began to study Union Pacific railroad schedules and court rumors about cash shipments. We would be heading north the following day, and at some point hoped to cross their tracks along the old U.P. line.

It was too pretty a day to spend it in the penitentiary. We hastened past the other exhibits—interestingly, the designers had bulked out the women's prison area with wall texts on significant Wyoming women, all of whom were too solid citizens to have done time—and soon we were back out in the free, fresh air. We moseyed over to Frontier Town, an Old West block of eight or ten wooden one-story buildings, including a blacksmith's, livery stable, saloon, marshal's office, and print shop. There was also a photographer's studio where one could get dressed up in period clothing and have one's portrait taken with an old bellows camera standing on a wooden tripod. We decided to have Davey and Mimi do this but found that another family, with twin boys, was ahead of us—it was the nice couple from our evening at the Overland Restaurant, Kathy and Bruce. Kathy went inside to help the boys, Keith and

Russell, select their costumes, while Bruce and Mary and I relaxed outside. It had been nagging at me how familiar he was, but there was a lot of trip talking still to be done. I told him we'd been to Niagara Falls, the Lincoln Home in Springfield, Illinois, Mark Twain's birthplace in Hannibal, Missouri, and Willa Cather's home in Red Cloud, Nebraska, and that I was driving my daughter a little crazy with all the cabooses we'd climbed into in museums and public squares. He said his family had seen Anasazi ruins, swum in hot springs, hiked in Utah slot canyons, and attended a few baseball games. It was a hot, lazy afternoon. As he talked I listened with one ear while I ran through assorted venues—*Booknotes?* . . . the Bread Loaf Writers' Conference? . . . a college symposium? . . . or perhaps he had a small part in a film I'd seen?

Maybe performing Shakespeare? Keith and Russell had by now found their cowboy outfits, complete with Stetsons and chaps, and Kathy and the photographer got them in front of the camera; Mary and our kids went in to watch. I kept up my part of the conversation, my mind gnawing away at possibilities. I'd grown up outside Philadelphia, Washington, and New York, I told him. He said he'd been raised in Virginia and still lived there. I'd gone to Boston University, I said. He responded that he'd gone to Miami University.

Suddenly it began to click: Virginia and Miami.

What was your major? I asked.

"Music."

Bingo. Now it became clear. I realized I was sitting on a bench in Laramie, Wyoming, with a Grammy Award–winning pianist and composer whom I'd seen on *Austin City Limits*, and whose LPs and CDs I owned, and I saw that the mental image that kept replaying in my head was definitely of him doing something, and what he was doing was playing accordion and singing with Bonnie Raitt in her stupendous *Road Tested* concert video. This other tourist dad was Bruce Hornsby, whose singles had gone to the top a dozen times, whose compositions had been recorded by the Grateful Dead, Willie Nelson, Huey Lewis and the News, Bob Seger, and Don Henley. He had recorded with Bob Dylan; Elton John; Bonnie Raitt; Crosby, Stills, and Nash; Stevie Nicks; Leon Russell; Chaka Khan; and the Yellowjackets.

Not that I let on as his boys said "Cheese" for the photographer and my kids began rooting through the costume trunks with Mary. Where I come from, that sort of thing isn't done. This is something people with the right sensibility in Los Angeles and Washington understand as well as New Yorkers: you keep your aplomb. If you encounter a television actor in L.A. or a noted senator in D.C. or any kind of celebrity in New York City, you don't lose it. You treat them like regular human beings and allow them their space and privacy. Maybe you'd say, "Love your work," but it is cooler to be pleasant and not let on.

I was happy enough to have made the recognition. And since we had been in college in roughly the same era, I could reply to his answer that of course, I'd heard of the Miami music program, for it was indeed well known even back then. The conversation drifted back to travel itineraries—Mary and I had made a strong case for the Great Platte River Road attraction for their boys if they were going through Nebraska, and Bruce and Kathy recommended Flaming Gorge if we had time for water play. The twins' costumes were back in the trunk, the photographer had finished the quick processing of their portrait and slipped it into a cardboard frame, and it was time to part. We all said good-bye, and Mary and I went inside to get our kids ready for their shot.

Later, we were back in the car and heading south of town to see what the terrain looked like on the almost entirely vanished military reservation of Fort Sanders, on Soldier Creek, now some scattered foundation stones and a few lonely buildings. There, in July 1868, an extraordinary assembly of army brass from the Civil War convened to come to the aid of former general Grenville Dodge, now chief engineer of the Union Pacific Railroad. Dodge had withstood a prolonged assault from his boss, U.P. vice president Thomas Durant, who had lengthened the railroad line in Nebraska and Wyoming with extraneous curves and was bleeding the enterprise dry in many other ways. Now Durant was trying to get Dodge (whose ethical and engineering standards were slightly higher) fired. He summoned the engineer to Fort Sanders for a meeting, but Dodge learned that some old army buddies happened to be close by in Colorado at the time. He called for reinforcements at the meeting. When Dr. Durant appeared at the appointed time, he saw that Dodge's supporters included Ulysses S. Grant, the favored Republican presidential candidate for 1868; William Tecumseh Sherman, army departmental commander in the West; Philip Sheridan, cavalry commander; and a retinue consisting of Generals Augur, Harney, Gibbon, Dent, Slemmer, Potter, Hunt, and Kautz. The sight of so many eagles may have made Dr. Durant protectively clutch his abdomen to protect his innards. If he was carrying a pink slip for Dodge, it never came out of the doctor's waistcoat pocket. Instead, Grant insisted that Dodge's straight lines and railroad grades be respected in the national interest. This bought the engineer a little respite.

Fort Sanders was also once the home of Calamity Jane, according to her own tales—she always said she was stationed there as an army scout in 1871–72, after accompanying the troops on an Arizona campaign. Thereafter she led a contingent up into Powder River country and participated in skirmishes with the Sioux and Cheyenne.

We pulled off Highway 287 into a restaurant parking lot. What could the soldiers of the army post have seen from that place? Smoke from the chimneys of Laramie, just north; the Laramie Mountains rising to the east, and the Red Buttes; the plains and future ranchland stretching south toward the

Colorado line with the high Rockies propped up behind; and rising westward as if in ascending stairsteps, Sheep Mountain, the Medicine Bows, and the white-topped Snowy Range, all lit in late afternoon light in extraordinarily clear, thin air. Perhaps a Union Pacific freight train was heading past the Red Buttes toward the ascent over Sherman Hill—like the modern-day yellow four-header we saw drawing a line of coal hoppers as we stood there in the Wyoming wind. We went inside the restaurant to inquire after dinner—the restaurant was called, appropriately enough, the Cavalryman—but the waiting list was long, the outrageous prices on a posted menu would have choked a railroad mogul, and the dining rooms were turned inward upon themselves with hardly any windows to frame that extraordinary view available just outside. And who wanted to look at a lot of spendthrifts forking down the beef?

We headed elsewhere.

Back in Old Laramie, we decided to patronize the Overland Restaurant again, and were no sooner seated than we found that the next table was once more occupied by the Hornsby clan. By now I'd told Mary who they were. Over the course of the meal, parents attended to their offspring and kept a pleasant conversation aloft about the long family trips we'd all taken as children. In some ways all four of us were trying to replicate agreeable experiences we'd had. During this, I thought about bringing up music but just wasn't in the frame of mind to puncture the easy anonymity, even though I could have with no difficulty as I knew we had acquaintances in common. But to tell the truth, as we all sat in that nice little restaurant down near the Union Pacific tracks, I wasn't in the mood. I was tired, and looking over at Bruce I could see he was tired around the eyes, too. As a matter of fact, we would learn later that he played more than seventy shows that year, from jazz festivals to bluegrass festivals, and he'd written music for films, appeared on television shows, and was about to release a double CD of live performances. I, by the same token, was on the crest of a long, uphill ramp: a frantic year finishing *Empire Express* with the Ambrose juggernaut rumbling down the tracks behind me; an exhausting six months of the publication process while writing and editing another book for Middlebury College, all while teaching; a wonderful, dizzying fall and winter when *Empire Express* was published, with a long book tour and a number of fast overnight trips to cities for television tapings, and an increasing number of lectures. We were both tired about the eyes.

And it had been a long day—ours had stretched from the Jurassic to the present. The kids had a date with HBO back at our motel; I had a date with the inside of my eyelids. We were hitting the road in the morning. We warmly thanked Kathy and Bruce for their company in Laramie, and they wished Mimi and Davey luck in school in September. Later, back home, in early winter, I sent a copy of my book to them, unveiled the scenario as it had unfolded to me, and we now exchange Christmas cards and letters. But that night in Laramie, it was just two sets of traveling parents herding their children back to motels.

In the morning, we aimed the Dodge north and west up the Laramie Plains. If we'd had the right CD with us, I know whose it would have been.

The sun had lofted above the low ridge of the Laramie Hills on our right, illuminating the white caps of the Snowy Range in the Medicine Bows on our left, as we rolled up the plains past the inevitable northward sprawl of the Laramie outskirts, at that point limited to prefab houses on lots carved out of the wide-open old buffalo grazing land. Then the two-lane concrete-and-tar-stripped road of Route 30/287 threaded through some hayfields and ranches before the landscape settled on bare, low-growth ranchland. There was almost no traffic in either direction. Laramie Peak stood backlit and blue, some eighty miles to our northeast but clearly visible at about 3,000 feet higher in altitude than our high plain. Soon all the Laramie Hills disappeared. Antelope placidly grazed in a dirt bike racetrack next to the road. A few unpleasant-looking gypsite ponds—they are said to be scooped out by high winds and filled through percolation—stood just to the east with their harsh white chemical beaches. The steel tracks of the Union Pacific ran parallel to us in sight just to the west.

In the spring of 1868, several thousand tracklayers followed the easy grade up the plains, working like a huge machine as the days warmed. "We are now Sailing," the happy contractor Jack Casement wrote to his wife back in Ohio, "& mean to lay over three miles every day." Hundreds of new men crammed onto wagons passed the Irish tiemen, railmen, and spikers, heading west to fulfill rock cut and grading contracts far down the intended line. A snowstorm surprised Casement's men in the second week of May, halting them for three days while 400 shovelers went out to clear off the embankments. "I never saw a worse storm," he wrote his wife. The Laramie Plains were famous for their fierce storms, as the railroaders would discover. A year later, a blizzard arrived and filled the great trough that was the Laramie Plains, blocking ninety miles of track and stopping all train traffic for three weeks. Hundreds of passengers were marooned at little whistle-stop Wyoming stations with no accommodations and scanty provisions. Jack Casement's brother and partner, Dan Casement, rode a big snowplow train across Wyoming and could hardly believe the ferocity of the elements. "Have seen a cut fill up in two hours that took one hundred men ten hours to shovel out," he telegraphed headquarters. Two hundred stranded passengers up in the town of Rawlins could not get the train crews to leave the station and their consoling bottles of whiskey until the passengers agreed to man the shovels;

they cleared a drift 1,000 feet long only to have the locomotive bog down again. Fifty of the angriest passengers then left on foot for Laramie, trying to follow the buried tracks and suffering terribly until they struggled in four days later. Miraculously, no one died. After years of contemplating the problem, the Union Pacific adapted the idea of its competitor, the Central Pacific Railroad of California, and in 1916 roofed over its tracks with miles of snowsheds.

High winds and drifts above Laramie have dogged the progress of automobile roads, too. Interstate 80 darts across the Laramie Plains and ducks through a notch in the mountains before veering off west; even with massive plows it has often been closed by untoward weather. Back in the old Lincoln Highway days, local road crews knew when to just give up and wait it out; weather and geography sometimes conquered back. When Alice Huyler Ramsey became the first woman to drive across the country in 1909, she wisely did it in summertime, so her sights on the Laramie Plains were akin to ours. All she had were old wagon trails, often impeded. For her driving stunt she was preceded across Wyoming by "pilot cars" driven by helpful men who knew the countryside. But this did not always speed her progress. "The land had stretched out into huge ranch country," she wrote, "some parts of which had fertile fields watered by irrigation ditches. They were crude water courses, across which our trail would pass from time to time, necessitating fording them. Naturally, to keep the water from escaping, they were constructed with a slight bank on each side of the ditch." Here, she noted, the pilot driver got bogged down. "Approaching," she continued, "he went down deep with the front wheels, which in turn rapidly rose out of the water on the far side; the rear wheels followed the others quickly into the ditch. Whether the bank was steeper than he judged, I'll never know, or why he deviated from the trail, I can't imagine. The rear wheels now came to a complete stop, almost entirely under water, as millions of bubbles issued from his active exhaust. The motor never stopped but it wasn't doing him one bit of good. That automobile was really stuck!" Ramsey's car negotiated the ditch at a shallower spot. Out came the block and tackle and rope, and pulling in reverse, she yanked him out.

Fifteen years after Ramsey rattled through, the Lincoln Highway was a well traveled though rutted gravel road through those parts, and the hamlet of Bosler at the Laramie River crossing was listed in a 1924 highway guide as having a population of seventy-five, with two schools, a newspaper, telephone company, express office, railroad station, tourist camps, and even a hotel. Now it is a wide place in the road where deserted houses far outnumber inhabited ones, and the playgrounds of the yellow-brick Bosler Consolidated School have been converted to an auto graveyard. Ahead, a thin line of green revealed itself to be Rock Creek—narrow, clear water reflecting the blue sky, fringed with greenery—meandering irresolutely around low hills toward an abandoned stretch of the original Union Pacific line, which was shifted ten miles west to its present course to avoid some undesirable river crossings and snow choke points. This, however, killed off the lively freight and cattle stock-yards town of Rock Creek quicker than Interstate 80 had killed little Lincoln Highway way stops like Bosler.

But while the trains still rumbled past Rock Creek's stores and saloons, banks and businesses, the first section of the No. 1 Overland Limited from Omaha paused at Rock Creek's division station in the wee hours of June 2, 1899, and moved on up the line, its engineer probably thinking of a breakfast pail in Rock Springs. A mile west of Wilcox, four passengers who had climbed aboard at that station pulled on masks and unholstered guns. One went forward, crawled over the tender, and dropped into the cab, pointing a gun at W. R. Jones, the engineer, and ordering him to halt the train just west of a bridge. Two more robbers were waiting there near a small signal fire.

The engineer and fireman were forced to walk back to the first of two rail-way mail cars, inside of which were two postal clerks sorting mail. The rob-bers shouted for them to open up but the clerks extinguished their lamps and took cover, refusing to open the door. They argued back and forth for some minutes before the robbers got impatient and set off an explosion with a dy-namite stick, rocking the car. They threatened to destroy the car. The clerks surrendered, but before the robbers could examine the mail the headlamp of the second section of the Overland approached through the darkness. One of the crew told the robbers that there were two carloads of soldiers on the next section. Frantically, the railroaders were hustled back onto the train, it was pulled forward a short distance, and the robbers set off a dynamite blast on the bridge. Though the structure remained standing, the rattled robbers de-cided to move forward again, but only after uncoupling the train behind the two mail cars and the express car.

The attenuated No. 1 Overland Limited pulled ahead another two miles, but by then the robbers were thinking about the two carloads of soldiers al-legedly down the tracks; instead of taking a long time going through the mail cars, they turned their attention to the express car. The attendant inside, E. C. Woodcock, refused to unlock his door. They attempted to blow out the door with more explosive, but as they had measured too lightly with the bridge they overcompensated with the door and blasted a huge hole in the side and roof of the car. Woodcock was stunned but uninjured. The robbers used more explosive on two safes, gained entry, and cleaned them out of about $60,000 in unsigned banknotes. Then they rode away from the wreckage and disap-peared into the night on mounts they had waiting for them. Lawmen found 100 pounds of dynamite at the scene the next morning.

According to a number of sources, Butch Cassidy and Flat Nose George Currie were credited with planning the robbery. The Wilcox heist was faithfully used by William Goldman when he wrote the screenplay of his film *Butch Cassidy and the Sundance Kid*, even to naming the hapless attendant Woodcock. The humorous scene, complete with the exploding express car, was on our minds that morning as we drove toward the crossroads town of Rock River (population 190, elevation 6,892). I had three books about Butch Cassidy on the seat beside me, the most authoritative of which was written a couple years before by Richard Patterson and published by the University of Nebraska Press. The literature around Butch Cassidy is often fanciful and seldom trustworthy, and another of my books, *The Wild Bunch* by F. Bruce Lamb, was an indispensable myth-buster. In Patterson's careful account of the Wilcox heist, the author pointed out what he calls "an interesting sidebar"— on board the Overland Limited that night in 1899 was a lawyer named Douglas Preston, who was on permanent retainer for Butch Cassidy. Preston was spotted by a friend who worked for the U.P. and who thought that the attorney acted very guiltily, and blabbered about having an alibi. As Patterson judiciously weighs the confusing claims and suspicions of several writers on the subject, one gets the impression that he believes the Wild Bunch to be the perpetrators and that among them were Flat Nose Currie, Harvey Logan, the Sundance Kid, and others, and that Butch Cassidy, one of the planners, rejoined them later to count and distribute the money before they were scattered into flight by a hard-riding posse, though nothing like it was depicted in the Goldman film. The chase moved northward past the town of Casper, there was a firefight between the lawmen and three of the bandits, and a sheriff, Joe Hazen, was shot in the abdomen and died the next day. With this death the train robbery made national news; Butch Cassidy's Wild Bunch became famous.

One of the safes they blasted was removed from the scene and eventually found its way to a little community museum in Rock River where it reposes, twisted shards of steel poking out into the air like an Alexander Calder sculpture, along with photographs of the dynamited express car—reduced to charred splinters, undercarriage, and wheels. Other holdings at the Rock River Museum include artifacts from nearby paleontological digs, which we would have liked to have seen also, but Rock River was mostly boarded up. A large, stately concrete canopy gas station, now closed, prompted images of the extinct Sinclair Oil dinosaur in my mind. The brick-and-sandstone-columned First National Bank building had been shuttered since 1927, and the only life seemed to be in some barefooted kids playing an obscure game out in front of a little country church. The museum had a delightful bright mural painted relatively recently on one wall—an outlaw squinting at a baleful dinosaur—but the building was as tightly shut and undisturbed as the old bank, despite posted hours on its door and in Laramie tourist handouts. We drank sodas from our ice chest and had to content ourselves with looking at the photos of the blasted safe and railroad car in my Butch Cassidy books before moving on.

North of Rock River we rolled through bleak flats of sagebrush and short-grass. Then we came in sight of majestic Elk Mountain, to the southwest and at 11,152 feet as much of a landmark as it had been for the passengers on the Overland Stage as they pulled out of the Rock Creek station. To our east was a low, undistinguished, red-colored bluff that actually had a distinguished, even dramatic history: it was Como Bluff, famed dinosaur digs for more than a century, which we would approach from a different direction later in the day. A few miles up the road and we slowed for a faded collection of low buildings, the centerpiece of which, heralded in old guidebooks and featured in the Ripley's Believe It or Not columns, was "the oldest house in the world," built entirely out of dinosaur fossils set in concrete. It sat behind an antediluvian picket fence, propping up memories of weeds. In the Lincoln Highway–era WPA guide it was billed as "The Creation Museum," one of those humble, idiosyncratic tourist attractions that have been erected by hopeful, uncapitalized but starry-eyed entrepreneurs everywhere to rake in the tourist dimes and quarters ever since the tourists were rolling by in horse-drawn wagons and worrying about Indians. In the Lincoln Highway era, people passing through paid a quarter to look at dinosaur fossils, artifacts, and relics. Now it had a "Ripley's Believe It or Not" sign in letters as faded as the words "Fossil Cabin" over the door, but no number of quarters was going to get us inside— it was closed tighter than a Union Pacific safe. Weather-faded drawings of Triceratops, Stegosaurus, and Apatosaurus were nailed on poles. Another sign read "Dinosaur Graveyard," but I guess even the grave attendant was dead. A blank-windowed, closed-up stone cottage stood a few yards away, with dinosaur leg bones leaning on either side of the front door next to some rusty old snow shovels and window screens, and signs read, BEWARE OF RATTLESNAKES and NO TRESPASSING. Empty, tumbledown pens and paddocks leaned behind the cottage, some adorned with antelope and elk horns, with empty little galvanized feed buckets wired on panels at lamb height: an old petting zoo. But the only things to pet here now had rattles on one end and fangs on the other.

A Union Pacific freight whistled by. It was all very theatrical, complete with weathered stumps whitening in the sun, wind-rattling metal signs, a rusty can blowing across gravel and weeds, and eerie-sounding gusts reaching us from the heights of distant mountains to the north. It only reminded us of the emptiness of our stomachs, it being several hours past noon.

We moved on, fighting the wind.

On July 19, 1885, a twenty-four-year-old Main Line Philadelphian named Owen Wister rode a rancher's wagon into the town of Medicine Bow,

Wyoming. His pedal extremities were still tender; he had been in-country all of sixteen days.

Wister had enjoyed many advantages in life. His father was a physician and his mother, a published writer and friend of Longfellow's and Henry James's, was the daughter of the famed British actress Fanny Kemble. When Owen was seven his mother inherited her parents' Germantown estate and they lived there with a retinue of servants and a circle of genteel society, their life punctuated by extended travels in Europe. Owen was talented at music and schooled at the exclusive St. Paul's boarding school in Concord, New Hampshire, where he pursued music and sports and discovered a love for literature: not only the classics, but contemporary writers like Walt Whitman and Mark Twain, whom his teachers thought too democratic to be considered seriously. He wrote poetry, essays, and literary reportage, and edited the school newspaper. He seemed forever in trouble with administrators for excessive frivolity and independence—traits, indeed, to which his stern father had objected for years. Wister entered Harvard in 1878, a member of the class of '82, and lost no time in continuing his musical and literary enthusiasms, contributing to the *Crimson* and soon becoming its freshman editor. Theatrical clubs and productions kept him busy his four Harvard years. In his freshman spring at an athletic meet he had his first sight of a junior, Theodore Roosevelt, slightly built with muttonchop sideburns and no mustache, and at the time occupied in a boxing match from which he would emerge with a bloody nose. He and Wister would become lifelong friends.

Wister spent a year traveling with his ailing mother in Europe after graduation. He had serious aspirations to follow a music career and studied in Paris, obtaining audiences in Bayreuth and Weimar with the great Franz Liszt, who warmly encouraged him. But Wister's father impatiently summoned him back home, having secured the graduate a post in a Boston banking firm. There, Wister made the best of a dreary bank teller's job by leading an active social life, and becoming a charter member of the Tavern Club, which included 100 Boston writers, artists, doctors, lawyers, and merchants, with William Dean Howells as president. Howells, then working on his novel *The Rise of Silas Lapham*, became a kind of mentor to Wister. The literary young man tried things out for himself with a highly autobiographical coming-of-age novel about a sensitive and artistic fellow whose parents failed to appreciate him. After months of labor he sent it to Howells, who advised him to hide it away: it was a "rebellious" work, shocking, profane, with excessive bad language and too many references to hard drinking. "So much young man never seems to have gotten into a book before," Howells wrote him, adding that it would be a grievous mistake even to publish it anonymously.

Wister took the advice and apparently destroyed the manuscript. By then, in the late spring of 1885, he had left his Boston job, returned home, and planned to enroll in the Harvard Law School in September. But suddenly Wister suffered a nervous collapse. It may have had literary roots in the recent rejection, or have been tied to the earlier reversal of a musical career. At any rate, his physician prescribed a rest cure—fresh air, physical activity, and contact with "humble folks in the fields." It was uncommonly good advice, and it would resonate down the corridors of American literature.

Wister's rest cure was to be at a Wyoming cattle ranch north of Laramie Mountain and not far from the ruts of the Oregon Trail. The notion of going West was not without precedent in his circle—several Harvard chums had done so, the most noteworthy being Theodore Roosevelt, attempting to mend his own tattered life after the deaths of his mother and wife and a humiliating defeat at the Republican convention. Patient Wister left Philadelphia on a westward-bound train, accompanied by two older women who were friends of his mother and who would keep their protective matronly eyes on him. His health improved with every mile he put between himself and home, and the literary young college graduate began scrawling impressions in a notebook.

"Here and there," read one of his entries, after the Union Pacific had borne him across Nebraska and into Wyoming, "far across the level, is a little unpainted house with a shed or two and a wagon. Now either a man on horseback or a herd of cattle. We've passed a little yelping gang of collies who raced us but got beaten. The sky—there is none. It looks really like what it scientifically is—space. The air is delicious. As if it had never been in anyone's lungs before."

The train deposited Wister and his companions at Rock Creek in the wee hours of the morning. "The remains of the moon is giving just enough light to show the waving line of the prairie," he noted. "Every now and then sheet lightning plays from some new quarter like a surprise. The train steamed away into the night, and here we are." They passed the rest of the night in a room in the depot; the walls were festooned with animal skins he could not identify. The next day they were conveyed fifty miles north by stagecoach. "I can't possibly say how extraordinary and beautiful the valleys we've been going through are," Wister wrote. "They're different from all things I've seen. When you go for miles through the piled rocks where the fire has risen straight out of the crevices, you never see a human being—only now and then some disappearing wild animal. It's like what scenery on the moon must be. Then suddenly you come round a turn and down into a green cut where there are horsemen and wagons and hundreds of cattle, and then it's like Genesis. Just across this corduroy bridge are a crowd of cowboys round a fire, with their horses tethered."

One of his first sights at the Wolcott ranch was the branding and castrating of seventy-nine calves. He started a beard, slept in a tent, bathed every morning in Deer Creek, and marveled at the profusion of game. "If I don't learn to shoot," he said, "it won't be the fault of the wild animals of these parts." A day later: "This existence is heavenly in its monotony and sweetness. Wish I were going to do it every summer. I'm beginning to be able to feel I'm something of an animal and not a stinking brain alone." And on July 14: "I'm a quarter of a century old today."

A few days later he and the rancher were heading south the ninety miles to the nearest railroad station at Medicine Bow, arriving there after nineteen hours of driving and a night on bedrolls in the mountains. Their purpose was to meet the midnight train to fetch a shipment of trout and bass fingerlings for stocking Deer Creek. Medicine Bow did its best to impress him and failed. "This place is called a town," he commented cynically in his notebook. " 'Town' will do very well until the language stretches itself and takes in a new word that fits." He had a lot of time to kill, so he walked around to make sure he didn't miss anything.

> Medicine Bow, Wyoming, consists of: 1 Depot house and baggage room; 1 Coal shooter; 1 Water tank; 1 Store; 2 Eating houses; 1 Billiard hall; 6 Shanties; 8 Gents and Ladies Walks; 2 Tool Houses; 1 Feed stable; 5 Too late for classification. Total 29 Buildings in all.

It was a tally that would one day be of use to him in his career to come.

The local storekeeper let Wister and Wolcott nap for a few hours on the counters of his establishment until the train came in. This, too, recorded in his notebook, would be useful one day. Back at the Wolcott ranch, riding, taking part in roundups and chores, hunting and fishing, getting brown as a bear and hairy as a cayuse, Wister repaired his soul and toughened himself for a return to Eastern society, and even Harvard Law.

But the West was now inside him. Through the subsequent space of years of law school and entry into a Philadelphia law firm, he went back to Wyoming at every opportunity, seeing also California, Oregon, Washington, and British Columbia. But it was Wyoming that spoke clearest to him. All the while he filled fifteen notebooks with what he saw and heard, and although most people have taken him at his word that he gave no thought to being a writer until autumn 1891, after a particularly vivid summer at Yellowstone and his usual central Wyoming haunts, Wister's journals were a writer's working notebook, a treasure house of impressions. Strangely, his family— his wife, whom he married in 1898, and his children—would not know of the notebooks' existence until after his death when his desk was opened and inventoried. Never before in his life, Wister would say, had he ever been able to sustain a diary, no matter how significant or even thrilling his experiences whether in Boston, Europe, or Philadelphia. "But upon every Western expedition," he said, "I had kept a full, faithful, realistic diary: details about pack horses, camps in the mountains, camps in the sage-brush, nights in town, cards with cavalry officers, meals with cowpunchers, round-ups, scenery, the Yellowstone Park, trout fishing, hunting with Indians, shooting antelope, white tail deer, black tail deer, elk, bear, mountain sheep—and missing these same animals. I don't know why I wrote it all down so carefully, I had no purpose in doing so, or any suspicion that it was driving Wyoming into my blood and marrow, and fixing it there."

The notebooks are a joy—and most so in the summer of 1891, when the beautiful desolation, the comforting isolation, the nobility and the cruelty of nature and one of its instruments, man, sent his pen to scratching with the highest eloquence.

"Reached here at 2:15 A.M. this morning," he noted at Chadron on June 10, 1891.

> Then up and breakfast and off on this train which takes from 6:30 A.M. till 7 P.M. to go 192 miles. It has one passenger coach behind a miscellaneous assortment of great freight and stock cars, and the engine is always just around the next curve and can be identified only by the jetting black smoke it shoots heavily up. I don't object to the gait of this train. It oozes along through the draws, and the sun blazes comfortably down. I sit on the back platform and find it great luxury to stare at the things I know so well. Not a house, not an animal. But the train goes so slow that the birds sing louder than the noise it makes.
>
> Every now and then a cracked, dry watercourse up through the mounds, and on its bank black, stiff, little pines whose trunks and boughs are twisted in leaning contortions, as if they had been lazily stretching themselves and caught forever midway. Now and then a stream, whose little green valley is choked with cottonwood and bushes. Then a bald station and "town," ragged cattle, wire fences, wooden mansions suggesting that firmness and permanence of stage scenery. Then more trundling, with occasional trestle work over a chasm along which a lonely trail winds below. We're too late, nearly an hour. But I don't seem to care. Sometimes the country makes you think of a face without eyebrows. The occasional windmills look like stage setting too. So do the solitary meaningless sheds one passes. All seem constructed to be shifted at a moment's notice. Then a man with a huge broad hat comes by driving a team slowly, and apparently from nowhere to nowhere.

A long trail ride with a cowpoke named Tisdale turned into a procession of deep shocks for Wister, for Tisdale was a base and cruel man. At one instance he "taught a lesson" to a horse insufficiently obedient by gouging out an eye (a scene Wister would later replicate in his great work). The Easterner grasped a sobering fact about the darker side of existence, and something clicked: he saw the urgency of chronicling both sides, the light and the dark.

"I begin to conclude from five seasons of observation," he confided to that last notebook,

> that life in this negligent irresponsible wilderness tends to turn people shiftless, cruel, and incompetent. I noticed in Wolcott in 1885, and I notice today, a sloth in doing anything and everything, that is born of the deceitful ease with which makeshifts answer here. Did I believe in the efficacy of prayer, I should petition to be the hand that once and for all chronicled and laid bare the virtues and the vices of this extraordinary phase of American social progress. Nobody has done it. Nobody has touched anywhere near it. A few have described external sights and incidents, but the grand total thing—its rise, its hysterical unreal prosperity, and its disenchanting downfall. All this and its influence on the various sorts of human character that has been subjected to it has not been hinted at by a single writer that I, at least, have heard of. The fact is, it is quite worthy of Tolstoi, or George Eliot, or Dickens.

Many years later, after Wister had become a novelist, he explained (in a memorial volume about his long friendship with Roosevelt) what happened next. In the autumn of 1891 in Philadelphia, "fresh from Wyoming and its wild glories," he had dinner at his club with a kinsman, Walter Furness, who was also passionate about the wide-open spaces. "From oysters to coffee we compared experiences," he recalled, as they darted from lived experiences to the waiting empty pages for posterity.

> Why wasn't some Kipling saving the sage-brush for American literature, before the sage-brush and all that it signified went the way of the California forty-niner, went the way of the Mississippi steam-boat, went the way of everything? Roosevelt had seen the sagebrush true, had felt its poetry; and also Remington, who illlustrated his articles so well. But what was fiction doing, fiction, the only thing that has always outlived fact? Must it be perpetual teacups? Was Alkali Ike in the comic papers the one figure which the jejune American imagination, always at full-cock to banter or to

brag, could discern in that epic which was being lived at a gallop out in the sage-brush? To hell with tea-cups and the great American laugh! we two said, as we sat dining at the club. The claret had been excellent.

"Walter," Wister blurted, "I'm going to try it myself—I'm going to start this minute!"

"Go to it," responded Furness. "You ought to have started long ago."

Wister went up to the library of the clubhouse, and by midnight a good slice of his first Western story "was down in the rough." When polished and sent to *Harper's Magazine*, its first recipient, the story, entitled "Hank's Woman," was accepted. So was Wister's second tale, "How Lin McLean Went West." *Harper's* sent him West in the spring of 1893 to write stories and articles and on his return he gave up the law. The next summer he went back to Wyoming.

A year later his first Western story collection was published, titled *Red Men and White*. He had by then issued two forgettable volumes, a burlesque of *Swiss Family Robinson* and a fantasy novel. *Red Men and White*, though, was solid and took hold—it was tough and realistic but also eloquent and literary, and nothing like it had yet been seen in print. But it was all preparation. More stories appeared in magazines and were bound in two subsequent collections. But slow down and look: if one peers back at Wister's journal titled "Frontier Notes, 1894"—writer's notes of stories, gossip, dialogue, and description for later use—one finds a passage taken from life that, with many of the earlier stories reworked and welded together into linked chapters, would coalesce into the great work of a lifetime.

> Fetterman Events, 1885–1886. Card game going on. Big money. Several desperadoes playing. One John Lawrence among others. A player calls him a son-of-a-b—. John Lawrence does not look as if he had heard it. Merely passes his fingers strokingly up and down his pile of chips. When hand is done, he looks across at the man and says, "You smile when you call me that." The man smiled and all was well.

The process of mining the riches in "Frontier Notes, 1894" and his stack of short writings would occupy Owen Wister during the last years of the nineteenth century and the first of the twentieth. Medicine Bow at midnight with its twenty-nine buildings and a train vanishing into the blackness would supply his opening imagery. A procession of no-accounts would unite and slide into the single pair of boots of a man called Trampas. And a few impressions of a kind of gentle, mute nobility he saw in several Westerners—including a Point of Rocks stagecoach station operator from Virginia and a cowpuncher and cavalry scout from Arizona—would assemble into an unnamed hero he called the Virginian. And from that he had his title.

Fenimore Cooper's Leatherstocking would have approved of Wister's romantic superman. And by then others had approached the West (or were about to) from other literary as well as geographic perspectives—a generation before, Bret Harte and his California mining camps; and more recently Alfred Henry Lewis (writing then as Dan Quin) and his dialect-voiced tales of Arizona cattle drovers. Others were on Wister's boot heels, like Andy Adams and his Texas cowboys. But as Wallace Stegner has noted, Wister's *The Virginian: A Horseman of the Plains*, published in 1902 and instantly, wildly popular, is the sourcebook, the established template for America's archetypal cowboy, its mythic horseman. Western literature high and low would follow its path. And so would Hollywood. In 1902 Gary Cooper was being weaned and learning to walk in Helena, Montana; William S. Hart was a successful thirty-two-year-old Broadway actor; Tom Mix, at twenty, was between service as an artillery sergeant and a drum major; Tim McCoy, at eleven, was the son of the police chief in Saginaw, Michigan; Henry Fonda was not even in the planning stage in Grand Island, Nebraska. Joel McCrea had a date with his parents in South Pasadena in 1905, John Wayne in Winterset, Iowa, in 1907, Alan Ladd in Hot Springs, Arkansas, in 1913, and Clint Eastwood in San Francisco in 1930. Bill Pullman, the latest to play the Virginian or his type, would first find voice in Hormel, New York, on the fifty-first anniversary of the book's publication. In its first year, *The Virginian* sold 100,000 copies. It was translated into many languages, went through multiple U.S. editions, and has always remained in print.

"WHEN YOU SAY THAT, SMILE!" We stood in Medicine Bow, Wyoming, seven miles north of the fossil cabin on Route 30, at the base of a large brown pyramid constructed of fragments of petrified wood. The plaque upon which these words were inscribed was dedicated to Owen Wister by the State Historical Landmarks Commission in July 1939, one year after his death. Nearby stood a one-and-a-half-story log cabin formerly used by Wister near Jackson Hole and moved down to this spot on the west margin of the highway; beyond was the abandoned U.P. train depot with its bright red Spanish tile roof, now the community museum. When he first came to Medicine Bow in 1885 Wister inventoried twenty-nine buildings, a generous count that included a water tank, coal shooter, and five "too late for classification." Medicine Bow may be about the same size now, though present buildings seem more substantial than the shacks, lean-tos, and false fronts of 1885. The most impressive structure now is the three-story Virginian Hotel, grandly built of rusticated stone block masonry in quasi-Mediterranean style with a pagoda tile roof. "Reasonable Rates," promised a display advertisement in the official Lincoln Highway road guide of 1924. "Free Road Information. A $130,000 Hotel.

Modern in every detail. The Virginian Hotel takes its name from Owen Wister's Novel, the manuscript of which was written here." That, of course, was a tall tale—the hotel was built in 1911, nine years after the novel was first published. The hotel thrived through the 1920s and 1930s on Lincoln Highway tourists eager to see the novel setting come alive; survived the gas-rationed 1940s; enjoyed the resurging, gas-guzzling, motor-crazed 1950s and 1960s; and miraculously lived past the shock of the interstate siphoning off summertime throngs. Indeed, in 1984 the Virginian Hotel was given a million-dollar refurbishment, and is probably brighter, spiffier, higher-polished, and more sumptuously Victorian than in any of its previous ages.

We had reserved the Owen Wister Suite, of course, for one night at the top Medicine Bow room rate of $85. We registered on a yellow pad and found our way upstairs to the second-floor lobby, with red-flocked wallpaper, ornate and polished Ionic wood columns, Turkish-style carpets, velvet-upholstered furniture, and prints and paintings in old gilt frames. Down at the end of the south hallway was our suite—two bedrooms with brass beds, armoires, and bed stands flanking a sitting room with chandeliers, Victorian armchairs, sofas, a rolltop desk, lace curtains, and fringed lamp shades. Photographic portraits of Owen Wister and Theodore Roosevelt hung side by side. There was a dressing room with tri-folded mirrors and the bathroom had a tub with clawed feet. There was one electric plug, in the sitting room.

The suite was a delight, but we deposited duffels and hustled downstairs to the coffee shop for a much delayed lunch, eating in a booth next to red wallpaper and wainscoting and under a stamped tin ceiling painted gold. There were a mural and many framed antique photographs on the wall. Grilled sandwiches and fried potatoes took on a distinctly romantic cast, helped as much by sharp hunger as by decor.

Afterward, we ambled (the only way people walk in Medicine Bow) back outside to examine the depot museum. Freight trains rumbled by at a few miles per hour, but one had to look both ways on Route 30 for quite some time before one saw some car traffic to avoid while crossing.

Inside the Medicine Bow Museum ("Showcasing Four Generations of Ranching Around Medicine Bow" read a sign), I met Barbara Weiser, the director, who was personally acquainted with three if not four generations in town. Past her counter, we entered the wonderfully disorienting jumble of community museums everywhere. In this place it included a branding iron collection, saddles, scythes, buffalo overcoats, a stuffed sage grouse, radios, crank telephones, a treadle sewing machine, and a blue-gray enameled Home Comfort stove. One room, dedicated to the Lincoln Highway, noted that the first car to sputter into town arrived in the fall of 1911—a Winston 6, though Horatio Nelson Jackson had passed through in 1903.

There was a room commemorating Wister and *The Virginian* in print and films. In a room dedicated to everyday life on the Plains were photos and documents about the non-everyday occurrence when "The Lone Bandit," Gentleman Bill Carlisle, also called "the last train robber," was the subject of a furious posse search in 1919 in Medicine Bow. Serving time at the Rawlins State Prison for three armed train robberies in 1916, Carlisle escaped Rawlins in a packing crate, changed clothes, found a gun, and boarded another U.P. train heading west from Rock River to scare up some cash. Shot in the hand, he escaped on foot. The train pulled up at Medicine Bow, word spread quickly, and citizens formed a search party. A typescript displayed at the museum entitled "Childhood Recollection by Hazel Robertson Morgan," typed all in capital letters, continues the story. "It so happened my father was home," Hazel Morgan recalled. "He was a contractor for hauling equipment from one oil field to another. . . . My father had purposely chosen to build our home in this locality so that he could use the broad prairie behind our house to store oil well equipment. As I remember there was a lot of big pipes, some of which were big enough for a man to crawl into. When my father heard of the train robbery, he ran into the house to tell my mother to stay in the house with all of us children. We did not have a lock on the door and I remember mother placing a chair under the doorknob to secure the door. All of us were scared to death, and it took quite a while before Daddy came back to our home. He told us it was Bill Carlisle, a well known bank robber of the day, who had held up the train. The posse broke up when they had searched everywhere it seemed possible for him to be. Bill had hidden himself in one of the big pipes that lay on the prairie behind our house." Carlisle was soon apprehended at Esterbrook, Wyoming, and served the rest of his sentence at Rawlins. After his parole in 1936, Gentleman Bill Carlisle bought a little tourist camp outside Laramie and for the sake of tourists capitalized proudly on his checkered past. In his four holdups on the Union Pacific, the Lone Bandit had stolen all of $731.22.

We were heading to a considerably older treasure field, Como Bluff, but before driving off we ducked into a low building down from the hotel, which housed the noted Little Dip Bar, named in honor of the Diplodocus dinosaur, one of many unearthed near Como Bluff. The Little Dip boasts a forty-foot-long bar carved from a four-and-a-half-ton jade boulder found near Rock Springs. The proprietor, Bill Bennet, was a retiring fellow with a large brush mustache and a curiously bemused expression. He sat on a stool behind his jade bar whittling away on something and in almost monosyllabic terms made it known that he could not serve sodas to our children at his bar, or indeed to their parents, unless we all sat in the table area. We did so and retrieved our sodas from him, wondering how many years it had taken him to perfect his strong, silent demeanor, or whether it came naturally. When our eyes adjusted to the dimness we began to see how he had decorated his establishment.

Murals had been painted on all walls and ceilings—of local scenery, historical events, and scenes and acquaintances out of the artist's life. The painter was Bill Bennet himself, and his work was primitive and spirited and bright. But what took my breath away were his extraordinary wood carvings—I got up to examine each and every one, and he watched me carefully out of the corner of his eye, taking little gouges out of a hunk of tree branch in his hand, as I drifted around his rooms. The fully three-dimensional carvings were mounted on walls and in a room partition made of dozens of empty aquariums. I found a wagon with an eight-mule team, a stump with three faces emerging from the rough bark, and many representations of Old West characters. The greatest was a wall plaque of famous bandits and lawmen, with near-photographic relief images of Butch Cassidy, the Sundance Kid, Bat Masterson, Tom Horn, Wild Bill Hickock, and others. These were all works of genius—done with three small, ordinary penknives and one gouging blade. He showed me his tools when I went over to effusively compliment his work. It was the best folk art I'd seen since Ogallala, probably the most accomplished of our whole trip, and we had seen much whether in galleries or community museums, all along the way. If he derived any pleasure from a stranger's compliments, it was hidden behind that brush mustache. By then the sodas were finished and the elder stratum of town had begun to trickle in for cocktails. We went off to look for dinosaur beds.

West on Route 30, north on Route 487 a few miles to a right on Marshall Road, and then we were bumping across cattle grates and following a rutted lane across an area called Greasewood Flats. The sky was immense and deep blue as befitting our altitude of more than 6,800 feet. Scores of ground squirrels fled the approach of our car and fanned off into the short brown grass and sage and a few prairie dog colonies appeared here and there but the citizens were mostly below stairs. Grazing cattle and antelope regarded us with intense curiosity as we passed, prompting my son to start a new antelope count: his total for an hour's drive, in fact, would be an extremely satisfying ninety-two. I knew my count would be considerably fewer; I was hoping to see at least a couple Apatosaurus-sized holes, for I knew that there were old dinosaur quarries in nearly every direction, most remarkably the Sheep Creek sites to our north and the Bone Cabin and Como Bluff quarries ahead of us, as Marshall Road curved east and then south past some dry creeks and vanished lakes toward Como Bluff. Drawing slowly nearer, we dipped down onto an old lake bottom where the soil color changed abruptly from buff to red and then briefly the exposed stratum of rocks became the roadbed. Cliffy outcroppings rose on either side of us and the mind began to manufacture bony images from the roughened rock. We pulled over. Two golden eagles launched themselves from a twisted old pine tree and climbed the sky.

In the summer of 1877, two Union Pacific employees discovered a large saurian bone eroding out of the bluff here. They were William Edward Carlin, the station agent at Como, and William Harlow Reed, the section foreman. They followed the bone with digging tools and eventually uncovered more, including a shoulder blade some four feet, eight inches in length and vertebrae joints some ten inches long and two and a half feet in circumference, and many large teeth and claws. Dollar signs seemed evident all over the ancient rock, and they followed Como Bluff for six or seven miles, finding evidence of many buried animals. Not that they knew it yet, but they were picking away at one of the richest dinosaur fossil beds in the world: exposed rock formed during the Triassic, Jurassic, and Cretaceous ages, from 65 to 245 million years ago. Carlin and Reed wrote (under pseudonyms to disguise their find) to a prominent paleontologist at Yale, Othniel Charles Marsh. "You are well known as an enthusiastic geologist, and a man of means," they said, "both of which we are desirous of finding—more especially the latter." They sought money for the bones they'd turned up, and regular employment to uncover more.

Marsh agreed as soon as a large wooden crate full of fossils arrived at Yale. He sent off a check for $75 and dispatched a representative to see things for himself. But he was an unpleasant employer, paying minimal wages and staying perpetually behind in his payments, all the while demanding more fossils. The two bone hunters worked through the frigid winter and spring of 1878, opening four quarry sites, two of which yielded rich and varied deposits, which were laboriously hauled to the railroad, crated, and sent off to Yale. The hard work and infrequent payments opened rifts by the end of the summer; Carlin and Reed stopped talking to each other, Marsh terminated the contract, and Carlin decided to go freelance. He began working for Marsh's bitter enemy, Edward Drinker Cope of Philadelphia, digging new quarries at Como from a tent camp in early 1879. Marsh was so infuriated that he instructed Reed to smash any fossils left in his pit when he moved to more fruitful quarries, lest they fall into the hands of his rival. For the next several years, then, separate digging gangs sent tons of bones off by rail to either New Haven or Philadelphia, eventually to be sorted and assembled at the Peabody or the Academy of Natural Sciences. In the race to publish descriptions, names, and classifications, both Cope and Marsh committed many errors that it would take decades for others to correct, the most famous being Marsh's error that Brontosaurus and Apatosaurus were different creatures, which he compounded by placing the wrong skull on the skeleton of the former. That mistake was not to be corrected until 1975.

Meanwhile, other institutions became players in those desolate plains around Como, including Harvard University and New York's American Museum of Natural History, whose legendary paleontologists, Barnum Brown and Walter Granger, discovered an enormous, nearly intact Apatosaurus at the museum's Bone Cabin Quarry, between Como and Medicine Bow, and shipped boxcarsful of other finds back to New York City.

The red bluff went east and west of where I stood in a notch in front of my car. Eroded holes gaped off to my left, quite possibly old quarries opened by Marsh, Cope, and the others in that decades-long dinosaur gold rush. The crumbly layers of rock and clay were solid beneath my feet. There is a marvelous phrase of John McPhee's from his magnificent book *Rising from the Plains* about Rocky Mountain geology along the 42nd Parallel in Wyoming. Evoking prehistoric pictures of this area, he writes "of a meandering stream, with overbank deposits, natural levees, cycads growing by the stream. Footprints the size of washtubs. A head above the trees. In the background, swamp tussocks by the shore of an oxbow lake." I thought about this picture and about many in my own mind's eye registered over years of wandering. As a child going to collections in Washington, Philadelphia, and New York City, I had stood below towering skeletons, many of which had shaken this Wyoming earth when they walked. As a young adult I had gone to see exhibits in New Haven and Cambridge. In my current Parental Era my children have stood next to me, their hands in mine, in the saurischian and ornithischian dinosaur rooms at the Museum of Natural History in New York, staring up at the full Stegosaurus; the skulls of Camarasaurus and Diplodocus; and for the longest time, the extraordinary Apatosaurus. It had been remounted with the proper skull since I was a boy, and it stood with its head twenty feet above us, up over the chandeliers, its shoulders fifteen feet high, and its impossibly long, curving, thirty-foot tail. Long after my children went off in search of new marvels, I stood, almost smelling the warm, leafy breath puffing down on me.

On Como Bluff I crouched down and put out my hand, thinking of the type of cutaway paintings one found in the old *Life* magazine when I was a boy: strata of soil and rock below my feet, with remnants of fossils layered beneath us, and now, inches below the palm of my hand, an up-tending muzzle of an Apatosaurus, its long neck reaching down through the red rock toward shoulder blades, backbone, ribs, pelvis, and long, long curling tail, and I held my hand down on the ancient mud, still warm under a declining sun.

Later we found ourselves back at the Virginian Hotel, full of hope about a dinner in the formal dining room, which we'd seen earlier lavishly decorated and set with silver and goblets gleaming atop white linen. But a waitress told us they didn't use the formal dining room except for special occasions and signified that our presence wasn't special enough to make exceptions, so we ate where we had before in the coffee shop, from the same menu, which was limited to food that could be fried on a stovetop or grilled below. It had slightly less allure than at lunchtime when we had it before. After we finished and Mary took the children upstairs, I paid the check and noticed through the front window that the parking lot was becoming clogged with pickups.

"Big night," explained the cashier. "We've got four Wyoming road crews as guests tonight!" They would fill all the other rooms in the hotel, he added, and the motel annex. He leaned forward. "In your suite," he said in a kind tone, "you may want to keep your bathroom door closed tonight. For some reason noise comes up through there from below."

I peeked into the bar. From checkered floor tile to stamped tin ceiling it was certainly authentic for Medicine Bow. The redbrick walls were covered with mounted animal heads and antlers; photos and paintings of cowboys, chuck wagons, and local prizefighter heroes from half a century ago; and pennants, tractor caps, and flags. Loving cups, stuffed jackalopes, bean pots, Radiolas, and other knicknacks stood on shelves above a highly polished slab of a bar. A wooden sign proclaimed "Home of the Virginian" below a five-foot wide panoramic photo of Medicine Bow taken in the Teens from atop the Virginian Hotel. The place was filling up with deeply tanned young men and women, the men still sweaty from their roadwork and the women glowing with vitality and smelling of shampoo. The hanging lights over pool tables were already lit, and someone was feeding quarters into the jukebox.

Upstairs, I realized that our delightful suite was directly above the bar and coffee shop. Tammy Wynette's voice floated out of our open bathroom door as if she were singing in the shower. Someone told a joke right below me and five or six people threw back their heads and roared. Cigarette smoke was circulating freely through the suite from below. Full-throated truck engines revved outside in the lot. Someone beeped a horn and laid down some rubber as he peeled out onto the road. "Maybe we could find a place to sleep on a store counter," I muttered, but my family missed the reference; the reply came back to me from outside as a Union Pacific freight engineer pulled down his air horn cord as he neared the grade crossing across Route 30 from the hotel. Then the Oak Ridge Boys began harmonizing in our bathroom.

"You know, we could just leave," I said.

"No, let's stick it out," replied Mary. She could sleep through anything, whereas I was the one prone to light sleep and working out the problems of the world while studying the ceiling.

"All right! Way to go!" someone yelled from downstairs.

There was no television for the children that night, of course—that no hotel rooms here had television Mary and I had minimized to them, stressing the authenticity of our antique rooms—so instead we sprawled in velvet chairs and lounges, turned the lights low behind fringed lamp shades, and listened to old-time radio shows from CDs played on my laptop. The Mercury Theatre would have resonated better in that old-fashioned room without the trucks, shouts, songs, and thumps from below, and Jack Benny would have been funnier without the riotous competition, but the old shows set us in the right direction for bed. There in the Owen Wister Suite, I stuffed towels around a bathroom radiator pipe that seemed to be the conduit for jukebox music, firmly shut the door, realized that it hardly muffled the din from below, and began to read to Davey before sleep.

"And ten," said he, sliding out some chips from before him. Very strange it was to hear him, how he contrived to make those words a personal taunt. The Virginian was looking at his cards. He might have been deaf.

"*And* twenty," said the next player, easily.

The next threw his cards down.

It was now the Virginian's turn to bet, or leave the game, and he did not speak at once.

Therefore Trampas spoke. "Your bet, you son-of-a-bitch."

The Virginian's pistol came out, and his hand lay on the table, holding it unaimed. And with a voice as gentle as ever, the voice that sounded almost like a caress, but drawling a very little more than usual, so that there was almost a space between each word, he issued his orders to the man Trampas:—

"When you call me that, *smile!*" And he looked at Trampas across the table.

Yes, the voice was gentle. But in my ears it seemed as if somewhere the bell of death was ringing; and silence, like a stroke, fell on the large room. All men present, as if by some magnetic current, had become aware of this crisis. In my ignorance, and the total stoppage of my thoughts, I stood stock-still, and noticed various people crouching, or shifting their positions.

"Sit quiet," said the dealer, scornfully to the man near me. "Can't you see he don't want to push trouble? He has handed Trampas the choice to back down or draw his steel."

Then, with equal suddenness and ease, the room came out of its strangeness. Voices and cards, the click of chips, the puff of tobacco, glasses lifted to drink,—this level of smooth relaxation hinted no more plainly of what lay beneath than does the surface tell the depth of the sea.

For Trampas had made his choice. And that choice was not to

"draw his steel." If it was knowledge that he sought he had found it, and no mistake!

Davey fell asleep soon after this scene—he had the ability to blot out the noise that came from the bathroom pipes and thundered through the bedroom floorboards. A loud horn outside. Another U.P. freight rumbled by across the street, setting off sympathetic vibrations and rattles in the window. Now Randy Travis had taken up his shift in the bathroom and someone below began a heel-stomping competition out on the dance floor. I tried to read a few more pages of the novel.

> They were now shouting for music. Medicine Bow swept in like a cloud of dust to where a fiddler sat playing in a hall; and gathering up fiddler and dancers, swept out again, a larger Medicine Bow, growing all the while. Steve offered us the freedom of the house, everywhere. He implored us to call for whatever pleased us, and as many times as we should please. He ordered the town to be searched for more citizens to come and help him pay his bet. But changing his mind, kegs and bottles were now carried with us. We had found three fiddlers, and these played busily for us; and thus we set out to visit all cabins and houses where people might still by some miracle be asleep. . . . Do you know the sound made in a narrow street by a dray loaded with strips of iron? That noise is a lullaby compared with the staggering, blinding bellow which rose from the keg. If you were to try it in your native town, you would not merely be arrested, you would be hanged, and everybody would be glad, and the clergyman would not bury you. My head, my teeth, the whole system of my bones leaped and chattered at the din. . . . Everybody was to come out. Many were now riding horses at top speed out into the plains and back, while the procession of the plank and keg continued its work, and the fiddlers played incessantly.

Sometime later—after I gave up reading and tried to sleep with my head thrust between two pillows, the noise rising and falling oceanically and punctuated by bellows and sharp laughter—I groggily lost my sense of where—or when—we were. I began to worriedly fixate that the high-spirited revelers in the saloon downstairs were going to start shooting their pistols at the ceiling and through our floorboards, and I wondered if putting one of our big duffel bags beneath the bed and my slumbering son would keep him safe until morning. Then some barely rational part of my consciousness cut through those dazed thoughts, reasserted our placement in a modern era when the cowboys downstairs in the saloon didn't pack steel, and acknowledged that besides the booming jukebox all the raucous noise was at least good-natured talking, laughing, and singing—no fights or obscenities, just letting off steam. They would not ventilate the Owen Wister Suite's floorboards or its occupants. I drifted in and out of some kind of dismal substitute for sleep, my teeth vibrating in time to the bass guitar on the juke, being slammed upward toward consciousness by one outburst or another, but on some level I realized that the noise had trained me to be able to make a reasonably accurate count of the saloon occupants below, and that there were fewer people than before. I sank for a while into a deeper place—I don't know for how long—but then Hank Williams echoed up into our rooms with his yodels spurring the last three barflies into coyote howls.

That was around 4:00 A.M.

At 5:00 A.M. all crews rose, thumping up and down the corridors on stragglers' doors and laughing at expressions of pain or dismay. By dozens they all clumped downstairs, revved up their trucks, and spun off down the highway, leaving us with only the freight trains. We settled in for a few grateful hours of sleep.

At 8:30, when we blearily reawakened, the parking lot was decorated with a small squadron of enormous, shiny, suburban-looking motorcycles, all pulling tiny two-wheeled trailers. By the time we got down to the coffee shop the bikers had finished their breakfasts and were posing in front of the Owen Wister murals for snapshots. These cycle jockeys were all from Ohio, all in their mid- to late fifties, portly men with Fu Manchu mustaches—several of whom had exchanged their cycle helmets for cowboy hats for the duration of their meal, which explained the tiny trailers—and their slimmer women companions in bright sweatshirts. As they laughed and posed for pictures, it made me pine for the likes of Marlon Brando and Lee Marvin in *The Wild One*.

Mary must have read my mind, or I hers. "What are you rebelling against, Johnny?" she asked me, saying the line from the famous film.

"Whattaya got?"

West on the Lincoln Highway at 11:00 A.M. Sagebrush and grass over light buff powdery soil. Elk Mountain tall to the south, and to the north, Fossil Ridge, Slate Ridge, the Medicine Bow River wandering somewhere and finding Difficulty Creek; a rose-colored cliff, five rounded hills, and the Freezeout Mountains. Mule deer and pronghorn antelope. An old Lincoln Highway motel complex—boarded, bereft, faded, forgotten, vanquished, and vacant, with this legend spray-painted on a wall: "NO TRESPASSING—SURVIVORS PROSECUTED."

The road crews were out laying white gravel over the old highway concrete in preparation for asphalt. They looked shell-shocked, poleaxed, gut-wrenched. They wielded tools and vehicles and orange flags with painful, exaggerated care.

From deep back in my own misspent youth in Boston taverns, blues clubs, and rock emporiums, a phrase swam up into my mind, and I uttered it aloud as we motored past these wan survivors on the morning after the night before: "I thought I was dancing—until I stepped on my hands!"

The Medicine Bow Mountains towered now to our south. Just north was the hillside sprawl, trailers and bungalows, of the old mining town of Hanna. The old original U.P. grade appeared. Then Interstate 80 snarled in after its brusque surmount of the Medicine Bows. Heading toward Rawlins and the Continental Divide, we had no choice but to join it as it swept over to engulf and devour Route 30, and us.

10

Crossing the Divide

From its mouth on the Missouri River to the western face of Scotts Bluff we had followed most of the length of the Platte River, some 350 Nebraska miles. So it was pleasant to greet the North Platte again, forty-odd miles west of Medicine Bow, where it flowed north from its faraway source in the Medicine Bow Mountains. Not the shallow, braided, island-choked and quicksand-prone watercourse we had grown used to, it was now a slow, deep catfish stream fringed by cottonwoods, willows, and purple sage. It was crossed here by the no-nonsense interstate, by an attractive little metal truss bridge from the latter Lincoln Highway days, and by a dilapidated old wooden bridge from the earlier twentieth century. We were pulling off the superhighway anyway and could not resist driving back and forth a few times on the little truss, which looked like it had been lifted from a model train set. On cue, a train whistle sounded off to the north, and we bumped toward it on a concrete road along the riverbank.

In early August 1843, buffalo hunter Kit Carson led the army exploring expedition of John C. Frémont across a willow-choked and ravine-cut pass from the Laramie Plains on their remarkable and influential journey toward Oregon and California; they forded the North Platte and followed it for a distance, camping somewhere near here. Over fires fueled by sagebrush they dried large amounts of thin-sliced buffalo meat taken by Carson and the hunters from ample buffalo herds. Suddenly about seventy mounted Indians charged over the low hills, shouting and brandishing weapons, only to halt at the surprising sight of so many armed whites and a formidable mountain howitzer. They made signs for peace. They were of the Cheyenne and Arapaho tribes and explained the warlike behavior from having mistaken the explorers for Shoshone, their enemies. "They had been on a war party," Frémont wrote in his report, "and had been defeated, and were consequently in the state of mind which aggravates their innate thirst for plunder and blood." The war party had been chased back all the way from Green River by Shoshone and Snake Indians and had a right to feel rueful. The pipe of peace made its rounds, and with an exchange of presents and a meal, the raiders rode off.

A couple of miles north of the highway, across a U.P. grade crossing, was the site of Fort Fred Steele. It was established in June 1868 to protect railroad workers and overland emigrants, decommissioned in August 1886, sold off at auction, and marked as a state historic landmark in 1914. In 1924 the town of Fort Steele contained a hotel, four businesses, and a few cabins, and was advertised as good for trout fishing and camping by Lincoln Highway tourists. By 1940 the highway had moved and only a few weatherbeaten log and board-and-batten buildings were left near a bullet-ridden red, white, and blue Lincoln Highway marker and a vandalized cemetery. Now there was little left but rubble, with cellar holes filled with bricks, charred wood, rusted bedsprings, cast iron stoves, and a large safe, presided over by stone chimneys. Only a powder magazine remained intact, south of the tracks and next to a graveyard whose military occupants had been moved to the Fort McPherson National Cemetery, leaving civilians to fend for themselves. We waded for a while in river water below the railroad bridge, waving to the engineer of a slow-moving freight, and were suddenly amazed to see a pelican, of all things, flapping and gliding back and forth over the river, clearly impatient to get back to brunch.

In 1909 Alice Huyler Ramsey, the first woman to drive across the country, had crossed the river here in her Maxwell. Learning down in Laramie that the wagon bridge over the North Platte had washed away during a flood and had not yet been replaced, she had to obtain advance permission from the local stationmaster to use the railroad bridge. She waited on the east bank while companions hoofed it across the bridge and down the railroad grade to the station, and more than an hour passed while the official telegraphed for clearance. Finally, when her friends waved her ahead, Ramsey had to drive onto the tracks, with her right set of wheels on the outside of the right rail, and her left set in between. It was necessary to keep the car going steadily in low gear to prevent the wheels from getting lodged between the ties. She bumped down three quarters of a mile and then she was out on the bridge—"open to the flowing water of the river below and no place for 'scared cats,'" she recalled. "It was one of those places where you don't look down—you keep plowing steadily on."

The original 1868 railroad crossing over the North Platte ran several miles to the south, and even with the army post established that summer of 1868, raiders from the Northern Cheyenne and Arapaho tribes continued to attack railroad builders. One young surveyor, Arthur Ferguson, was sitting in his tent when a bullet buzzed through, which he ascribed to Indians—four railroaders had been killed and scalped a few days before. Still, the Iron Horse was opening up more country to settlement and exploitation. "The time is coming, and fast, too," he wrote in his diary, "when in the sense it is now understood, THERE WILL BE NO WEST."

This might have been disputed at the end-of-track just a few miles away across the sagebrush plain, where a new Hell on Wheels town had grown up overnight. It was called Benton. The mild Massachusetts editor Samuel Bowles of the *Springfield Republican* passed through and made haste to get out. It was, he sneered, "a village of a few variety stores and shops, by day disgusting, by night dangerous, almost everybody dirty, many filthy and with the marks of lowest vice, averaging a murder a day, gambling and drinking, hurdy dancing and the vilest of sexual commerce."

Benton, one Cheyenne newspaperman commented, "like the camps of the Bedoin Arabs, is of tents, and almost a transitory nature as the elements of a soap bubble." It would last about as long. As the Union Pacific crews pushed away from the dirty little town it began to fade—and the businesses unscrewed their portable buildings and folded their tents and moved after them. Months later a returning visitor found only rubble and wreckage, a few skeletal chimneys, and—of course—a weed-choked cemetery, all of it coated in bitter white alkali dust.

Fifteen miles west of the North Platte on old Route 30 was Rawlins, the seat of Carbon County, a town of wide, right-angled streets shaded by big trees; mansions and kit-built bungalows; a handsome stone railroad depot; and block upon block of humbly scaled, relic Art Deco architecture. It was home to the Wyoming Territorial Prison (now a tourist destination with a high emphasis on its active days' record of executions) and also the current state penitentiary. The town was named for General John A. Rawlins, a lawyer in peacetime from Galena, Illinois, which was also the hometown of General Ulysses S. Grant: Rawlins served as Grant's adjutant in the Civil War and was fiercely loyal and protective. After the war, when Rawlins was going downhill with tuberculosis, Grant asked his former officer U.P. chief engineer Grenville Dodge to take Rawlins on his summer 1867 tour of the railroad construction sites and camps and on his reconnaissance of a route from the Laramie Plains westward across the Continental Divide. Concerned about the lack of water in the area, Dodge ordered his men to fan out across the sagebrush hills west of the North Platte; he and Rawlins were the ones to find a fresh, cold spring gushing from a rock. Rawlins tasted the water and called it "the most gracious and acceptable thing" he had ever found. He told Dodge that if the engineer were to name anything after him on this journey it should be a spring—not a pass or a mountain but a place like this. Dodge replied that this miraculous place in the desert would be called Rawlins Springs, as it was. It would prove to be a godsend to the railroaders, and Rawlins became an important U.P. division point and distribution town, though its early years were pockmarked by the usual lawlessness of frontier settlements.

One of its most notorious episodes was set in motion by a train-robbing gang composed of "Dutch" Charlie Burris, "Big Nose" George Parrot, Frank

James, and Tom Reed. In the summer of 1878, the outlaws plotted to derail a Union Pacific train for its payroll shipment. East of the mining settlement of Carbon, they pulled out spikes and looped wire around the rail, lying in wait for the train so they could yank the rail and dump the train. Fortunately, a section crew passed on a handcar, noticed the setup, and hurriedly respiked the rail. The outlaws fled into Rattlesnake Canyon. Two deputies, Bob Wooderfield of Rawlins and Tip Vincent of Carbon, pursued them to Elk Mountain and found their camp. Wooderfield bent over the campfire, found the ashes still warm, and straightened, exclaiming, "We're hot on the trail!" Instantly he was shot through the head and killed. Vincent, shot in the knee, tried to flee on his horse but was killed with another shot. They were the first Wyoming lawmen to die in the line of duty (which, given the raucous early years of railroad hell towns and general outlawry, seems miraculous); their bodies were not found by a search party until nine days later.

The outlaws fled north toward the Hole in the Wall hideout and then Montana; by then their trail had grown so cold that no one had any idea who had killed the deputies or where they'd gone. Two years later, though, in 1880, the bandits were drunk in a Montana saloon and began boasting and joking about the botched rail job and their escape after shooting the deputies. Someone overheard and telegraphed Carbon County sheriff Robert Rankin, who immediately left for Montana. Rankin surprised Dutch Charlie and Big Nose George and arrested them; the accomplices escaped. The prisoners were taken back to Wyoming by train, but as the train stopped at the town of Carbon, an angry mob boarded and seized Dutch Charlie Burris. Off the train, they forced him to stand on a whiskey barrel beneath a telegraph pole and strung him up; the sister-in-law of Deputy Wooderfield was given the honor of kicking the barrel out from under him. "This will teach you to kill my brother-in-law," she shouted.

Big Nose George Parrot, whose real name was George Manuse and who truly had a large, hooked nose over a bristling, droopy mustache, was jailed in Rawlins, tried, found guilty of murder, and sentenced to be executed on April 2, 1881. He spent the ensuing months in the county jail with shackles riveted to his ankles, but he somehow obtained an old case knife and began to work on the shackles. On the night of March 22, he swung the shackles on Sheriff Rankin and knocked him unconscious, but the sheriff's quick-thinking wife slammed the jail door shut. A blacksmith was summoned to fix new bolts in the shackles. Later that night, five armed men pushed their way into the jail and forced Mrs. Rankin to give up her prisoner. Outside, his neck in a noose with the rope thrown over a telegraph pole crosspiece, Parrot was put atop an empty kerosene keg and it was kicked out from under him—but the rope broke and he fell, prone and still alive. The vigilantes then found another rope, sent it over the crosspiece, and prodded Parrot to climb a twelve-foot ladder leaning against the pole. Since his hands were not tied, when they yanked the ladder away, he clutched the pole frantically. However, the weight of his shackles dragged him down, and he finally choked to death.

The saga was far from over. The vigilantes left the body dangling for some hours before it was cut down and given to the town undertaker. Instead of burying the body, however, the undertaker was persuaded to donate it to a young doctor in town, John Eugene Osborne, who had only just received his medical degree the previous spring from the University of Vermont at Burlington. Dr. Osborne had been hired by the Union Pacific Railroad as a surgeon and moved to Rawlins, where he divided his time between ministering to railroad employees, private practice, and side businesses in a drugstore and the sheep business. Upon taking possession of Big Nose George Parrot's corpse, Dr. Osborne made a death mask, and apparently began an autopsy—a process that he stretched out over a year by keeping the body in a strong salt solution in an old whiskey barrel. Two other local doctors, Dr. Thomas Maghee and Dr. Lillian Heath, assisted in various dissections. Dr. Osborne tanned skin from the body's thighs and had them made into a pair of shoes; he used skin of the chest to make a medical bag. He offered some of the skin to Dr. Heath, but she asked for a souvenir bone instead, and was presented with the top of the skull, which Osborne had sawed off. Rawlins must have been quite a place in those days.

Dr. Osborne finally tired of the pathological project and buried the whiskey barrel with the remains behind his office. His medical practice took a back seat to livestock raising as Osborne became one of the largest and richest sheep owners in Wyoming, while he also pursued a political career. Between the years 1883 and 1920, he was a member of the Wyoming Territorial Legislature, mayor of the city of Rawlins, governor of Wyoming, chairman of the Wyoming delegation to the Democratic National Convention, elected to one term of Congress, a member of the Democratic National Committee between 1900 and 1920, and was first assistant secretary of state during the first term of President Woodrow Wilson. Thereafter he returned to Rawlins to tend his sheep and assume the chairmanship of the board of the Rawlins National Bank. And then this pillar of society who had served his adopted city and state, and his country, so well; who had made a pair of shoes and a medical bag from the skin of a lynched murderer; and who had sentimentally presented the top of the skull to a pretty young female physician in town, died a wealthy and revered man in 1943.

Seven years later, on the morning of May 11, 1950, workmen excavating for a new store building on West Cedar Street in Rawlins uncovered a whiskey barrel near the rear of the Union bus depot and the alley running between Fourth and Fifth streets. The foreman pried open the barrel and inside found a number of bones, including a skull with the top sawed off. A crowd had gathered to get a look-see, and one of the throng, a local merchant, remembered that old Dr. Lillian Heath had her souvenir skull segment. He ran off to fetch it. The local coroner pronounced it a perfect fit—and declared the bones to be those of Big Nose George. Also inside the barrel were a pair of shoes, some rotted, unidentifiable clothing, and a bottle of Lydia E. Pinkham's Vegetable Compound. The shoes had been constructed with glass nails. The whereabouts of the medical bag commissioned by Dr. John E. Osborne, however, were never ascertained, though the death mask he had cast had been displayed in the Rawlins National Bank for many years.

The silver-colored death mask and the shoes now reside in the Carbon County Museum in a churchlike little building in Rawlins, where we saw them. The mask even replicated Big George's bristly mustache, and the two-toned shoes looked like a size 6½ narrow—Dr. Osborne must have had tiny feet. The shackles were there, too, along with a filigreed gold watch and winding key presented to Mrs. Rosa Rankin, wife of the sheriff, with this inscription: "Presented to Mrs. Rosa Rankin by the Board of County Commissioners of Carbon County for bravery in preventing the escape of Big Nose George from jail. March 22, 1881." There was also the relevant page from the Carbon County jail register, pronouncing a finish to Big Nose George Parrot: "Went to jine the angels via a hempen cord on telegraph pole front of Fred Wolf's saloon."

Relics from the days when Rawlins's Front Street was nothing but mostly one-story false-front stores, and when people arrived in town either by railroad or by horsepower, crowded the museum—saddles, spurs, bridles, branding irons, a buggy wheel wrench, a bull nosering, barbed wire, firearms, and the handcuffs used by Sheriff Rubie Rivera on the Lone Bandit, Bill Carlisle, when he was captured in 1916. There was also a photograph of them in Rawlins: the Lone Bandit was jug-eared, rabbit-toothed, and skinny. Another infamous Wyoming figure was represented by a pair of well-worn spurs with figurative designs incised in the metal and tooled leather—they had been owned by Tom Horn, the murderous enforcer who worked for big ranchers against rustlers, sheepherders, and homesteaders, and was finally tried and hanged in 1903 for killing the fourteen-year-old son of a sheep rancher. Many artifacts from the displaced Plains culture were there, too—scores of arrowheads and spear points, a Sioux war bonnet, clubs, and baskets; Santee Sioux moccasins of white deerhide with flower designs, colored stripes, and cut-glass beads; exquisite fringed and flower-figured leather arm gauntlets; an Arapaho bone scraper used to make buckskin by scraping the hides and massaging the animal brains into the skin to make it soft and pliable; and an Arapaho cradle board constructed with a willow stick frame covered with a flour sack and decorated with beads.

Joyce Kelley, the museum director, showed us around, giving us a brochure for a self-guided walking or driving tour of Rawlins and telling us about an upcoming Rawlins event, the annual Gathering of Cowboy Poets and Red Desert Western Arts Roundup later that July. She seemed to be most proud of her museum's little sheepherder's wagon built by James Candlish, a Canadian blacksmith. It featured a double captain's bed with drawers, a single bunk above, with stove, storage, a dry sink, and lamp within arm's reach. Nearby was an interesting firearm called a Nightherder, used at lambing camps: it was designed to fire a blank cartridge every half hour to frighten off coyotes and other predators. The director also pointed to her extensive Lincoln Highway display, including guidebooks, signposts, and artifacts from travelers. There were mud hooks used on cars in the Teens, a carbide lamp employed in the days before electrical headlamps, and an Atlas Cool Drink water bag; the flax bag was soaked in water before filling, and exterior evaporation would cool the water inside. Locally, when the Union Pacific shifted its right-of-way to the south, the state of Wyoming took over the abandoned roadbed and scraped it down to get an additional width for the Lincoln Highway. This raised new problems. "It has been impossible," warned the 1924 Lincoln Highway guidebook, "to get rid entirely of the old steel left by the railroad and some care must be used or one is apt to pick up an old spike or bolt." It was good advice for anyone who followed a railroad grade too closely.

The next morning the sun was trying to break through a cloudy sky as we pressed westward on the interstate, rising past eerie, eroded gray sandstone protrusions to dreary sagebrush flats, with mountains rising faraway to the north and south. Parallel just yards north of us was the thin, weedy, untraveled concrete ribbon of the Lincoln Highway, while about a mile south was the Union Pacific roadbed, crowned with red Sherman granite ballast. A long freight train pulled by two engines, consisting of half coal hoppers and half tankers, passed us heading east at Creston Junction. We were now on the Great Continental Divide Basin, at an average of about 7,000 feet, still lower than the surrounding hilly and mountainous terrain. Everywhere else in the Rockies, the Continental Divide marks the line where water flows either east, eventually to the Atlantic Ocean, or west, eventually to the Pacific, but the basin here was so flat that it was difficult to tell when one reached that divide, and in any case it was so dry that water had little opportunity to flow anywhere. The interstate and the old Lincoln Highway followed the lead of the Union Pacific, which skirted the southern edges of the Red Desert, where early explorers and emigrants described almost hallucinogenically shifting colors of the dry, sandy soil. Frémont's 1843 expedition penetrated only a short distance onto that desert basin before giving up and striking due north toward the Sweetwater Valley and South Pass. Grenville Dodge's railroad surveying parties suffered terribly as they struggled to mark the best line across the basin, which included scouring the wasteland for springs, seeps, or creeks, and produced nothing but dry defiles—"shallow graves of deceased rivers," Dodge called them—or undrinkable alkali lakes. At the place where a marker noted the Continental Divide, elevation 7,178 feet above sea level, we could see the curvature of the earth; at the horizon were the fluffy white crowns of huge thunderheads. Just south of the interstate was a famous Lincoln Highway monument, dedicated to its high priest, Henry Joy: a statue behind an iron fence, with four concrete Lincoln Highway markers at the corners, said to be the least visited of any official historical marker in the state, though the din of nearby interstate truck and car traffic was deafening.

At Wamsutter, an old wool-shipping center from the days when hundreds of thousands of sheep wintered on the desert, finding the self-cured bunch grass to be nutritious, the old shipping depot was visible. But aside from a large billboard on the highway painted in bright circus colors and proudly proclaiming, "Gateway to the Red Desert," now Wamsutter had more natural gas storage tanks than dwellings, even trailers. Back in 1914 when Effie Price Gladding made her transcontinental driving trip, the region startled her with its bleakness. At a grade crossing her party waited for a long freight train to move from blocking the roadway. "The train conductor came along," she recalled, "and we exchanged greetings. 'It's good to see you,' said the conductor, 'you motor people are about the only signs of life we fellows see out here on the desert.'" Later they encountered a large caravan of emigrants and it suddenly seemed as if half a century slipped away. "They looked foreign and were evidently in search of new farms and homes," Gladding wrote.

> They were drinking, and watering their tired horses at a small station on the railway. There were plenty of little children in the caravan. One woman dandled a tiny baby. A little farther on we came to a second and smaller camp. These people were traveling from Kansas to Washington. "There is good land there still that can be taken up by homesteaders, fine fruit lands," said they. One man had seen the land and was acting as guide for the others. Their wagons were drawn by horses and burros. The children were sweet, cheerful little people, but the whole party looked somewhat underfed. I would have liked to give them all the luxury of a hot bath in a big tub to be followed by a substantial supper. They had their water with them, having hauled it from the last point where water was to be had.

As they drove along, Gladding's party constantly saw the remains of former camps by the roadside. "Old tin teakettles, pieces of worn-out campstools, piles of tin cans," Gladding noted. "These are mute and inglorious monuments to the bivouacs of other days. These immense plateau states are very dependent upon canned foods, and all along tin cans mark the trail."

Buttes stood off to the northeast as we passed Red Desert Station, while, just yards to the south beyond the railroad, bluffs rose. Little redbrick buildings were appearing every ten miles or so, guarding natural gas pipelines. A few miles farther we came to Tipton, a negligible town, only one ranch visible, and we pulled off the interstate onto rutted dirt roads at Exit 152, on a quest to drive near the site of one of the biggest and headline-grabbing scams of the Gilded Age, at a sandstone mesa called Table Rock.

In mid-1872, two prospectors appeared in San Francisco reporting that they had found a fabulous "diamond field"—worth at low estimates $50 million—at a secret location somewhere out in the Western wastes. Diamonds and other gems could be picked off the ground. The miners exhibited a pile of diamonds and rubies that San Francisco experts estimated being worth $125,000; samples sent to Louis Comfort Tiffany in New York brought an even higher figure. A handful of prominent bankers eagerly organized a mining company. They also sent some of their number out to the diamond fields with the two prospectors.

Under great secrecy, the party was conveyed to a remote Union Pacific station in Wyoming Territory, where they transferred to mules and went off on a two-day circuitous ride, the last stage of which was conducted blindfolded. At the crest of a mesa the blindfolds were removed and the investors were invited to try their luck in the dirt. Within minutes, especially with the prospectors pointing out likely places, they began to turn up diamonds, rubies, emeralds, and sapphires. They returned to California in high excitement, the corporation quickly grew an eminent board of directors, including the Civil War generals George B. McClellan and Benjamin F. Butler, by then a Massachusetts congressman, and a renowned mining expert, Henry Janin, and established handsome offices in San Francisco and New York. Twenty-five seasoned businessmen invested $2 million in stock. The House of Rothschild became the company's European agent. Not only Louis Tiffany but Horace Greeley was said to be involved. Envisioning millions per year, with the world gem market moving from Amsterdam to San Francisco, the directors bought out the two prospectors' claim for $660,000.

The reports hit newspapers nationwide in August 1872; rumors put the location of the fabulous find in Arizona or Colorado. The directors were swamped with pleas from investors who wanted in, necessitating the employ of a squad of correspondence clerks. Before winter came to the high country, the company sent a party of fifteen businessmen out to see the fields for themselves. "The implements used by them seem to have been ordinary jackknives," reported the *San Francisco Bulletin*, "an improvement on the boot heels of the original locators. If so much wealth can be turned up by such primitive means, what might be accomplished with shovels and pickaxes?" Were it not for the lateness of the season, the nation might have seen a repeat of the frenzied migrations of the Gold Rush, and an extraordinary tumult on Wall Street, if the incorporators decided to go public.

Dramatically, the bubble burst. The renowned government geologist Clarence King, who had been leading surveys along the 40th Parallel (including southern Wyoming), read newspaper accounts while vacationing in Hawaii, and rushed back to America to confer with his scientists. They pieced together enough topographical information to pinpoint where the diamond fields might be located, and entrained for Wyoming. King's party reached Table Rock in early November. Sure enough, rough diamonds, rubies, and sapphires almost presented themselves in anthills and rock crevices. But then King's cook, a German, kicked up a diamond and rushed over to the chief. "This is the bulliest diamond field as never vas," exclaimed the German. "It not only produces diamonds, but cuts them also." Unmistakably it had been shaped by a lapidary. Further investigation showed that gems had been salted all over the mesa, just under the earth's surface, by the simple expedient of poking holes with a sharp iron rod, dropping in a diamond, and covering the hole with a quick twist of a boot heel.

King's party got back to San Francisco in late November; he simultaneously informed the corporation and summoned the wire services. "Wherever a printing press ran," the rueful Asbury Harpending (one of the original investors) later said, "the world knew the story of the diamond fraud." Reporters and detectives finally traced the story to Europe, where the previous year the two prospectors had secretly purchased bags of raw, industrial-grade diamonds and lesser gems for use in the scheme. Gleeful headlines continued for weeks, until finally the public turned to a new fancy, the developing Union Pacific Railroad scandals involving congressmen, senators, and even a vice president. But it would be a long time before the eminent men who had been suckered by two grizzled prospectors could show their faces again. "It has been from the first the most adroit and skillfully managed affair in the annals of fraud," pronounced the *New York Times*.

We felt our way across the desert following ruts for miles and once or twice dead-ending at a turnaround next to a Texaco natural gas well. Then we circled one mesa around its eastern shoulder—passing badger burrows off in the sagebrush and scaring off numerous antelope—and climbed the road through badlands toward Table Rock, a remarkable square-shaped, flat-topped prominence. Flashes of lightning lit the dark sky behind it—a robust thunderstorm was sweeping up toward us from the southwest. Were it not for the lightning and the well-advertised danger of flash floods and washouts in these parts, it would have been fun to hike up with the family to the summit of

Table Rock just to poke anthills and crevasses for the fantasies of leftover diamonds and rubies. But the greater worry was how quickly we had gotten ourselves out in the middle of nowhere: it just seemed dangerous to take one's wife and children up there. Mary was avid, but fortunately for my peace of mind the rapidly approaching storm settled the issue. We took a good long look at Table Rock, while a little family of antelope took a good long look at us.

Finally we rolled on down the hill past cliffs and draws and arrived at Bitter Creek Station on the U.P. line. It was merely a couple of trailers supervised by barking hounds. Somewhere near here in June 1867, the year before the railroad pushed through, a U.P. surveying team was attacked by a force of 300 Sioux warriors, who ran off the whites' livestock and pinned them down on a ridge. The chief surveyor, Percy Browne, a promising and popular young engineer, took a gut shot and lay in agony the whole day under the merciless sun until the Sioux tired of the standoff and disappeared. Browne's companions bore him on a litter for miles until they found help from a wagon train, but he soon died. Browne's recommended line along Bitter Creek would be adopted, and Grenville Dodge would name a pass in his honor.

Back on the highway, a procession of yellowish gray buttes stood just north of us across sagebrush flats. Behind us, we saw that the fast-moving dark storm had rushed northwest, filling gulches and draws, probably visible back in Rawlins. Such roaring winds and pelting rains were terrifying to unprotected travelers on horseback or wagons. Both the Pony Express and Ben Holladay's Overland Stage had passed by here after 1862. A driver experiencing such a tempest here on the edge of the Red Desert would have pressed westward, his eyes peeled for the sight of a sudden upthrust of gray cliffs—more than 1,000 feet high and a mile long—called Point of Rocks, where the shelter of a stagecoach station waited. We neared the dramatic wall, seeing that they truly had a "strange moth-eaten appearance," as the WPA guide had foretold, being heavily pockmarked with rounded caves and holes scoured by winds and rain. A shabby trailer park clung to the base of the picturesque rock, and beneath wide, homemade signs just off the highway exit ramp was a shanty quickstop advertising a Laundromat and cut-rate fireworks.

On the south side of the interstate, beyond four deserted old trackside buildings, were the remnants of the Overland Stage station—stonewall ruins of the horse stable, and one stone building of several rooms stabilized with a roof resting on big timbers. Inside, a dirt floor, hearth and chimney, minimal graffiti, and window cavities framed a view of Point of Rocks with interstate traffic zooming heedlessly by. Here Owen Wister had met a soft-spoken Virginian who performed his managerial duties, made the travelers comfortable, tended his ailing wife in the next room, and made a profound impression upon the young writer. Outside the station, wagon ruts trailed away into the distance. We walked along them beneath tall eroded rocks until distance and the hot wind swept away the roar of traffic and we were left with only the wind and silence. Then we returned to climb a pockmarked ledge until we could look down from its heights into the stagecoach station, and then westward along Bitter Creek past more prominences toward Rock Springs, some twenty-five miles away.

In September 1850, on the return leg of his army reconnaissance of the emigrant route from the Missouri River to the Great Salt Lake, and his investigation of the budding Mormon colony there, Captain Howard Stansbury and his party followed Bitter Creek eastward from its mouth at Green River, led by the famous mountain trapper and guide, Jim Bridger. From the time they entered Bitter Creek Valley they spied pockets and seams of glossy black rock amid the other eroded, crumbly horizontal strata of "escarpments, rounded into fantastic forms of bastions, buttresses, and turrets, by the action of the winds and the rains," and "in many cases quite beautiful." It was the evidence of "quite abundant" bituminous coal, Stansbury noted, that would make any occupation of that hot and dry valley practicable. Bridger told him that he had used the coal for years "and that it burned freely, with a clear, white blaze, leaving little residuum, except a small white ash." Later that year the Overland Stage Company established a primitive rock "hotel" for travelers next to a spring at the base of a prominent butte, about fifteen miles from the mouth of Bitter Creek. The Pony Express had a relay stop there during its brief existence. It would not be until 1867, as Union Pacific surveyors staked a line from the Laramie Plains to Green River, that the bleak alkali and sagebrush flat with its unpleasant-tasting mineral spring attracted any serious consideration for anything other than a way stop. Surveyors made ample note of the extensive coal deposits in their field reports to chief engineer Grenville Dodge, and Dodge, during his own reconnaissance with General John Rawlins, noted coal seams and hoped to get himself a personal interest in mining enterprises as the rails advanced farther into Wyoming.

The next year, 1868, as the U.P. line edged out onto the Continental Divide basin, two brothers from Scotland, Duncan and Archibald Blair, appeared and took over operation of the stage station. But their primary interest was the black rock and the ability to ship it out on the railroad. They opened a coal seam some distance west of the Overland station and imported laborers to work it. The workers lived in tents or in dugout shelters scooped out of the hillside in a rude cluster of habitations they called Blairtown. At the same time a Missouri businessman named Thomas Wardell was scouring the length of the U.P. right-of-way in Wyoming for coal deposits. Wardell's method was more orthodox than wildcatters'—he leased promising lands from the railroad and then operated in partnership with the U.P., receiving hefty discounts for freight rates and supplies. Already making a fortune with mines at Carbon, between Medicine Bow and Rawlins, Wardell sent scouts across to Bitter Creek. They struck pay dirt east of the stage station. The outcrops of coal could be traced for miles. Wardell sent an urgent telegram to Bevier, northeastern Missouri, where he had made his first big money in coal, to immediately dispatch a force from his mines there. Finding shelter by burrowing into the steep banks of Bitter Creek, the experienced miners were dramatically productive in their first shaft, Number One Mine, separating some 365 tons from the seam during their first short season in 1868. (That mine would operate for forty years, its production growing to 2,000 tons per day.)

The mining camp was called Rock Spring—the "s" came much later—and productivity rapidly increased. Wardell imported even more mineworkers in waves from England and Wales. Mines Two, Three, Four, Five, and Six opened up, and Rock Springs became the largest coal producer west of the Mississippi River. The Blair brothers' independent mining enterprise a few miles away at Blairtown was squeezed out of existence with nondiscounted freight charges and the high cost of supplies.

Other nationalities trailed in to find jobs, and many of the miners brought their families, or started them. The first school was organized in the dugout home of the teacher. There were six pupils, each of whom spoke a different language. The raw camp became a town, although not an outwardly organized one: houses were planted haphazardly, and the winding footpaths connecting them and the mine entrances became dirt roads standing at crazy angles from one another, whose baffling crookedness remains, memorialized in concrete and asphalt, to this day. Initially residents bought sulfurous water hauled from the rock spring for twenty-five cents per barrel, but mining disrupted underground water patterns and the spring dried up; by then the U.P. ran a daily water train from Green River. It wasn't until the late 1880s that the town laid a water main to draw from that clear-running stream; no septic system was even contemplated until the 1920s. Until then, Bitter Creek grew increasingly bitter as it passed through town; springtime floods were typhoid time.

Work in the early years was brutal and primitive. Miners hacked at coal seams with picks and shovels, loosening the rock with black powder and fuses. Mules hauled the heavy chunks out in little carts, the loads were transferred to wagons and lugged down to the U.P. tracks, though this last effort was made easier when a spur line ran up to individual mine entrances. This being long before the days of electricity, miners depended on headlamps affixed to their caps, burning cotton wicks rooted in lamp oil reservoirs, which gave them a wan yellow light and clouds of smoke. As a coal miners' history of Rock Springs reports, "Old timers tell of witnessing the fantastic flickering light of many oil cap lamps, as the men swarmed across the snow-covered ground to the mines in the early hours of the winter days. Since there were no street lamps in Rock Springs, the men had to wear their cap lamps to work, in order to find their way through the streets. They left home and returned thence in their pit clothes."

Accidents underground were commonplace, though it is recalled that most people who died in Rock Springs died after fights. The first undertaker opened his business in the late 1870s—prior to that, burials were rudimentary. The town history says that "at one time the miners had had to ransack the town to find 'a woman who could pray' to assist in the burial services." There was no doctor there until 1880 and no hospital until 1893. The first physician was Dr. Edward Day Woodruff, who was hired by the coal company and the Union Pacific and began work in November 1880. His memoirs were published in a 1931 pamphlet by the State Department of History. They detail an exciting life and career that got off to a rousing start when the nearest physician, who lived in Rawlins—and had to be Dr. John E. Osborne, the expert-to-be on the corpus of Big Nose George Parrot, since the dates match exactly—learned of this new competition. "He came right to town," recalled Dr. Woodruff, "hunted me up and started to raise hell. He said I was poaching on his territory, and I'd have to get out or he'd have me run out. That made me pretty mad. I told him I didn't consider it consistent with professional ethics for one doctor to treat another as he was doing, that this was virgin territory open to all comers." Dr. Osborne tried to have Dr. Woodruff scared out of Rock Springs by paying thugs to make "the tenderfoot doctor's" life miserable, but Woodruff packed his own steel and warded off all threats and attacks until he proved himself and was accepted. And Dr. Osborne had cadavers, politics, banking, and sheep to occupy him.

As early as the 1870s, Rock Springs was being called a melting pot. It was said that some fifty-seven nationalities had poured in for work that would help fuel the Gilded Age, but of course the image of national identity, culture, language, and religion melting into a smooth homogenous whole was as much of a myth in Rock Springs as in most places. Irish, Welsh, French, German, Polish, Swedish, Danish, Finn, Italian, Yugoslav, and Greek, and on and on, they kept a largely clannish existence with their own congregations and social clubs. But what did bring them together was the men's hard, dirty, dangerous work and their collective interest as workers.

The first clash between labor and management came in 1875. By then some 500 Rock Springs miners earned a decent wage of $6 to $10 a day, depending on the tonnage extracted. But work patterns were irregular; there was little activity in the summers and in the other season the week was limited to about three days. As winter approached that year, the company issued an increased production quota of 25 percent, which the miners refused to meet under the current system. They struck, closed down the mines, and when threatened, counterthreatened to set the mines afire. The corporation answered by calling in the U.S. Army to restore order, and by importing some 300 Chinese laborers hired from sources farther west. This was done in a pact with the Union Pacific Railroad, which pledged to keep the Chinese employed in the slow summer months on their trackbeds. The strikers were fired; there being no other work in town, they left.

Labor unrest thus addressed, the mines went back in operation. Only about fifty white workers were left on the job, and although cowed they chafed under strict rules forbidding them from harrassing the Chinese. Racist incidents were not unknown, however, as, over the next decade, the workforce in the Rock Springs mines increased at a ratio of two Chinese to one white. Resentment grew until finally the situation was ready to explode. On August 28, 1885, labor leader John L. Lewis wrote from Denver to warn the mine managers of the "grave" tensions. He said there was a storm brewing and that if grievances were not addressed an outbreak was inevitable. "There is nearly seventy-five of our men lying idle at Rock Springs at the present time, while the Chinese are flooding in there by the score." Lewis also wrote the Union Coal home office in Omaha. "For God's sake do what you can to avoid this calamity," he begged. "The pressure is more than I can bear."

The letters had barely been opened, much less acted upon, when, on September 2, 1885, a fight broke out between a white miner and two Chinese deep in Number Six Mine. Others rushed to the scene brandishing picks, shovels, and drills for weapons, and the ugly fight in a tunnel room ended with several white men severely cut and bruised and four Chinese terribly wounded, one fatally. The whites poured out of the mine, calling for reinforcements; they picked up revolvers, shotguns, and rifles and stormed off to town. Their number out in the streets swelled after dinner, and all saloons were hurriedly closed. Some sixty or seventy whites, all armed, followed by about the same number of stragglers, headed for Chinatown to clean it out. They sent an advance team to warn the Chinese that they had an hour to get out of town, and soon they could see the Chinese leaving their houses carrying bundles, most running out into the badlands. The mob grew impatient, and charged Chinatown early, shooting and shouting. A great many panic-stricken Chinese fled across the steep-banked, foul Bitter Creek and the railroad tracks, taking to the hills, spurred by a hail of bullets over their heads. The whites set fire to all the houses and stores in Chinatown, cornering a number of Chinese in the burning buildings, who were later found, either shot or burned to death. "All night long," reported the *Rock Springs Independent*, from whose harrowing story this has been condensed, "the sound of rifle and revolver was heard, and the surrounding hills were lit by the glare of burning houses." At least thirty Chinese were killed, a figure that seems an underestimate, given the hundreds who disappeared into the hills without food, water, or adequate clothing. September nights on the high desert are cold. Some estimates of the deaths run as high as fifty, but there is no way of knowing. Those were the days when such accountings were careless, as with casualty figures reported during the building of the first transcontinental railroad in the late 1860s; after one Sierra avalanche, newspapers reported that precisely three white men had perished and "twenty or thirty Chinese."

We stood before a display case on the second floor of the beautiful Richardsonian Romanesque old city hall building, now the Rock Springs Historic Museum, staring into the reflection of light within a pale green celadon rice bowl from a vanished Chinatown. It was surrounded by other artifacts—a porcelain vessel decorated with enameled red, blue, and gold peonies set above a golden basket; a white porcelain Four Flowers serving bowl hand-painted with chrysanthemum, prunus, tree peony, and lotus; wine cups, sauce bowls, and spoons; and olive brown ceramic vessel fragments once used to store condiments like bean curd, dried mushrooms, and shrimp paste. Though these particular objects were unearthed among the ruins of houses constructed on the ashes of the 1885 conflagration, a deeper archaeological dig would have discovered similar remnants of that ancient and refined culture in the lower strata—along with, possibly, more human remains.

Rock Springs's Chinatown was reborn. The Chinese government telegraphed a formal protest to Washington and sent a delegation to Wyoming—U.S. troops arrived to suppress the rioters, Union Pacific train crews were instructed to pick up any refugees they spied wandering in the desert, and what survivors as could be rounded up were sent to temporary haven at Evanston, Wyoming. Authorities then needed to discourage subsequent strikes, death threats, and mayhem directed at the refugees as well as Chinese working at other labors in the West. Survivors willing to go back to work at Rock Springs had armed escorts—soldiers were posted in the town until 1899 when they were suddenly needed in the Philippines. Chinese who resumed mine work would evade further deadly dangers in living quarters rebuilt by management, but would nonetheless be the victims of racism, hostility, and harassment, still being paid at a lower rate than white men and enduring the cruelties of gangs of boys in the streets and in their private homes for decades. No one was brought to justice for the acts of September 1885. A grand jury indicted a token number of the boldest and most blatant sixteen men for murder and arson, but could not summon sufficient evidence to pursue the case. The panel issued an equivocal finding denouncing the crimes as well as the stain on "the fair name of our Territory," but widening blame to the U.P. and the coal corporation.

Scattered across ten blocks of downtown Rock Springs, straddling eight Union Pacific Railroad tracks and overshadowed by the high, dramatic dark walls of Pilot Butte, are many remnants of the dramatic growth and wealth of the city between 1880 and 1940, but no trace of Chinatown. The enclave dwindled as the workers aged, retired, and died without replenishing their numbers—not many Chinese women ever lived in Rock Springs—and cultural interchanges were limited to patronizing excursions down to the quarter during Chinese festivals, when, for some years around the turn of the century, a huge ricepaper and bamboo dragon was paraded in the streets, held aloft by a long line of Chinese men. In the 1920s as a sentimental gesture, a handful of old men were repatriated to China with a sinecure after a "let bygones be bygones" banquet with speeches by coal and railroad executives and United Mine Workers officials.

If there are no explicit reminders of the immense Chinese contributions to the growth of that city beyond the exhibits in the museum and the Chinese flag that flutters from poles commemorating the fifty-seven nationalities in Rock Springs, a self-guided walking tour setting out from the city hall museum with a twenty-eight-page guidebook and gatefold map shows what some of that coal money paid for. There are rusticated sandstone edifices, big redbrick or brownstone mercantiles, churches, banks, markets, hotels, clubs, halls, and theaters. One stone building on North Front Street, embossed with the legend "Labor Temple 1914," was owned by the United Mine Workers and contained the office of Douglas Preston, first criminal attorney in Wyoming and on retainer by Butch Cassidy. Another address down North Front once identified a beer garden where Martha "Calamity Jane" Cannary worked in the early 1890s, leaving behind legends about emptying both six-shooters while "shouting verbal oaths well tarnished."

The walking tour takes one past marvelous, massive architectural trophies from the Gilded Age through the Art Deco years, and past the delightful big metal triumphal arch adjacent to the train tracks and spelling out in neon letters, "Home of Rock Springs Coal." Given the location, one continually gapes at the sight of these huge piles of stone, particularly the old city hall and the federal post office building, and wonders: what is holding them up? Underground, mines followed the coal seams regardless of what was above until the entire town was underlaid by the multilevel tunnels, hundreds of miles of them, each shored up with timbers. Photographs from the early era make mine construction and use of the wooden supports appear haphazard, improvised, claustrophobic, and extremely dangerous. Not surprisingly, cave-ins occurred. One famous episode in 1878 involved the ground giving way beneath a little rock family house: "all that saved the family," reports the Rock Springs walking tour guidebook, "was the carpet securely tacked around the edges. It supported them until friends reached through the doorway and helped them to safety." Many larger structures could not be built until holes were drilled through the mine layers and foundations were anchored to solid rock. When the post office building was built in 1911 it was necessary to drill sixteen holes, and pipes were inserted into each hole, and down these pipes concrete was poured until the tunnels were filled; whether the architects, engineers, and contractors were ever able to get a good night's sleep after such an improvisation is unknown.

The people of Rock Springs assumed that black gold was forever, but the West has a way of reinforcing that all enterprises are impermanent. By the late 1950s all but two of the underground mines in Sweetwater County ceased operations, and Rock Springs's Number Eight Mine, the Union Pacific's last coal mine in Wyoming, closed down in August 1962. Diesel fuel had already been elected king on the rails, though a huge open pit mine east of town would keep power plants burning for years to come.

Decades later, the uncertain ground beneath Rock Springs continued to subside. In the 1980s so many buildings sank into shafts that aid from a federal abandoned mines program funneled millions of dollars into the shafts, sometimes relocating structures. All the while we were in town I found myself wanting to tiptoe, imagining warrens of pitch-dark shafts and tunnels below one's feet, filling with inky, alkaline groundwater, a nightmarish McElligot's Pool outlet in which floated rotting old pick handles, work gloves, boots, lost miners, and memories.

Rock Springs is anything but a played-out old boomtown. Year-round residents seemed to enjoy a prosperous town of nice neighborhoods, open businesses, vigorous schools, hospitals, libraries, and cultural and recreational centers, though one could not ignore pronounced mall sprawl and heavy traffic, still a sign of some kind of vitality. Much of this prosperity had roots in an earlier big oil squeeze in 1969 and the construction of the nearby coal-fired 2,000 megawatt Jim Bridger Power Plant. One can imagine a dismissive expectoration of tobacco juice from the old mountain man onto the polished shoes of whoever named the plant. But thousands of workers poured into the area to work on the construction or to find jobs in the burgeoning petroleum and natural gas fields of southwestern Wyoming, the trona mineral mines, and the open pit coalfields that produced some 10 million tons annually. The boom was entirely too much for the city to handle. Population doubled, crime soared, and corruption festered in many municipal and county levels including the police department, making Rock Springs too reminiscent of the Hell on Wheels towns of the late nineteenth century. But the city lived through its difficult growing pains and now seems intent on enjoying its rewards. With its placement on the interstate, proximity to heritage and historical tourists tracing the nearby California/Oregon Trail and the Mormon Pioneer Trail, and its nearness to the popular Flaming Gorge recreational lake, Rock Springs's home-grown prosperity is increased by passersby in their SUVs and

RVs. The motels and restaurants were well populated for a summer in which gasoline pump prices were nudging perilously close to $2 per gallon, and late one Monday afternoon traffic down Springs Boulevard from our motel was the slowest we'd encountered since Cheyenne, or perhaps North Platte.

At the long stoplight at Springs and Stagecoach boulevards, I found myself gripping the steering wheel and craning my neck to the deep blue sky above the city, populated only by clouds and the high, ice-white trails of long-distance jets. It was easy to imagine all that open air suddenly disturbed, with alert citizens of Rock Springs pulling over in their cars and perching on running boards to see the sight of a giant dragonfly-like device whirring through the atmosphere of June 1931, hovering overhead, and then alighting out at the airfield. It was an autogiro—short wings, propeller, and an ungainly overhead rotor like a helicopter—and the pilot who gracefully emerged and theatrically removed his—*no, her!*—goggles and cap was none other than Amelia Earhart. Chewing gum—this attempt at the first transcontinental autogiro flight was sponsored by Beech-Nut Chewing Gum, and that name was stenciled on the Pitcairn PCA-2's fuselage—she flashed her trademark toothy tomboy grin, gamely answered local reporters' questions, and posed for photos with the assembled, awestruck crowd.

Earhart had an eventful time getting as far as Rock Springs. In April 1931 she had set the official altitude record of 18,415 feet for the Pitcairn autogiro at the aircraft's plant at Willow Grove, Pennsylvania. On May 29 she took off from Newark, New Jersey, aiming to set the first transcontinental autogiro flight. It was no picnic—the autogiro had a very limited range, so she had to stop at least once every two hours to refuel (seventy-six times in the round trip), and its top speed was only 80 miles per hour—if there were no headwinds. In long trips an automobile could do almost as well; at one time during the flight, a Nebraska man in a car paced her for 100 miles and nearly beat her. The autogiro also handled poorly. She crashed it three times that summer, walking away from each accident unscratched. "I'll never get in one of those machines again," she vowed once. "I couldn't handle it at all."

Not that she would say such a thing against the craft in public. In a few weeks Hearst's *Cosmopolitan* magazine would be on at least some newsstands or bedside tables in Wyoming, that issue for August having a signed article entitled "Your Next Garage May House an Autogiro." "These days I am having exciting experiences in the air with my new autogiro," Earhart wrote. "The excitement lies not in the flying alone—the fun of that one gets used to—but in the fact that the craft is new and unknown." She eagerly looked forward to a time when autogiros "will be rubbing shoulders with automobiles in our garage-hangers of tomorrow. At least, it is no fantastic prediction to foresee the time when perfected autogiros will be as commonplace about country estates as touring cars and motor boats." Good-sized backyards or

large, flat roofs were all one needed to take off or land the windmilling gad-
gets. "Perhaps when the ships are more nearly perfected, fishing and hunting
trips, week-end excursions and commuting (with reservations) by air will
prove the easy means of travel aviation enthusiasts predict. Addicts of golf
and country clubs, and those who go to beaches and resorts, will benefit
when and if flying becomes such mobile transportation."

Unfortunately, after she gulped down a sandwich, refueled, leaped into the
air over Rock Springs, and disappeared off to the west, the transcontinental
publicity stunt for Beech-Nut set no record. When she arrived in Oakland,
California, she learned that another pilot had beaten her by a week. But she
turned around and flew back to New Jersey, flying 11,000 miles in some 150
air hours. A year later, after she returned to more conventional aircraft, she
became the first aviator to duplicate Lindbergh's solo flight over the North Atlantic, making landfall outside Londonderry, northwestern Ireland, and giving the citizens of Rock Springs an occasion to relish their brief contact with
celebrity.

Our path had taken us away from the California/Oregon Trail back at Scotts
Bluff in Nebraska, and we had missed more than 150 miles of its length, including such sites as Fort Laramie, the Guernsey trail ruts and Register Cliff,
Ayres Natural Bridge, and Independence Rock. But I could not have us miss
South Pass. A pilgrimage to that significant landmark some sixty or seventy
miles away was worth such a backtracking detour, and a pilgrimage it would
certainly be.

Route 191 headed north from Rock Springs along the valley of Killpecker
Creek, named long ago by disappointed, thirst-quenched soldiers. Pilot Butte
stood stolidly to the west, succeeded by the striated Badlands Hills. Abandoned U.P. short lines seemed to thread off eastward into every dry creekbed
of the Leucite Hills, soon dead-ending at closed-down coal mines. Black Rock
Butte and Emmons Cone, an extinct volcano, were visible east of the hills.
Then the road left the creekbed and crossed bleak alkali flats populated only
by sage and greasewood, though eight miles to the east the parched edge of
the Great Divide Basin was marked by the enormous Killpecker sand dunes,
pale in the morning light, with the bluish Jack Morrow Hills rising beyond;
the dunes are second in size only to those found in the Sahara. We saw a red-
tailed hawk hovering at the roadside as if studying a menu, intent on a
ground squirrel breakfast. Old ruts crisscrossed through sagebrush. A sheep
wagon abandoned out on a rise had a bleached white canvas top. All seemed
hopeless for life. But then, ten or twelve miles farther, an astonishing explosion of vivid green—it was Eden, literally. Eden (population 220, down 36
from the 1940 census, altitude holding firm at 6,590 feet) is surrounded by

extensive irrigated alfalfa and clover fields, and many cottonwoods, all nourished by deep artesian wells. A large number of horses and ponies grazed in sight of a tavern and the Oregon Trail Memorial Church. Ranchers stacked green hay bales in fields. In the year 1849, famed New York newspaper editor Horace Greeley, working on a travel memoir about his cross-country stagecoach journey, spent an August night in a log house near there; he was told it was the only house between Fort Laramie and Fort Bridger.

The Wind River Mountains were visible to the northeast, perpetually snowcapped. We drove four miles more and came to the crossroads settlement of Farson (population 325, altitude 6,580), consisting of some trailers, a gas station, a café, and a handsome big brick building. We stopped for BLTs at the Oregon Trail Café, where a man with more keys on his belt key ring than there are doors in Farson chatted up the highly made-up blond teenaged counter girl. Afterward we walked over to the brick two-story Farson Mercantile, "Home of the Big Cone," as a sign told us. The big cone referred not to any of the extinct volcanoes one could see along the road but to a truly impressive scoop of locally made ice cream that, if eaten too quickly in the heat and altitude, guaranteed a headache.

Across Route 191 was the meandering Big Sandy River—a trickle, three feet across. We stared down at rocks and sunlight reflecting off the bit of water. "Whoever named it the Big Sandy River," said Mimi, "must have been a midget." Not far away was a concrete marker and medallion noting that the Pony Express had a station at this crossing of the Big Sandy. And a little distance away was the bronze plaque marking the June 28, 1847, conference between Jim Bridger and Brigham Young. Bridger, who had been credited with discovering Great Salt Lake and its broad valley on the far side of the Wasatch Mountains, obliged Young and his pilgrims by giving directions to the valley and answering questions about whether a large colony of Latter-day Saints could sustain itself there. Bridger doubted it—and was slightly irritated by Young's stubborn faith that it would be the Mormons' promised land. "I'll pay you $1,000 for the first bushel of corn you can grow in that valley," he scoffed, never thinking he would have to pony up.

From Farson, Route 28 headed arrow-straight to the northeast toward South Pass, with lazy Little Pacific Creek wandering nearby. The Wind River Range was rising tall, jagged, and snowy as we drove through low sculpted buttes, one of which had a microwave tower on it, and we were again in sight of the Killpecker sand dunes back south. This stretch of highway was more littered with water bottles and soda cans than any road we had traveled since we left Vermont, which seemed shameful since this was the California/ Oregon Trail, traversed by so many patriotic heritage tourists.

Near a sign marking the line between Sweetwater and Sublette counties, we passed a yellow schoolbus towing a sailboat. It was driven by a bearded man in sunglasses and a nautical cap. A little farther we pulled off to look at the polished granite marker set some yards off from the highway to note "the Parting of the Ways," where emigrants turned either to Oregon or California. As has been explained in many guidebooks, most pungently in Gregory Franzwa's *The Oregon Trail Revisited*, this spot was the false Parting of the Ways, the real one being ten miles closer to Farson. Here, in deep and thick sagebrush, two wagon wheel ruts passed by, continuing westward toward the Big Sandy, more desert, and Green River, brightly visible in light orange-colored soil through the clumps of dull green sage. Another rutted wagon trail detached and headed off southwest; it was made not by emigrants or forty-niners but by the iron-wheeled freight wagons and stagecoaches connecting South Pass and the Union Pacific Railroad station at Green River in the three decades between 1870 and 1900. It was still remarkable to walk in those old ruts, whether toward the railroad or toward paradise in California and Oregon. The state of Wyoming erected a sign correcting the error of the engraved stone marker, for some reason electing not to simply move the stone to the correct spot.

No sooner were we back on the road than we had crossed into Fremont County, and pulled over again at the South Pass Vantage Point, crunching a lot of glass shards of the Budweiser brown variety. We stood buffeted by winds, gazing across the approach to South Pass, surprising even with preparation because it seemed so low and flat and wide—twenty miles wide—a saddle between the craggy high Wind River Range and Pacific Butte and Oregon buttes. We walked down the wagon ruts for a while until the sound of the stiff wind at our backs drowned out the occasional truck passing a mile away.

Robert Stuart had come this way, eastward, on October 22, 1812, on his return journey from Astoria to St. Louis and discovered this easy way over the Continental Divide. "We set out at day light, and ascended about 3 miles," he wrote, "when we found a spring of excellent water, and breakfasted; 5 more brought us to the top of the mountain, which we call the big horn, it is in the midst of the principal chain; in scrambling up the acclivity and on the top, we discovered various shells, evidently the production of the sea, and which doubtless must have been deposited by the waters of the deluge." Not until 1824 did other white men pass through, a legendary pack train party of Rocky Mountain Fur Company trappers including Tom Fitzpatrick, Jedediah Smith, Jim Clyman, and Jim Bridger. Captain B. L. E. Bonneville had taken his wagons and party through on July 24, 1832.

The Reverends Marcus Whitman and H. H. Spalding had brought their brides, Narcissa Whitman and Eliza Spalding, through the pass on July 4, 1836, pausing at Pacific Springs to pray and dedicate the Western lands as a "home of American mothers and the Church of Christ." They went forth to preach and multiply at the Walla Walla mission in Oregon, the Whitmans being rewarded by the Cayuse Indians in 1847 with rifle balls and toma-hawks. Their dramatic story, related in the pages of *American Heritage* maga-zine when I was a boy, and later most memorably when I discovered the sublime historian Bernard DeVoto, had cemented South Pass in my mind long ago. Consider his paragraph (in *Across the Wide Missouri*) about what then happened as the Whitman-Spalding Party resumed their journey across South Pass and a party of trappers and Indians galloped down to welcome them: Eliza Spalding, "in heavy boots and swathed by yards of skirt" and rid-ing sidesaddle, "tall, naturally thin and emaciated by travel and illness, dark-haired, sallow under tan, frightened and appalled by the uproar of hospitality. And Narcissa Whitman who was neither frightened nor appalled—she was delighted. A smaller woman than Eliza but by no means emaciated, the pe-riod's ideal in womanly curves, blue-eyed, tanned now but memorably blond. Men always remembered her face and red-gold hair. Men in fact remembered Narcissa, and though she was dedicated to God's service she was charged with a magnetism whose nature no one could mistake."

And then the Pathfinder, John C. Frémont, went through on August 7, 1842, finding the ascent no greater trouble to surmount than walking up "the Capitol Hill from the avenue at Washington." His published travel jour-nal encouraged tens of thousands of emigrants to follow, but not all got the full distance and many despaired at the punishment of the long miles and the loss of everything familiar. When the Donner-Reed Party surmounted South Pass, nooning at the summit on Saturday, July 18, 1846, it unwisely pressed past Pacific Spring down to the Dry Sandy, where cattle drinking from stand-ing pools were quickly poisoned—only the beginning of a series of tragic mis-fortunes for that group. And there is the heart-wrenching testimony of Lavinia Porter. "I would make a brave effort to be cheerful and patient until the camp work was done," she wrote, journeying with her husband and brother from Kansas to California in 1860. "Then starting out ahead of the team and my men folks, when I thought I had gone beyond hearing distance, I would throw myself down on the unfriendly desert and give way like a child to sobs and tears, wishing myself back home with my friends and chiding my-self for consenting to take this wild goose chase."

We drove slowly up toward the pass. Young Samuel Clemens had ridden through with his brother, Orion, in an Overland stagecoach in 1861, having only a few dark moments during that jouncing, twenty-one-day journey to Carson City, Nevada. He found South Pass "more suggestive of a valley than a suspension bridge in the clouds," and the stage conductor told them that at the Continental Divide, streams ahead of them would finally empty in the Pacific, while those behind were just beginning a long journey eastward. Clemens's imagination soared backward out in that direction, following the Sweetwater to the Yellowstone to the Missouri down to St. Louis and then

joining the Mississippi, the stream he had memorized all the way to New Orleans, until it entered "into its rest upon the bosom of the tropic sea, never to
look upon its snow-peaks again or regret them." They rolled "whirling gayly
along high above the common world," agog at "dissolving views of mountains, seas and continents stretching away through the mystery of the summer haze." And down off that wide, grassy saddle on the desert floor, "always
through splendid scenery but occasionally through long ranks of white skeletons of mules and oxen—monuments of the huge emigration of other days—
and here and there were up-ended boards or small piles of stones which the
driver said marked the resting-place of more precious remains. It was the
loneliest land for a grave! A land given over to the coyote and the raven—
which is another name for desolation and utter solitude."

At the summit of South Pass, 950 miles west of Independence, Missouri,
and standing at an altitude of 7,525 feet above sea level, we followed a dirt
road to South Pass City, an agglomeration of log cabins, outhouses, and the
dilapidated Oregon Butte Trading Post and Rock Shop, which was closed. The
thirty restored structures of the South Pass City State Historic Site were
sprawled up the hillside. The Carissa Gold Mine nearby, with its ramshackle
ore conveyer and mine house, had been discovered by soldiers from Fort
Bridger in 1865 and enjoyed a brief rush until 1867. We went looking for
two famous stone markers standing yards from each other. The first stone
was placed near the wagon wheel ruts by Ezra Meeker, first president of the
Oregon Trail Memorial Association, in 1906 when he launched his attempts
to retrace and preserve the trail: "Old Oregon Trail," it said, with year dates
subsequently corrected.

The second stone was dedicated to Narcissa Whitman and Eliza Spalding,
the first white women to cross the Divide. I could not help but think of a
friend, by this writing departed, Arthur King Peters, the elegant and erudite
author of *Seven Trails West* (1996), a wonderfully written and concise celebration. With exquisite photos, drawings, and paintings, it told of Lewis and
Clark, the mountain men, the Santa Fe, California/Oregon, and Mormon
trails, the Pony Express, the overland telegraph, and the first transcontinental
railroad. At that moment Arthur was with his wife, Sarah, up at his ranch
in Wilson, Wyoming, near Jackson Hole and overshadowed by Rendezvous
Mountain. But in a lunch in a Swedish restaurant back in New York City, before I took my family West on this journey, Arthur had regaled me with anecdotes about his research; together, we had entertained his charming research
associate Cynthia Henthorne with tales of explorers and railroaders until all
the other patrons had gone back to work. I'll not forget one of his retracings
of the Oregon Trail. "I was at South Pass with an old friend," he told us, "and
stayed down at the car to change film in my camera. My friend went up to the
Whitman-Spalding monument and he stood there for a long time; I noticed

his shoulders were shaking. Finally I was ready to go over. He walked off to look at the mountains by himself and I found myself standing by the little stone marker, thinking about Narcissa and Eliza and that extraordinary passage into the unknown they had taken, and finally to such a tragic end. And then I found my shoulders shaking, too, and realized I was crying."

We drove east across the Divide, turned around at a wide place in the road, and headed back over. Across the Sweetwater River, we passed a lone cyclist in a bright yellow shirt, near the half-roofed ruins of an old log cabin.

David, Mimi, Mary, and Davey the day we left Orwell, Vermont.

Rose Donahue, taken about 1907, the year her father died.

Charles and Rose Haward and their grandson, 1951.

Famed railroad curator Don Snoddy and William Thompson's scalp, at the Union Pacific Museum, Omaha, Nebraska.

Grenville Dodge, a reckless man but able chief engineer for the Union Pacific Railroad.
(Library of Congress: US262-15959-815738)

Karl Bodmer, *Teton Sioux Woman*. Watercolor on paper.
(Joslyn Art Museum, Omaha, Nebraska)

On the Lincoln Highway, the horseless carriage requires some assistance.
(Nebraska Department of Roads)

The Great Platte River Road Archway Monument, Kearney, Nebraska.

Kool-Aid advertisement, early 1940s.
(Adams County Historical Society Archives, Hastings, Nebraska. KOOL-AID is a registered trademark of KF Holdings.)

Distinguished Union Pacific junketeers pose at a triumphal arch at the 100th Meridian, 1865.
(Photo by J. Carbutt; Union Pacific Museum)

A pioneer family poses before their dugout sod dwelling, tucked into a Nebraska hillside.
(Nebraska State Historical Society)

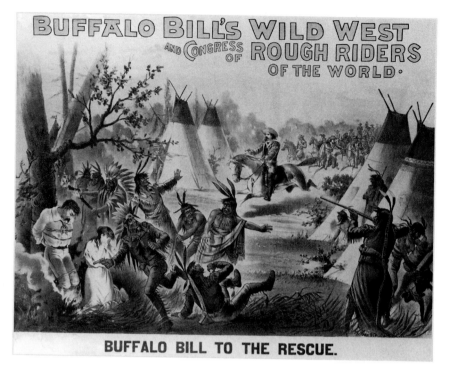

BUFFALO BILL TO THE RESCUE.

Riding into mythology, Buffalo Bill effects a daring rescue in this color lithographic poster for his celebrated Wild West Show, circa 1887.
(Buffalo Bill Historical Center, Cody, Wyoming; Gift of the Coe Foundation; 1.69.108)

(Left) "Enemies in '76, Friends in '85": Sitting Bull and Buffalo Bill pose in the studio of William Notman, 1885.
(Buffalo Bill Historical Center, Cody, Wyoming; P.69.2125)

(Above) Davey, Mimi, and Mary stand at Scotts Bluff, Nebraska, on the pioneer trail.

William Henry Jackson, *Approaching Chimney Rock*. Watercolor on paper.
(Scotts Bluff National Monument, Gering, Nebraska)

This classic drawing of a Hell on Wheels railroad town saloon, called The Big Tent, was in short-lived Benton, Wyoming.
(Author's collection, from J. H. Beadle, The Undeveloped West, *1873)*

Butch Cassidy's gang poses in a studio at Fort Worth, Texas, 1900. *Left to right:* Harry Longabaugh (the Sundance Kid), Will Carver, Ben Kilpatrick, Harvey Logan, and Butch Cassidy. *(American Heritage Center, University of Wyoming; 7742A)*

The notorious Union Pacific train heist outside Wilcox, Wyoming, in 1899, was memorialized in William Goldman's script of the film *Butch Cassidy and the Sundance Kid.*
(American Heritage Center, University of Wyoming; 27593)

Death mask and shoes taken from desperado Big Nose George Parrott, displayed at Carbon County Museum, Rawlins, Wyoming.
(American Heritage Center, University of Wyoming; 26662)

Owen Wister, author of *The Virginian*, template for America's archetypal cowboy.
(Library of Congress, LC-USZ62-37313)

The upper Laramie Plains of Wyoming became a magnet for eager dinosaur fossil hunters after railroad foreman W. H. Reed's discoveries in 1877.
(American Heritage Center, University of Wyoming; 29810)

The cover of *Harper's Weekly* on September 26, 1885, graphically depicted the Chinese massacre at Rock Springs, Wyoming. *(North Wind Picture Archives)*

The Arch postcard, Rock Springs, Wyoming.
(Author's collection)

Amelia Earhart poses with her autogiro and citizens of Rock Springs, Wyoming, during her tour for Beech-Nut chewing gum, 1931.
(Sweetwater County Museum, Green River, Wyoming)

Castle Rock and temporary and permanent bridges,
Green River, Wyoming, 1868.
(Photo by A. J. Russell; author's collection)

Jim Bridger, whose backyard was the whole West.
(American Heritage Center, University of Wyoming; 2614)

Stone charcoal kilns mark the
entrance of the ghost town at
Piedmont, Wyoming, site of two
famous Old West abductions.

At the Golden Spike Ceremony on May 10, 1869, chief engineers Samuel Montague and Grenville Dodge shake hands.
(Photo by A. J. Russell; author's collection)

THIS TRAIN

STOPS

20 Minutes for Supper at the

Golden ▐ Hotel

PROMONTORY, UTAH.

FIRST-CLASS MEALS, 50 CENTS.

THE GOLDEN SPIKE

Completing the first Trans-continental Railroad was driven at this point May 10, 1869. Don't fail to treat yourself to a first class meal at this celebrated point.

T. G. BROWN, Prop.

10 MILES OF TRACK, LAID IN ONE DAY. APRIL 28TH 1869

(Left) At desolate Promontory Summit, this Golden Hotel promotional flyer made boards and canvas seem palatial, and poor fare palatable. *(Author's collection)*

(Above) Chris Graves at 10-Mile Sign, Promontory, site of the famous Central Pacific building triumph in 1869.

Chuck Sweet, Bob Chugg, and Chris Graves grin through the desert dust at Oasis, Nevada, after a long day of following the original railroad grade.

Samuel Clemens began writing under the name Mark Twain in Virginia City, Nevada, where he stood for this portrait in 1864.
(Mark Twain Papers, Bancroft Library, University of California, Berkeley, California)

In 1948 Bing Crosby *(right)* posed with Elko mayor Dave Dotta in the denim tuxedo presented to the singer who owned a large ranch nearby.
(Northeastern Nevada Museum, Elko, Nevada)

An 1860s passenger train rolls through Palisade Canyon in Nevada.
(J. J. Reilly photo; Union Pacific Museum)

Desert magic: Artist Arlene Halford sculpts a mule out of old, rusty baling wire found near her Bishop, California, campsite, while Mimi looks on.

Chinese railroad workers and a foreman pause their handcar on the Central Pacific line.
(E. & H. T. Anthony photo: Bancroft Library, University of California, Berkeley, California)

C. C. Nahl, *Sunday Morning in the Mines*, 1872.
(E. B. Crocker Collection, Crocker Art Museum, Sacramento, California)

(*Left*) Mary, Mimi, and Davey at the Golden Gate Bridge, 2000.

(*Below*) David, Mimi, Mary, and Davey reach Pacific tidewater at San Francisco's Mariners' Museum, 2000.

11

Green River to the Rim

C astle Rock towers over the winter scene—a light snow has fallen, emphasizing the rock's striations, and the snow has softened the outlines of railroad ties, iron rails, bridge girders, and unfinished stone abutments. It's December 1868 or January 1869. Andrew Jackson Russell, the New York portrait painter who has been photographing views along the Union Pacific Railroad's construction for the better part of a year, has planted his big wooden box camera on its tripod and is facing southeast, focusing across the temporary timber railroad bridge and the uncompleted permanent span over Green River, which is frozen with a broken white crust. A work train obligingly halts on the bridge and chuffs, its black coal smoke being carried away by the stiff wind from the south, as three men perch on its pilot, the engineer and fireman look out from the cab, and at least four or five others lean out of train windows or hang out over stairs. On the temporary bridge two men pump their handcar into the foreground and pause, while others clamber onto the hewed stone on either riverbank. One man walks out to midpoint on the bridge and puts his back to the wind—by his overcoat and jaunty hat and proprietary stance he seems like an official. And Russell darts across from his little darkroom wagon with a glass plate wet with collodion and silver nitrate solution and sets it into the back of his camera. He takes one last look at the composition. It is perfect. He removes the lens cap, counts, caps, and then races off to the developer, while the railroaders go back to work and Castle Rock stands out against the cold blue sky.

I have spent many hours with a stereopticon viewer, staring into the images captured by A. J. Russell (and his counterpart for the Central Pacific Railroad, Alfred A. Hart) along the right-of-way. There is something masterful about Russell, whether one is thinking of his painterly composition, his eye for detail, or his ability to capture life as well as wild terrain vividly, vibrantly. It is his view of Green River, Castle Rock, and the permanent and temporary bridges, with that locomotive and handcar and water tower and raw, new buildings, and twenty railroad men perfectly poised as if placed there by a painter, that remains my favorite. I went to some expense to have a print made directly from Russell's glass negative, now owned by the Oakland Museum after many years' entombment at the American Geographical Society. Castle Rock was in my mind's eye that morning as we left Rock Springs and followed Bitter Creek Valley.

We moved west between steep sandstone bluffs of the massive Pilot Butte, scored by wind and water erosion, dirt bikes and ATVs, and past high striated buff-colored mountains. The Union Pacific tracks swung along south of the highway; older railroad bridges had been left standing below and alongside more modern ones. We neared the Green River. In the old days before the railroad, wagon teams would smell the clear water far off and get agitated. The fording place for wagon emigrants was some forty miles north, and William Swain's journal entry from 1849, made at that higher Green River crossing, reflected a common experience for travelers depending on horses, mules, or oxen, and having to cross desert stretches with no water or forage. "Before we reached the river, our cattle became aware that we were nearing water and showed signs of great impatience. . . . When the stream was actually in sight, we found it necessary to unyoke the teams and let them loose to prevent them from stampeding with the wagons. The approach was down quite a steep mountainside. As soon as they were free, they rushed pell-mell down into the river. It was a beautiful, clear stream, and they stood in it drinking and cooling their feet for a long time. It required a good deal of urging to get them out, drive them up the hills, and reyoke them."

We rounded the edge of White Mountain, gaping up at abrupt, red, and rectangular Pulpit Rock to our southwest (which Brigham Young did not pass by nor preach near, contrary to oft-repeated myths), and then the two fancifully headlike Kissing Rocks to the south. To the north was Sun Bonnet Girl Rock, a hard one to decipher as anything resembling a human form, and Sunset Rock, which the settlers at Green River city named because it was the last rock to catch the red rays of the setting sun, regardless of the time of year. Then, the columnar Giant's Club. And then an exit lane took us in a long, lazy curve down toward the city of Green River (population 12,711, altitude 6,100 feet). I felt like I could smell the water, too, and suddenly, to my family's excitement, Castle Rock hove into view—unchanged since Russell's indelible 1869 image but for the city streets draped at its base and sprawling across the far bank of the Green River into residential neighborhoods. We drove down into the city streets, our eyes locked on Castle Rock. I thought about the first time I had seen these monumental rock formations, driving west on the interstate nearly three decades before in a White Rabbit hippie bus with nine other passengers and a dog. We drove through one dark late-summer night and the shadows of the buttes hulked over the road like sleeping dinosaurs. I was thinking about the fur trappers' rendezvous at Green River and about Jim Bridger, from boyhood tales, and about the one-armed explorer John Wesley

Powell, whose famous expeditions down the Green River and eventually through the Grand Canyon of the Colorado had made history.

I thought about Powell now, too, as we followed little signs through the streets and over the Union Pacific tracks, through a neighborhood of cottages and bungalows, and over a narrow 1910 steel bridge to a tidy park draped across Expedition Island. We parked and walked down to the river's edge on the west bank of the island, where rocks were perfect for skipping across the clear water with its greenish bottom. Swallows vaulted and swooped over us. Across the river were nearly white cliffs. A girl crossed a red pedestrian truss bridge toward another park. A U.P. engine hooted from the yards, invisible to us but not far behind us. The railroad crossing was also out of sight but just around a slight bend. It was not the permanent bridge depicted in A. J. Russell's winter 1868–69 photograph: a week before the May 10, 1869, Golden Spike Ceremony to be held in Utah's Promontory Mountains, chief engineer Grenville Dodge came through western Wyoming to Utah and was horrified to find the subcontracted stonework of all bridges in the area so dangerous that he refused to let trains be shunted off the temporary tracks onto the permanent. In Utah, lazy and corrupt masons had stacked bridge abutment stone without mortar, hiding it beneath one course of dressed and mortared stone, collecting their pay, and disappearing. "I do not believe the masonry will hold up a truss," Dodge reported, "and the placing of heavy sandstone on the piers will only help crush them." All would have to be rebuilt. The same was true at Green River and along Bitter Creek—the contractor used local sandstone that was hardly of structural quality. "All went well until warm weather came," one engineer complained, "when the masonry dissolved, the bridges became unsafe and the road was put back on the temporary crossing." Already the entire nation was beginning to turn its attention toward the much anticipated "wedding of the rails" in Utah, and bridges, culverts, and embankments were all giving way. When A. J. Russell had taken his historic photograph of Castle Rock and the temporary and permanent bridges over the Green River, he was recording a disaster just waiting to happen.

John Wesley Powell and the nine intrepid men who would follow his lead into the uncharted landscape downriver might have appreciated knowing about the condition of those bridges, with mortar powdering out and friable stones vibrating loose. According to Powell's journals and the superb biography of him written by Wallace Stegner, *Beyond the Hundredth Meridian* (1953), as trains bearing U.P. executives and notables were gingerly taken over the temporary bridge in May 1869 toward the Golden Spike Ceremony, Powell and his men were camped below while they caulked boats and packed provisions. (Subsequently some readers have interpreted this as "beneath" the bridge rather than "below," or downstream, and indeed there is room for either interpretation.) Why the sojourners would pitch tents beneath those rumbling work tracks, in danger of hot falling cinders raining down from the locomotive stacks, is a mystery. Then again, Powell's men chose the campsite while their leader was elsewhere seeing to the boats and supplies, and they did not always think ahead. In one instance, a month after the full expedition pushed off into the river current, at a campsite on a sandbar on June 16, the cook lit his fire too near some dead willow trees and they ignited. Soon all the brush was ablaze and sparks flew everywhere in the stiff wind, settling on clothing, hair, and beards as the men were driven at a desperate run into their boats, which were swept through rapids almost a mile downstream. Powell himself had climbed to the canyon rim to look around, and could only watch helplessly from far above at this mishap rising out of carelessness, which cost them some important gear but luckily resulted in no injury.

But that was in the unforeseeable future when the ten adventurers stepped into cold water below the train bridge and pushed their boats into the current. Major Powell waved his hat at the gawkers of Green River who assembled to watch them embark. Over and over in my mind I freeze them all in that moment, especially Powell, the New York–born geology professor from Illinois, waving with his one good arm—his right arm having been amputated above the elbow after being shattered by a minié ball at the battle of Shiloh in 1862—heading off, underfunded and underequipped, into the unknown. Powell envisioned a leisurely ten-month sojourn, taking copious notes and instrumental readings and adding hugely to scientific knowledge. In reality, once they disappeared around the bend they relinquished control to geology and could only react to powerful currents, undertows, rapids, falls, swamping waves, rolling boulders, deeply cut canyons, punishing portages in 100 degree heat, and the unceasing echoing roar of rapidly falling water. It was like living beneath an express train, twenty-four hours a day, under constant peril—their boats repeatedly swamped and battered loose of caulk, their rations lost overboard or wetted until spoiled, their scientific instruments lost or shattered, their field notes swept away. One man quit after losing all his gear in a bad dunking that cost them one of their boats; he walked out to the nearest trading post in northeastern Utah. And with that last report of their whereabouts, Powell's men continued downstream toward the junction with the Little Colorado.

Rumors of their death would flash across the continent three times, as they paddled past granite escarpments, lava extrusions, polished marble walls, fanciful landforms, waterfalls, water-worn arched caves and amphitheaters, side canyons leading to cataracts, and cliff-clinging Moqui ruins with time-worn steps ascending from the river. They contended with rattlesnakes, scorpions, mosquitoes, spoiled food, almost no game to be had or fish to be caught, a procession of accidents each day, on and off the water, and a smoldering but growing resentment among the men. Surrounding terrain ranged from badlands to the exquisite pastoral scenery of Glen Canyon to the terrifying, claustropobia-inducing, and nearly unclimbable mile-high cliffs of the Grand Canyon. Tragically and ironically, Powell knew on August 28 that they were within two days of emerging to the first outpost of civilization when three of his half-starved, ragged, and exhausted men refused to endure even one more rapid (which looked to be the worst they'd encountered) and resigned. They said they'd rather climb out of the Grand Canyon and strike overland to escape. They would get up over the rim but not much farther, being riddled with Indian arrows. Two days later, on August 30, 1869, Powell's last two boats with him and his five remaining companions were swept into view of a Mormon family out seining the river for their bodies or the expedition's debris, and they were finished. They had been in motion between May 24 and August 30, ninety-eight days and 900 miles, and, as Wallace Stegner has said, they unlocked a vast territory of the West. Powell's articles in *Scribner's Monthly Magazine*, his official reports, and his journals, all brought meaning and context to that unlocking.

Staring out over Green River toward the white cliffs, seeing the swallows swooping and hearing magpies chattering, in my imagination I let Powell and his men unfreeze at that moment when the current seized their boats; he finished the arc of his salute and sat down, and they disappeared around the bend.

Back on the east side of town, we went past the handsome Neoclassical red-brick Union Pacific depot with its red-tiled roof and colonnaded entry at trackside, completed in 1910. It had been commissioned by U.P. chairman E. H. Harriman to recognize the increasing importance of Green River in the passenger and freight business. Unfortunately Harriman died a year before the building was completed, but his wife attended the dedication ceremony. She was overheard saying that the grand, sprawling structure was "the most expensive outhouse in the west." The birth of the railroad's involvement here was hardly grandiose or well planned. After Grenville Dodge's surveyors staked their way down the Bitter Creek Valley to the confluence with the Green River, but before the tracklayers and railroad town surveyors arrived, speculators laid out a town on bottomland by themselves just north of the Overland Stage crossing, after buying what appeared to be clear title from the Overland's Ben Holladay. When the U.P. got to Green River in October 1868 they beheld a tent and adobe town with residents gleefully rubbing their hands at the expectation of becoming a winter terminus. This annoying usurpation gave the railroad no chance to profit from land sales, so the track-layers just hurried by over the temporary bridge and pressed westward. Thirteen miles west on Black's Fork River, the U.P. founded the terminus town of

Bryan; in a matter of days the population of Green River plummeted from 2,000 to less than 100. But Green River had the upper hand, with a ready water supply that Bryan lacked. By 1872 the repentant U.P. moved its division terminus back to Green River; with vast freight yards, machine shops, storehouses, roundhouse, coaling facilities, and stockyards, the resulting employment and business building Green River into an important city, with brawn and vitality.

Near the base of Castle Rock, towering 650 feet overhead, we found the eventful story of Green River from the fur trappers' era to the present told in the attractive, well-curated galleries of the Sweetwater County Museum, where director Ruth Lauritzen showed us around.

There were extensive Native American exhibits, primarily on the Ute and Shoshone tribes; mountain men and the beaver trade, explorers, railroaders, emigrants, ranchers, coal miners, the Chinese of Rock Springs, and desperadoes were also carefully but compactly covered. One in the last class we had encountered before: rabbity Bill Carlisle, the Lone Bandit, whose deeds we had tracked in Laramie, where as an old man and parolee he had operated a tourist cabin camp; Medicine Bow, where he robbed a train and hid in construction pipes; and Rawlins, where he did hard time. Here we found Bill Carlisle humanized—he had run away from his orphanage at thirteen to work as a cowboy, and in 1916 robbed his first train outside Green River. "I held up my first train because I was desperate—hungry, cold, jobless, with a nickel in my pocket," he told authorities. "While I was standing at the station waiting to jump a freight, the Overland Limited came through. I stood looking at the warm, comfortable coaches with their glowing lights and well-dressed people. . . . the idea of holding up the train came to me on the spur of the moment." That first heist gained him $52, the launching of a mediocre, unrewarding, and short-lived career in crime.

The museum had more artifacts and many more photographs in storage than it could ever hope to display, even in larger surroundings. Among the holdings I found many treasures, including representative railroading stereographs from A. J. Russell, portraits of bank robbers, Chinese miners, Amelia Earhart and her autogiro, and many from the Lincoln Highway era, which made Green River light up with new anticipation.

The county museum has some nice documentation of how that venturesome travel era began to change for the better, starting on the afternoon of August 3, 1915. Then, the official Lincoln Highway Caravan drove into Green River on its transcontinental promotion tour from New York City to San Francisco's Panama-Pacific International Exposition, where the battered and dusty automobiles would go on exhibit as proof that the journey was possible. The lead car into Green River was a 1915 Stutz, painted white with a red, white, and blue hood and blue wheels, the Lincoln Highway colors. Behind were the pride of the Packard, Studebaker, and Oakland lines. The party took many snapshots and movie film, stayed a day, and then gaily rumbled off to the west: "San Francisco or Bust." They did not bust. And within a few years the local Lincoln Highway promoters here in town were boasting about their $50,000 shale gravel road; their avoidance of dangerous railroad grade crossings; their 286-foot-span bridge over Green River; their four hotels, garages, auto supplies, tires, and repair shops; their safe speed limits of 15 miles per hour; their free campgrounds, and their beautiful scenery. Green River became a desirable way stop for auto tourists out to see the fabled and flamboyant West.

West of Green River was broken terrain with sandstone hills and bluffs. The Lincoln Highway split about twenty-five miles past Green River, one of several interesting compromises from its earliest days. The North Branch went northwest across sage flats and low hills following Ham's Fork and the Union Pacific branch line tracks (they headed to the Utah Northern and Oregon Short Line railroads) toward the towns of Granger, Opal, and Kemmerer, past old coal diggings and dinosaur and fossil fish beds and finally the Idaho line. The South Branch, which we and millions of other motorists followed on Interstate 80, paralleled the Oregon/California Trail, the Mormon Trail, the routes of the Overland Stage and Pony Express, and not far away, the U.P.'s original main line toward Utah. Work crews were resurfacing the interstate below Wildcat Butte and Church Butte, the latter's tall, eroded black and blue sandstone cliffs truly resembling a cathedral, as all of the guidebooks, emigrant journals, and railroaders' diaries had reported. Shortly after, just before the highway crossed the waters of Black's Fork, we took advantage of the exit at Lyman, noting excitedly that we could see snow on the blue Wasatch Range ahead and that we were nearing the western edge of Wyoming, with Utah ahead. Bluffs to the north of us had a remarkable green stratum across the middle. Lyman (population 1,896, elevation 6,695) was a healthy town with an abundance of water, green hayfields, and sheep pastures as kind to the eyes as gardenlike Green River.

A couple of miles west, we gained famous Fort Bridger, named after the extraordinary trapper and guide Jim Bridger. Not only was the name prominent in nineteenth-century history, it had also once figured in the imagination of a boy who had grown up reading history, eventually had written some, and was now driving our car. Who could have blamed me for slowing the Dodge as it approached the parklike fort itself—palisaded log walls, white-painted stone buildings, and figuratively decorated tipis, across from the pebbly little Black's Fork River.

I can't remember when I first became fascinated by Jim Bridger, who was born in Virginia on St. Patrick's Day, 1804. It may have been in the late 1950s and early 1960s in the pages of the *American Heritage* magazine my maternal grandfather, Charles Haward, had given me a subscription for. It may have been in the line of young-readers' histories published by Random House, the Landmark Books series. I mowed grass and did other household chores to earn the money to purchase the books that now sit in my son's room. The fascination could only have been fueled by the knowledge I learned at some point during a summertime stay at my grandparents' place—the Hawards' two-acre Circle H Ranch in Kansas City—that Jim Bridger had lived his declining years in a stone farmhouse he built himself not even two miles southwest of where my grandparents' house would be built. He had even been buried there. Not until I was working on *Empire Express* and researching the writings of Union Pacific chief engineer Grenville Dodge did I learn that between Bridger's death in 1881 and the year 1904, his bones had rested on that nearby slope, now overlooking the quick bustle of Kansas City's Highway 435 and its shopping malls.

Dodge wrote a passable biography of Bridger, his onetime scout (it was at least more reliable than Dodge's autobiography). "In person he was over six feet tall," Dodge recalled, "spare, straight as an arrow, agile, rawboned and of powerful frame, eyes gray, hair brown and abundant even in old age, expression mild and manners agreeable. He was hospitable and generous, and was always trusted and respected." Dodge owed him a high debt since it was Bridger's idea to investigate the branch of Crow Creek in eastern Wyoming that led to Dodge's famous "magic route" across the Rockies, between the future towns of Cheyenne and Laramie. In 1904 Dodge had Jim Bridger exhumed and reburied in a proper city cemetery, Mount Washington, below an elaborate obelisk. So Bridger figured in both my life and imagination. It was pursuing full-scale biographies like Gene Caesar's *King of the Mountain Men* and Shannon Garst's *Jim Bridger, Greatest of the Mountain Men* that got me into trouble with frowning public librarians in Port Washington, New York, in the early 1960s, as I trespassed into the adult wing. Fortunately, I argued my case well and obtained visitation privileges into the adult stacks years before I would have been officially eligible. So in a reading sort of way, Jim Bridger had guided me into new and amply rewarding terrain.

He had memorably done this for contemporaries across his vast backyard in the West. It has been said that Bridger carried a photographic image of the West's topography in his mind and knew where every trail led, what every pass and cutoff gained and what they lost, where every drinkable spring and stream hid behind hills, where one could live off the land or be killed by it; in his youth and middle age he was a fierce and canny warrior, adopting the ways and wiles of plains and mountain warfare and earning the respect of Native peoples. Even my hallowed WPA guide to Wyoming weighed in on Jim

Bridger, who made and followed the trails for fifty years: "The whole central and northern Rocky Mountain area was his range. Men who came to the frontier convinced of their own superiority often had to be led through the wilderness like children by Bridger."

He established his small fort with Louis Vasquez on Black's Fork in 1843 with stores, a blacksmith shop, and a supply of iron. Emigrants who expected facilities along the lines of Forts Laramie and Leavenworth were in for a letdown. "The buildings are two or three miserable log cabins, rudely constructed and bearing but a faint resemblance to habitable houses," wrote one traveler, Edwin Bryant. Another, Joel Palmer, sneered that "It is built of poles and dabbed with mud; it is a shabby concern. Here are about 25 lodges of Indians, or rather white trappers' lodges, occupied by their Indian wives. They have a good supply of skins, coats, moccasins, which they trade for flour, coffee, sugar, etc." There Bridger delighted in joshing inexperienced emigrants, telling them he remembered back to when Pikes Peak was just a hole in the ground; in one of his oft-told yarns, he had once come upon a petrified forest. There petrified birds sang petrified songs, and when he tried to jump across a gorge and found he had underestimated it, he was saved by landing on petrified air.

He was married three times—his first wife, the daughter of a Flathead tribal chief, died in childbirth in 1846; his second, a member of the Ute tribe, suffered a similar fate in 1849; the concerned father then sent his two children to board at a St. Louis parochial school. His third wife, a Shoshone and the daughter of Chief Washakie, bore him a third child and then raised all three.

The trading post thrived for ten years. In July 1846, the Donner-Reed emigrant party camped near there "in a beautiful Grass bottom," resting four nights and three days. James Frazier Reed met Bridger and pronounced him one of the few honest and dependable mountain men in the region. But Bridger's fort began to languish after that summer, as emigrants began following "cutoffs" that considerably shortened their journeys west. Also, Utah colonists there in western Wyoming set up competing posts. Giving up on trade, Bridger moved his wife, two daughters, and son to the farm he bought in Missouri in 1855, and formally went into the scouting and guiding business for the government. Commendably, though he was gone for immense stretches of time, Bridger maintained support of his family and made sure his children received a good education.

Fort Bridger was taken over by Mormon settlers and traders after 1853 and acquired several stone houses and a formidable stone wall 14 feet high and 400 feet long, but the works were burned out by the Mormons themselves when they fled in advance of U.S. colonel Albert Sidney Johnston during the federal attempt to corral Brigham Young's faithful. Rebuilt by Johnston's troops, Fort Bridger became a guard station for the Overland Stage, a reassurance for surveyors and tracklayers of the Union Pacific, and a base for troops detailed to the Sweetwater mining district, the Indian agencies of Wind River and Uinta, and other missions. The army abandoned Fort Bridger in 1890 and the buildings were auctioned and taken elsewhere. When in recent years the state of Wyoming marked the fort as an official historic site, planning historical and cultural exhibits, shops, and demonstrations, and began to rebuild it, the architects were guided by photographs taken in 1868 and 1869 by Andrew Jackson Russell, photographer of the transcontinental railroad.

We circled Bridger Butte—the Donner-Reed Party had camped at its base the day after leaving the fort—then rejoined the interstate. Somewhere below the cliffs, Effie Price Gladding, the indefatigable auto tourist on her transcontinental drive in the summer of 1914, had an adventure. "We came upon two draught horses feeding peacefully by the roadside," she recalled.

> As they saw us, they immediately came into the road and began to trot just ahead of our machine. First we drove gently, hoping that after their first fright they would turn aside into the great plain which stretched for miles, unbroken by fences, on each side of the road. But no, they trotted steadily on. Then we drove faster, hoping to wear them down and by the rush of our approach to force them off the road. Once they were at the side of the road we could quickly pass them and their fright would be over. To our disappointment they broke into a wild gallop and showed no sign of leaving the road. They were heavy horses, and we were sorry to have them thundering so distressfully ahead of us. Then we dropped into a slow walk and so did they. But as soon as we traveled faster, they broke into a gallop. For ten miles they kept this up. We were quite in despair of ever dropping them, when suddenly we came to a fork in the road. To our joy they ran along the left fork. Our route was along the right fork and we went on to Fort Bridger glad to be rid of the poor frightened beasts.

I hope those draught horses found their way home. I would have been happier trying to get around them on a little country road instead of battling tractor trailers for a bit of asphalt, but thankfully we would not be on it for long; it was time to retrace a stretch of the original Union Pacific right-of-way, abandoned in 1902 in favor of gentler grades and a straighter route along Antelope Creek and the convenience of two tunnels more than a mile long at a place called Altamont. We were looking for a ghost town, site of the most famous kidnapping in railroad history. At the Leroy Road exit, twelve or thirteen miles past Fort Bridger, we pulled off and picked our way generally southward. Sagebrush hills rose east and w st, and the country was broken by the circuitous and sunken Muddy Creek. None of the dirt roads we followed were marked, and despite an excellent DeLorme map, with a scale of one inch equal to four miles, we got lost. "This is maddening," I said from the front passenger seat, as we drove raising immense clouds of dust. "I can see Piedmont marked on the map but I can't figure out how to get there!" "They put ghost towns on the maps?" Davey asked. "Believe me," I replied, "they put everything on this map—water towers, spillways, gravel pits, cemeteries, and definitely ghost towns. But this place is like a maze." Abruptly we came up against a railroad grade—but it was a modern one with rails shining in the bright sun. "Ah," I said, "I see where we might be, but let's be sure." I had Mary stop the car and I got out to fetch something out of the back. "Release the secret weapon," said Mimi.

Back in the car, I booted up my Mac laptop and plugged in a yellow dingus and set it on the dashboard. It was a toy I'd acquired from DeLorme just days before we left home, an Earthmate global positioning system receiver. For about $100 it allowed a user to connect with several of twenty-four orbiting satellites that beam identity-keyed low-power radio waves down to Earth, accessible from anywhere on the globe. With mapping computer software one can find, in less than a minute, one's exact placement on a highly detailed map on the screen; information about latitude, longitude, elevation, traveling speed, and direction are superimposed. One can even have a window on the map show the number, numerical identity, and position of the satellites guiding one from overhead. Across nine states we had frequently used it to "watch" the Dodge's progress speeding down a state highway marked by moving green arrows on the map, or crawling down a dirt road and crossing creeks or brooks whose names we discovered by looking down at the computer. We tracked our elevation at Hannibal, Missouri, Grand Island, Nebraska, and Sherman Summit, Wyoming. We had not yet used it, however, to find our way out of a labyrinth like the one of unmarked dirt roads and unpopulated terrain around the valleys of Muddy, Antelope, and Little creeks.

"This is so cool," said Davey "We're finding a *ghost town* with a *computer and satellites!*" In a minute we were an X on a Wyoming map, I zoomed in to the highest magnification in about six clicks, and there we were in yellow terrain broken up by numerous creekbeds between Bigelow Bench and Tom's Draw, not far from Dog Spring. With such aid we turned around and slowly but resolutely threaded back through the tangle until we were following deeply cut and winding Muddy Creek.

"We're now driving on the original railroad grade from 1868," I announced. "This is the way people came to get to the Golden Spike Ceremony.

This is the way Robert Louis Stevenson rode the U.P. in 1880. Uinta County took possession sometime after it was abandoned in 1902 and just scraped it down a little to widen it." Copper green and orange-striated hills stood above the road. Ahead, a marker rose surrounded by guardrails: it was the Muddy Creek Campsite where Brigham Young had stopped with his flock. Down at a wide place at creekside we could see four cars and a trailer, with a knot of men and women walking carefully through brush: historical pilgrims following the Mormon Trail. Beyond the marker and down the grade, the surprising sight of a green meadow and a haying operation in progress, then an old ranch, then the ruins of a log cabin. And then to our left appeared four immense stone beehives, three intact and one fallen partially to rubble. "We're here," I told my family. "This is Piedmont."

Under the bright sun and majestic blue sky and in a stiff wind, a dozen or more derelict wooden buildings were scattered across the sandy, high-grass slopes on either side of the creek, roofs mostly intact but long bereft of windows and doors. An iron-fenced cemetery with bleached white stones and the anomalous blossom of plastic flowers stood at the top of a long, hill-climbing path that began near where our car idled on the old railroad grade, just beyond those four stone beehives. They were charcoal kilns constructed by an entrepreneur named Moses Byrne and his brother-in-law, Charles Guild, the first settlers of Piedmont. Byrne and family had earlier run the Overland Stage station on Muddy Creek, not far south, and Byrne named the new town after the birthplace of his wife and his sister-in-law in Italy. Piedmont began as a tent town when the Union Pacific grading crews followed survey stakes down from Bridger and began to climb a steep grade to a 7,400-foot summit dividing the basins of Muddy and Aspen creeks. Test wells there at the base located ample pure water and it became a watering stop and collection point for wood fuel for the U.P. locomotives. The U.P. planted a big water tank and constructed a roundhouse for helper engines. Moses Byrne brought in extra wood for his kilns—four to produce charcoal, which would then be shipped by rail to new iron factories in Utah, and one to cook lime for plaster. Piedmont became quite a bustling little village, with stores and businesses run by Byrne and Guild and a number of other settlers, and four saloons, and ranches nearby. I have a late-nineteenth-century map of Piedmont, which shows those four saloons, a livery stable, railroad section workers' houses, hotels, cabins, a schoolhouse, warehouses, chicken houses, pump houses, wells, and water towers.

It had a lot of life to it back then, but now it was just wreckage in the big landscape, soaking up the sun. We went inside the stone kilns, dry and empty and well constructed except for the collapsed one, each about twenty feet high. We walked around the houses and saloons, seeing nothing inside but broken glass and rubble, and then followed the cemetery path uphill to read the stones. A recently dead cow lay across the path baking in the late morning sun. We skirted it, trying not to breathe the indelicate air. "Ah, me," said Mimi. "The pox!"

Above was the ornate iron fence of the Byrne family cemetery, which had a sign saying it had been used between 1870 and 1931, though there were several markers dated as recently as 1996. Moses (1820–1904) was there beneath an urn-shaped monument with a faint plastic floral spray laid against it, next to his wife, Catherine (1829–1902), with their children, grandchildren, and great-grandchildren reposing nearby, including Joseph C. Byrne (July 18, 1917–June 30, 1918), beneath a sweet little marble marker, a life-size dove now mostly worn away.

One son of Moses and Catherine is not present, and there's a story there. Back when the Byrnes were still running the Muddy Creek stage station, a hunting party of Sioux came by and kidnapped their two-year-old son, Eddie, right out of their yard. The Byrnes did not find out until it was too late to pursue the raiders, and they gave up hope. Two years later, another Indian party rode up. It was led by Chief Washakie, the famous chief of the Shoshone. Without a word of explanation, Washakie leaned down from his pony and handed Moses Byrne his boy, now four years old. A history of Piedmont adds that Eddie Byrne grew up to be mayor of a town in Idaho.

Washakie became a regular patron of the Byrnes' and Guilds' stores after they moved to Piedmont, as was most of his tribe. The chief insisted that his people accept credit only if they fulfilled their promise to pay on demand; the Piedmont history says that "Washakie contended that a man's word was his law, and when a brave promised to pay for something he had received, he paid or was killed." Once, some of his men broke into the Guilds' house and demanded whiskey. Mrs. Guild whisked her son out a back door when they weren't looking and he fetched help from Moses Byrne and his ranch hands, who came galloping to the rescue and drove the Shoshone off. Soon Chief Washakie appeared to apologize to Mrs. Guild. He insisted she give him the names of the men who threatened her—she knew their identities since they were customers of the store—but she refused to turn them in, knowing they would be severely punished. Washakie later presented Mrs. Guild with a pair of moccasins and a purse as recompense for her fright.

One could look anywhere along the hills and creeks of western Wyoming for a place where Washakie, the charismatic and forceful leader of the Eastern Shoshone, had ridden, but Piedmont, with the wind whistling through empty window cavities, across an abandoned railroad grade, and past old settlers' tombstones, was good enough. Born sometime between 1798 and 1804, as a young warrior he was said to make his enemies quake with fear—

but as a diplomat dealing with antagonistic tribes, with the U.S. Army, with emigrants, railroaders, Mormon colonists, and settlers, Washakie showed immense gifts.

In many ways he can be compared to Spotted Tail, leader of the Brule Sioux in Nebraska, who could be cooperative with the encroaching whites and their cavalry forces, but who could be pushed only so far. When, in the years following the Civil War, the railroad and pressure from settlers and fortune hunters sparked the Indian Wars, Washakie was amenable to resettling with his people along the Wind River in Wyoming provided the federals would provision the Shoshone and protect them from the Crow and Sioux, but the leader derided the white man's ways and the proposal that the Shoshone adopt them—especially farming. "God damn a potato," he said to a federal Indian agent who rhapsodized about the quiet delights of raising one's own vegetables. The 1868 treaty signed at Fort Bridger sealed the move to the reservation. Although the Shoshone continued to enjoy buffalo hunts out in the Big Horn basin, and even aided the U.S. Cavalry in actions against the Sioux and Arapaho, reservation life took a terrible spiritual and cultural toll. Then, in 1878, the government proposed to settle the Northern Arapaho tribe—the bitter hereditary enemies of the Shoshone—on Washakie's Wind River reservation. Washakie, then in his late seventies but still the undisputed leader and spokesmen of his people, could not contain his bitterness and issued a scathing denunciation, angrily eloquent and morally irrefutable. "The white man, who possesses this whole vast country from sea to sea," he said,

who roams over it at pleasure and lives where he likes, cannot know the cramp we feel in this little spot, with the underlying remembrance of the fact, which you know as well as we, that every foot of what you proudly call America not very long ago belonged to the red man. The Great Spirit gave it to us. There was room for all His many tribes, and all were happy in their freedom.

The white man's government promised that if we, the Shoshones, would be content with the little patch allowed us, it would keep us well supplied with everything necessary to comfortable living, and would see that no white man should cross our borders for our game or anything that is ours. But it has not kept its word! The white man kills our game, captures our furs, and sometimes feeds his herds upon our meadows. And your great and mighty government—oh sir, I hesitate, for I cannot tell the half! It does not protect our rights. It leaves us without the promised seed, without tools for cultivating the land, without implements for harvesting our crops, without breeding animals better than ours,

without the food we still lack, after all we can do, without the
many comforts we cannot produce, without the schools we so
much need for our children.

I say again, the government does not keep its word!

The government went ahead with the Arapaho resettlement at Wind
River. The two tribes lived coldly and diffidently beside each other, with exceptionally little interplay and almost no intermarriage. Washakie died in 1900
there; in a final irony, the old chief was given an official burial with full military honors for his many conciliations and services to the U.S. Army, which I
suppose in some minds smoothed out the many betrayals. Today there is a
magnificent bronze statue of Washakie in the House of Representatives wing
of the U.S. Capitol. But there along Muddy Creek in southwestern Wyoming,
Washakie had shown kindness and fairness to the struggling families, enforc-
ing amity and honest dealings among his people, and restoring a lost boy to
his parents, thereby leaving another kind of monument for those who can
sense it among the ruins.

The abduction of Eddie Byrne along Muddy Creek was not the most famous
kidnapping in railroad history. For that we must look across the old right-of-
way and the battered, derelict old houses, saloons, and stores and reanimate
them with people and light in the night. On Thursday, May 6, 1869, just two
days before the planned Golden Spike Ceremony at Promontory Summit, a
Union Pacific engine arrived in Piedmont pulling a full complement of passenger coaches and one special Pullman palace car with a sumptuous
walnut-paneled, velvet-draped interior. The former presidential car, it was
built for Abraham Lincoln but never used by him until it was hooked up to his
funeral train for the long mournful ride from Washington to Springfield, Illinois. On this trip it was used by the vice president of the Union Pacific, Dr.
Thomas Clark Durant, that genius of manipulation called the "first dictator of
the railroad world," who ruled his corporation behind figureheads and helped
empty the railroad's treasury to the point where it teetered on bankruptcy
two days before its job was ceremoniously done. The stop at Piedmont was
unscheduled, and hundreds of travelers' faces crowded the windows, peering
out into the night past the flames of torches to see a sullen mob of 300 laborers who had stopped the train. They knew about the plutocratic passenger in
the next-to-last car and planned to use him as hostage to gain two months'
back pay that they suspected would otherwise never be theirs. In my imagination I peer at Durant's face with one question on my mind: is he nervous and
apprehensive about mob violence and financial ruin, or does a small smile

play on his face at the beginning of what might be his last brilliant scheme before the Union Pacific board of directors throws him out onto the cobblestones of William Street in Manhattan, as they most certainly will? Durant's palace car was shunted to a sidetrack with both switches locked. Another westbound train full of dignitaries pulled up and it, too, was stopped. Angry conductors were met at gunpoint and scurried back into the cars. No more traffic would go through until a minimum of $200,000 was produced by the railroad company.

Whatever Dr. Durant's thoughts were as he stood shoulder to shoulder with ruffians in the little telegrapher's office at Piedmont, I know those of the recipients of his eastbound and westbound calls for help. In Ogden, chief engineer Grenville Dodge reacted as the military man he was and immediately wired Fort Bridger for a company of infantrymen to rescue the executive. The message flashed up the wires of Weber and Echo canyons and over the Wasatch Rim into Wyoming and then stopped dead at Piedmont station, where an agent sympathetic to the strikers—he probably hadn't been paid, either—handed the message to the mob. They dictated a response: send the army and the hostages will suffer. "Men are getting desperate," Dodge told his wife, "and I do not blame them much." Meanwhile, in Boston, Union Pacific president Oliver Ames panicked and dashed off a wire to Dodge. "There seems to be no relief," he wrote, "and we feel that the vortex out there will swallow all that can be raised out of our securities—and then perhaps the mobs on line of Road will stop the trains and the next thing we shall hear is that the trains have been stopped and passengers robbed to pay starving men." Desperate as he was at this financial and public relations nightmare, Ames was powerfully suspicious. The mob in Piedmont was made up of woodcutters and haulers of a railroad tie contractor called Davis & Associates. Ames knew that the "& Associates" stood for Dr. Thomas Clark Durant. "Durant is so strange a man that I am prepared to believe any sort of rascality that may be charged against him."

Dodge agreed. It was only a matter of time—hours, perhaps—before thousands of disgruntled workers who had lived on empty promises for weeks or months shut down the national railway before it could open for business. Ames released a token first payment of $50,000 cash, which was gathered and bagged in U.P. offices east of the roadblock, and the men were mollified enough to release the dignitaries' trains, even Dr. Durant's palace car. Whatever the secret thoughts hidden behind those velvet curtains, there would have been a sigh of relief. Celebrants rolled on again, heading for the Utah canyons. There, at Devil's Gate, snowmelt-swollen waters would wash away the supports of a bridge, halting them again and delaying the Golden Spike Ceremony another two days, to May 10, 1869.

Piedmont held us as long as it could. But we had a hankering to see those famous canyons in a new state, the big-basin valley that opened up beyond them, and the great lake. Friends awaited us. We found our way back to I-80 and sped westward through hills and breaks to the Bear River Valley, yet another place of ghostly coal and rail towns. Bear River City, site of a furious gun battle in 1868 between outlaws and vigilantes, was a municipality whose half-life was measured in weeks. The signs for Evanston appeared, putting me in mind of its similar history with Rock Springs—railroad facility, coal mining, Chinatown, and even a recent spate of Hell on Wheelish headlines now subsided—but with all due respect to Evanston, for our purposes that summer it was redundant. With the rush of truckers and tourists, we swept upward toward the Wasatch Rim, zoomed over the Utah border, and began to plummet down the other side.

PART IV

12

Through the Canyons to Paradise

In August 1861, the retired Mississippi River pilot Samuel Clemens and his brother, Orion, sailed in the Overland Stage over the high rim of the Utah territorial border and down into Echo Canyon, the narrow, twenty-four-mile-long declivity that ran northeast to southwest through the tall red rock of the Wasatch Mountains. By mid-afternoon in August the walls begin to glow and take on extra dimensions. "It was like a long, smooth, narrow street, with a gradual descending grade," Clemens would write under the name Mark Twain in *Roughing It* (1872), "and shut in by enormous perpendicular walls of coarse conglomerate, four hundred feet high in many places, and turreted like mediaeval castles. This was the most faultless piece of road in the mountains, and the driver said he would 'let his team out.' He did, and if the Pacific express trains whiz through there now any faster than we did then in the stage-coach, I envy the passengers the exhilaration of it. We fairly seemed to pick up our wheels and fly—and the mail matter was lifted up free from everything and held in solution! I am not given to exaggeration, and when I say a thing I mean it." Sam and Orion jostled together and clung to doorposts as they peered out and up at the cliffs and formations sculpted by wind and rain.

We knew how they felt, ourselves zooming down a smooth high-speed highway at 70 miles per hour when a more meditative pace would have allowed us to study those figured canyon walls that had overlooked so much history and so much toil. But we were part of the frenzied stream of interstaters taking curves and descents immoderately in a rush toward the overbuilt suburbs of Salt Lake City down in the Great Lake basin or for the desert straightaways beyond, when one could really let 'er rip. I had driven (and in some places walked) through the canyons a number of times before, tracing the original Union Pacific Railroad route. Thanks to guidebooks and comparative photographic then-and-now studies by the Salt Lake City software engineer John Eldridge, I could point out the dramatic formations and name them as Mary drove us down Echo Canyon toward its confluence with the Weber River Canyon. There was red and gray Chicken-Cock Bluff, named by government explorer Howard Stansbury in September 1850; immense Castle Rock, towering more than 1,000 feet over the valley floor; at its base, needle-shaped Sentinel Rock; a few quick miles later, the capped perpendicular column called Winged Rock; the grayish crowd of projections called the Kettle Rocks; the shroudlike Hood Rock; the slim, gray, horizontal Hanging Rock, a natural bridge; Steamboat Rock, Pulpit Rock, and Mummy Rock, all readily identifiable as popular landscape art subjects from the stereographic era to the hand-colored postcard time of the Lincoln Highway. Just parallel and above the highway a westbound Union Pacific mixed freight, its bright yellow engines passing in and out of sunlight, passed an eastbound train and periodically disappeared. It seemed to pace us downhill. Then, as Echo Canyon opened into the larger Weber (pronounced Wee-ber) Canyon, we left Interstate 80 and joined I-84, plunging northwest through the quickly narrowing lower Weber Canyon toward Ogden on the lakeshore, passing beneath the weird grouping of rock figures called the Witches. The red walls rose as high as 4,000 feet above our heads, and we marveled that engineers had somehow threaded tracks and a big-traffic highway through that narrow slot. We paused for a few minutes at a rest stop below Devils Slide, two parallel stone reefs jutting forty feet up and out of the canyon side; it was such a squeeze there that were it not for a chain link fence one could almost stretch out one's arms to brush the sides of groaning, squealing boxcars and the speeding tractor trailer traffic. Incredibly, twenty miles downriver the canyon narrowed even more at the famous old bottleneck called Devil's Gate, and then we emerged from the mouth of Weber Canyon as if from out of a cannon, rolling across the outskirts of Ogden. Our heads spun from the tumultuous pinball transit of the Wasatch Mountains, the fast-motion imprint of those red-figured walls embossed behind our tired eyeballs. We'd traveled sixty miles in perhaps forty-five minutes.

It had not always been so easy.

In 1846—fifteen years before Sam Clemens's stagecoach had rushed down the Overland route along Echo Creek—the canyon was choked with boulders, 50- or 70-million-year-old Tertiary and Cretaceous debris fallen over great time from the cliffs above. These posed no serious barriers to a slow rider on horseback like the heroic mountain man James Clyman, who rode through on June 4, 1846, guiding a small party returning home from California. Even less did it obstruct the imagination of another member of that group, a twenty-three-year-old lawyer named Lansford W. Hastings, who was a piece of work.

Hastings, originally from Ohio, had been casting about for something to do with his life. Following the early impulses of the Great Migration, he had been led by experts across the empty West to the new American enclave in Oregon. He found it less desirable than California, to which he drifted and instantly recognized as a paradise in the making. He decided to help it out. He imagined a rosy future for himself as a California booster; the personal benefits might be limitless. Some historians have speculated that Lansford Hastings hoped to split California off from decadent Mexico and install himself as emperor; his ambitions beyond self-promotion and personal gain are obscure. "Whatever was in his mind," Bernard DeVoto commented acerbically in the brilliant *The Year of Decision: 1846*, "he did not have quite enough stuff. Put him down as a smart young man who wrote a book—it is not a unique phenomenon in literature—without knowing what he was talking about."

Based on his one journey and an encyclopedia's worth of supposition, Hastings went East to publish a traveler's advisory, the *Emigrants' Guide to Oregon and California*, published in Cincinnati in the spring of 1845 and an instantaneous best-seller. "Here perpetual summer is in the midst of unceasing winter," he would enthusiastically write about California; "perennial spring and never failing autumn stand side by side, and towering snow clad mountains forever look down upon eternal verdure." As DeVoto and subsequent trail historians such as Dale Morgan noted, Hastings's public relations efforts on behalf of California over Oregon were remarkable. His book was studied by thousands of wistful Easterners, and carried in the saddlebags and wagon boxes of many of the California-bound emigrants who assembled in jumping-off towns like Independence and St. Joseph, Missouri, in the spring of 1846. It was passed along and pored over by many of the 12,000 Mormons who had begun heading west in February of that year, and was even read by their new leader, Brigham Young, as he huddled in Nauvoo, Illinois, recovering from persecutions and violence that had taken the life of their beloved Latter-day Saints founder, Joseph Smith. But beyond his flowery rhetoric Hastings's best-selling guide was short on specific information and criminally inaccurate about realities out on the trail. "Those who go to California," he advised, as if wiggling fingers vaguely and dismissively in the air,

> travel [across Wyoming, and] from Fort Hall [present-day southeastern Idaho], west southwest about fifteen days, to the northern pass, in the California mountains; thence, three days, to the Sacramento; and thence, seven days, down the Sacramento, to the bay of St. Francisco, in California. . . . Wagons can be as readily taken from Fort Hall to the bay of St. Francisco, as they can, from the States to Fort Hall; and, in fact, the latter part of the route, is found more eligible for a wagon way, than the former. The most direct route, for the California emigrants, would be to leave the Oregon route, about two hundred miles east of Fort Hall; thence bearing west southwest, to the Salt Lake; and thence continuing down to the bay of St. Francisco.

As established in 1841 by the Bartleson-Bidwell Party, the California Trail followed the Bear River in western Wyoming in its great arc around the northern edge of the Wasatch Mountains in present-day Idaho to Fort Hall, and skirted north of Great Salt Lake before traversing the Utah desert. It was long and roundabout but it was proved and it was safe. But here was Hastings with his improvement that promised to save emigrants up to 450 miles—a powerful inducement, as it would get them to the final hurdle—the California Sierra—before early, deadly snow. But Hastings had seen nothing of Utah, and his cutoff glossed over the Wasatch Mountains, the hazardous bogs west of Salt Lake, and eighty miles of desert, not to mention the dry washboard of Nevada, the Forty-Mile Desert beyond the Humboldt Sink, and the Sierra Range. He hurried his book off to press. Thousands of emigrants jumped off in the spring of 1846, taking on faith his advice as to time, distance, and survival. Then he compounded folly. That spring, after obtaining sketchy impressions of the trans-Utah route from John C. Frémont in California, Hastings and Clyman set off on horseback. They survived the hazards that would prove so disastrous to ox teams and heavy wagons, then they picked their way eastward through the network of Wasatch canyons to Fort Bridger in Wyoming, where they parted. Hastings set out to advertise his "Hastings Cutoff" as a shortcut to California, recruiting an agent to intercept all wagon trains their side of South Pass: Lansford W. Hastings, Esquire, would personally guide all comers. One of several emigrant parties located by the agent was that headed by James Frazier Reed and the Donner brothers.

Hastings's first customers were members of the Bryant-Russell Party, a mule train burdened by no wagons whatsoever, led out of Fort Bridger on July 20 and into Utah by Hastings's associate, James M. Hudspeth. He improvised their way through a maze of Wasatch canyons with moderate difficulty to the Salt Lake Valley and sent them on their way toward Nevada. Hastings himself left Fort Bridger around the same time as the Bryant-Russell mule train, but he was guiding the first wagon train, named for George W. Harlan and Samuel C. Young of Michigan. This group of some forty wagons entered Echo Canyon and toiled downward with their wagons, shoving aside a multitude of rocky debris. Hastings left them to this slow work while he explored alternative routes and found new customers in the Swiss-German immigrant party led by Heinrich Lienhard. He got them started toward Echo, and then galloped ahead to regain the Harlan-Young group, which by now had descended Weber Canyon with James Hudspeth and on his assurance were bent upon squeezing through Devil's Gate. Their experiences in narrow Weber Canyon were not publicized for a year. "The [Weber] canyon is scarcely wide enough to accommodate the narrow river which traverses it," they related, and there was no room for roads between its waters and the abrupt banks. . . . Three times spurs of the mountains had to be crossed by rigging the windlass on top, and lifting the wagons almost bodily. The banks were very steep . . . so that a mountain sheep would have been troubled to keep its feet, much more an ox team drawing a heavily loaded wagon. On the 11th of August [1846] while hoisting a yoke of oxen and a wagon up Weber mountain, the rope broke near the windlass. . . . The faithful beasts . . . held their ground for a few seconds, and were then hurled over a precipice at least 75 feet high, and crushed in a tangled mass with the wagon on the rocks at the bottom of the canyon.

They pressed westward across the lower Salt Lake Valley and the desert, suffering badly and losing many livestock and some wagons. Meanwhile, the four-wagon Lienhard company followed the Harlan-Young route down lower Weber Canyon. By then Hastings and Hudspeth had nosed through better alternatives, but the Swiss-Germans were stubborn and could not be dissuaded.

"On August 6 we ventured upon this furious passage," Heinrich Lienhard wrote in his diary,

up to this point decidedly the wildest we had encountered, if not the most dangerous. We devoted the entire forenoon and until fully one o'clock in the afternoon to the task of getting our four wagons through. In places we unhitched from the wagon all the oxen except the wheel-yoke, then we strained at both hind wheels, one drove, and the rest steadied the wagon; we then slid rapidly down into the foaming riverbed, full of great boulders, on account of which the wagon quickly lurched from one side to the other; now we had to turn the wheels by the spokes, then again hold back with all the strength we had, lest it slip upon a low lying rock and smash itself to pieces.

Later, Lienhard wrote that the Hastings Cutoff would more accurately be named the "Hastings Longtripp."

This would certainly prove accurate for the Donner-Reed Party, which had already succumbed to advertising and by the time it reached Fort Laramie was so determined to save 450 miles with the Hastings Cutoff that James Frazier Reed ignored the warnings of his old friend James Clyman—with the Hastings book Reed had the proof in the print. The party continued on to Fort Bridger, negotiated the Wasatch Rim down to Echo Canyon, and followed the

Harlan-Young-Lienhard wagon ruts through boulder fields past the Weber River junction. Just four miles downstream on the Weber, however, they found a letter from Lansford Hastings—it had been inserted in a split stick and placed atop a large sagebrush. The Hastings Cutoff promoter advised that the lower Weber Canyon was too difficult for wagons, but if the emigrants would send a messenger ahead to find him, he would ride back and show them a different route to the Salt Lake Valley. Reed followed Hastings, found him, and noted the new and unbroken route, while the rest of the party rested in Weber Canyon, losing four days. When Reed reappeared and the party set out, on August 11, the route proved better than taking on Devil's Gate—but the path was still arduous and alarmingly time-consuming. From Main Canyon to East Canyon to willow-choked Little Emigration Canyon, where all members had to cut brush and trees for two days to make it passable, the Donner-Reed Party with its sixty-six wagons struggled over steep gaps using winches and teams, attained Emigration Canyon, and finally fought its way over a high obstruction later named Donner Hill. The mouth of Emigration Canyon was just beyond, and the Salt Lake Valley. Hastings's cutoff through the Wasatch had taken the party fifteen days—four days of waiting for Hastings's new directions and eleven of backbreaking struggle. Ahead were more mountains, mud flats that would mire wheels and imprison ox legs, and more than eighty miles of desert—far more than they had been led to believe—before they would reach water at the base of Pilot Peak near the Nevada border. Ninety years later, in the 1930s, the whitened bones of horses and oxen and debris of abandoned wagons and jettisoned possessions littered the route of the Donner-Reed Party. Much of their tragic path is clear to this day. And the circuitous road they broke from Weber Canyon to lower Salt Lake would be the path taken by the Mormon pioneers in July 1847 to their Zion on the lakeshore.

"Thence bearing west southwest, to the Salt Lake; and thence continuing down to the bay of St. Francisco." Lansford W. Hastings would continue to fail at most of what he would undertake—though not with the terrible consequences of starvation, death, and cannibalism visited upon the Donner-Reed emigrants in the Sierra snows in 1847. As a rancher in Arizona, during the Civil War he led efforts promoting annexation of the territory by the Confederacy. After the war he schemed to create an enclave in Brazil for ex-Confederates, but died at forty-seven during a voyage to the fanciful colony, a new bastion for chivalry, mint juleps, and slavery, the prospect for which expired with him.

In my mind's eye was the image of those yellow Union Pacific locomotives gliding down in sight of the interstate, beneath those high, figured red cliffs, but it took no stretch to superimpose that of the clanking, smoking, steaming

Iron Horse throttling down the same passage, 132 years earlier. After the summer of 1864, Echo Canyon and Weber Canyon became the path of the Union Pacific Railroad into Utah, following extensive explorations the length of the Wasatch and the Uinta ranges. Samuel Benedict Reed and a surveying party ascended the canyons with the blessing of Brigham Young, who instructed all the faithful colonists along their route to sell them supplies. "The scenery is magnificent," Reed wrote his wife back home in Illinois, "mountains composed of granite and gneiss towering four to five thousand feet almost perpendicular above us. The deep narrow gorge in which the river runs is only about 300 feet wide and is the wildest place you can imagine."

The graders and tunnelers and tracklayers would not congregate until late in the year 1868, when the transcontinental race between the Union Pacific and the Central Pacific had heated up to a heart-choking pace. The federal government and even some snoopy journalists were beginning to wonder if the enterprise might be rotten; it was actually worse than they expected. The U.P. was by then nearly broke and besieged by creditors, though those company directors tied in with the railroad's construction arm, the Crédit Mobilier, were considerably richer after voting themselves galactic dividends. The C.P. was stretched to its limit, too, being within 100 miles of the Nevada-Utah border and similarly unable to stop for fear of vanishing in a puff of steam and brimstone.

But the rush was on, and the Latter-day Saints were in the theoretical position to profit. First with the Union Pacific, and later with the Central Pacific, Brigham Young contracted to supply graders to the railroaders, passing the contracts to his sons and to trusted Mormon elders, who in turn used the church hierarchy to enlist their faithful into the labor. And they were desperate for cash: Utah had endured two successive summers of drought and two plagues of grasshoppers. Earlier Young had hoped the railroads would build into Salt Lake City, but engineering problems kept the two companies from considering a southern route around the lake; the need for money and trade was so acute that Young swallowed his disappointment at the bypass and made the best of it. The farmers in the Salt Lake Valley swarmed up into the canyons with their plows, teams, picks, and shovels. Three tunnels were necessary—one at the head of Echo and two small ones down in Weber—and for a time the Mormon work gangs even served as diggers and blasters. Other Mormon legions graded lines up the east shore of the lake and into the Promontory Mountains over the north shore, and across the desert to the Nevada border. The two railroad companies were so eager to win federal approval for their continuous track lines into Utah that they graded in parallel, often crossing each other's routes. It was a high-stakes gamble worth millions: whoever won the government's approval would reap all the benefits.

Many Mormons were apprehensive about the influx of "gentiles"—with their worldly, wicked ways—into insular Utah. They had heard about the Hell on Wheels towns of Wyoming and Nebraska, and sure enough, when the U.P. worked its way over the territorial line and established its winter quarters on the Wasatch Rim, predictions came true. The town of Wasatch, altitude 7,000 feet, had become the winter headquarters of the Casement brothers' tracklayers, and with a population of some 1,500 cold and restless souls, was getting fairly wild. The traveling journalist J. H. Beadle visited in January 1869 and stayed a week, during which time the thermometer ranged from 3 to 20 degrees below zero, never rising to zero. "During my stay," he wrote,

> the sound of hammer and saw was heard day and night, regard-less of the cold, and restaurants were built and fitted up in such haste that guests were eating at the tables, while the carpenters were finishing the weather-boarding—that is, putting on the sec-ond lot to "cover joinings." I ate breakfast at the "California" when the cracks were half an inch wide between the "first siding," and the thermometer in the room stood at five below zero! A drop of the hottest coffee spilled upon the cloth froze in a minute, while the gravy was hard on the plate, and the butter frozen in spite of the fastest eater.

This was another "wicked city," he noted. "During its lively existence of three months it established a graveyard with 43 occupants, of whom not one died of disease. Two were killed by an accident in the rock-cut; three got drunk, and froze to death; three were hanged, and many killed in rows, or murdered; one 'girl' stifled herself with the fumes of charcoal, and another inhaled a sweet death in subtle chloroform."

When the track advanced to Echo City, near the Weber confluence, the town swelled enormously. Often the sound of gunfire bounced off canyon walls in the clear, cold night air. Holdups, murders, hangings, and unex-plained disappearances became common. Beneath a trapdoor in one saloon putative lawmen found a large hole in which seven unidentified bodies lay amid the tin cans, empty liquor bottles, and other refuse.

Managers were so frantic for progress that safety shortcuts and resultant disasters were commonplace. Track was laid on frozen ground in places; later on in the spring, with the thaw, chief engineer Grenville Dodge saw an entire section of track slide off the grade under the weight of a locomotive and sup-ply cars. And bad luck dogged everyone. One night between Wasatch and Echo City, a drover was hauling a sled of freight and hastened his team to hurry across a grade crossing before an oncoming westbound supply train arrived. "It was a beautiful moonlight night in midwinter," recalled U.P. pay-master E. C. Lockwood, "and the snow was very deep. One of the sleds was about to cross the track when the runners settled down in the snow and the sled box containing the freight rested squarely on the rails. Our engine struck this sled box, tearing it to splinters, and scattering over the snow great quantities of baby shoes, destined for the Mormons at Salt Lake City. These tiny shoes caused us great trouble as they threw our cars off the track, right and left. Fortunately the engine remained on the track and was dispatched to the end of the track in Weber canyon for a wrecking crew, which was brought back in short order and the track cleared, for under no circumstances must the track laying be delayed. All this occurred about two o'clock in the morning; by seven o'clock the track was entirely cleared."

Another accident involved a supply train of some sixteen flatcars, which was on a relatively level segment at the top of Echo Canyon when the last four cars became unhitched; the main part had pulled about a half mile ahead when a trainman happened to look back to see the four trailing and fully loaded flatcars, which were now finding the advantage of the grade. He yelled at the engineer to "go like Hell"—and, with throttle open and whistle screaming, the work train fled before the maverick cars. Two brakemen were aboard the pursuing cars but they were fast asleep. Courting derailment at such high speed, the engineer blew the signal to open switches and clear the track as his train rushed through the night. The trainman had by then worked his way to the back car and ordered some workers to hurl stacked ties down onto the track behind them to stop the runaways. Sure enough, when they struck they catapulted high in the air. Somehow the two brakemen landed in snowbanks and were unhurt.

Disorder was riding into Utah on wheels of iron. The village of Ogden, settled in 1849 and chartered in 1851, lay in its path, between the mouth of Weber Canyon and the lakeshore. It had grown slowly, enjoying its solitude, and one does not think of it as aiming to be a junction of anything if it involved outsiders. William T. Coleman, interviewed in his ancient years about those early days, talked about this. "The emigrants," he says, "parked along the street in their covered wagons, and all the people were afraid of them and what they might do. The grocers would hide what little money they had in coffee cans and anyone who had any valuables would get them out of sight. They would let the people stay so long, then the sheriff, Will Brown, who lived on what is now 25th Street, just over the hump of the hill, where he could see all over town because there weren't many houses to obstruct the view, would gather a posse, and drive down the street and fire several shots with their rifles, and that was a signal for the emigrants to pack up and start moving again." Given how much the city police, and Mayor Lorin Farr, worked to discourage gambling, profanity, drinking, and other moral infractions in these years, it is no wonder that in his capacity as mayor and bishop Farr was looking, in 1867, with a dubious eye toward the railroads approaching from the east and the west. "The time is not far distant," he said gloomily, "when the police will have plenty to do." Indeed. By the next year officer Thomas S. Doxey reported "that there was some loose women in our town that would bear watching," and another officer, C. F. Middleton, took six men into custody for "fast driving and loud hollowing in the streets."

Meanwhile, the tracks of the Union Pacific moved out of the portal of Weber Canyon and over the plain, and were spiked past the log and adobe houses with their struggling little shade and fruit trees, to the place where a depot would stand. It was Sunday, March 7, 1869, but the celebration was postponed until Monday so that the faithful could participate. On the day Brigham Young would inaugurate a new company to build a railroad between Salt Lake City and Ogden—given a merry prelude by a brass band beneath welcoming banners—Ogden speakers marked the arrival of the "national highway" to Utah. There were residents still living in the 1930s to tell Dale Morgan and his WPA researchers of that day "when the entire populace gathered around in Sunday finery to see the iron monster. Suddenly the engineer blew the whistle, yanked a steam valve, and announced he was going 'to turn the train around.' A wild scramble for safety ensued and many ran pell-mell through a nearby slough in their fright, ruining their Sunday clothes. Some terrified children were not found until evening." An artillery salute punctuated the affair and, with it, any lingering hopes for solitude. As Morgan commented in the 1941 Ogden history, the city "awoke to an immediate lusty life such as had been experienced by no Utah city since the arrival of the first Mormon immigrants."

Ogden had become known as "the Junction City" when, from 1869, transcontinental passengers and freight changed trains from the Union Pacific to the Central Pacific (later the Southern Pacific). The future would be even busier: nine major railroads would run through town until consolidation and "progress" took their toll late in the twentieth century. Ogden was a junction for me and my family, too: we had been out on the road from Vermont for a month, on our own resources, but were about to relinquish some of our independence in return for company and local expertise out on the road. For long stretches we planned to follow the original Central Pacific grade—long since abandoned—across the northern Utah mountains and unpopulated desert into barren and remote high valleys in Nevada. The nearly state-wide Humboldt River Valley, path of the Central Pacific as well as stagecoach lines and the old California Trail, would ultimately convey us to within sight of the Sierra Nevada, of old barring the way to Eldorado for some covered wagon emigrants like the Donner-Reed Party. This junction at Ogden signified a big change. For the next several weeks we would be in close company with a vigorous, intensely gregarious history buff from California who seemed to know everyone with a story to tell along the way. As colorful as he was knowledgeable, he was a retired bank vice president named G. J. "Chris" Graves. I had known him for eight months—but had never set eyes on him.

Back in January, I'd received a letter on yellow parchmentlike stationery at my college office, embossed "BRASS EAGLE FARM, Newcastle, Cal.," in brown, antique letterhead that carried a printed watermark depicting an eagle with arrows in its talons. "Raw Wool and Woolen Products/Rare Books/Historical Research/Antique Farm Equipment," it advertised at the bottom. It was a nice fan letter, thanking me for writing *Empire Express*. The writer obviously had an encyclopedic knowledge of old iron rails and California history, and he wasn't afraid to share it with me. For the cost of postage Chris Graves offered to send me chunks of original pear-shaped iron rail laid in the 1850s for the Sacramento Valley Railroad (the first commercial line in California) and rail laid by the Central Pacific during original construction in 1864. When the companies had converted to steel rails in the 1870s the original iron was often recycled to other venues such as timber harvesting and quarrying, with the occasional rail being discarded in the weeds where, branded and dated by the original rail mill, it lay waiting for somebody to find. One day the UPS truck arrived with a heavy little box; cushioned inside were three pieces of iron. Fabricated in Massachusetts and New York State, the rails were taken by clipper ship to San Francisco, thence to a river schooner to Sacramento, and finally by flatcar to the railhead. Irish, Spanish, Italian, and English laborers had wrestled them into place onto the grades that had been built by the Chinese under the direction of James Harvey Strobridge, the Vermonter in charge of Central Pacific construction. My hands trembled with excitement as I pulled the rails out. Set on a windowsill to absorb the sun's rays, their warm, dappled, and scarred surfaces conveyed some kind of tangible energy from the distant past, and from the people who had placed them there.

In his accompanying letter, Chris commented on this. "It's fun to touch and hold something that is so close yet so far away," he said.

To think that those founding Central Pacific fathers like Theodore Judah, Leland Stanford, and the Crocker brothers had something to do with, or even rode on, that rail, and now you own that rail—well, it's fun. A few years ago the Catholic Church, in an effort to raise funds for Father Junipero Serra's sainthood cause, sold off the flakes of bones that fell out of Serra's casket. Yep, for $25 each, you could own a piece of Father Serra. These puppies were encased in plastic, and look really nice. And should you think that's strange, in 1890 the folks in Truckee, California, trotted up to the Donner Party site, dug up the timbers from the Murphy cabin, put 'em in bottles, and sold 'em for $1 each. Proceeds from the sale were used to build the Pioneer Monument at Donner Lake. So feel good about your rail—no bones, no blood, no cannibals. Just good iron from New York.

Over the ensuing weeks, Chris and I fell into a regular correspondence about history and historiography; about people, anecdotes, and artifacts. My correspondent was a member of a number of historical organizations associated with Western rails and trails: the Native Sons of the Golden West, the Sacramento Corral of Westerners, and, when he could stand the "awful politics," the California-Oregon Trail Association. With his "bride of thirty-eight years," Carol, he had walked the California Trail from Thousand Springs in northern Nevada to the Truckee Golf Course, nearly 400 miles over a period of five years. He had traced every inch of the original Central Pacific Railroad grade from the California-Nevada line down to Old Sacramento, and most of it, give or take a few yards, across the width of Nevada.

Following his retirement from the bank in 1995, he'd been freed to do as he pleased—tending large gardens around their home, inspecting flea markets and antiques fairs, rambling around historical spots of California and Nevada, and devoting long hours to idiosyncratic private research projects in state, county, and local libraries and archives. He seemed to love to roam. "Next week," he once wrote me, "my seven-year-old grandson and I travel to Death Valley. Today I bought him a jackalope to hang on his bedroom wall. His mother, my daughter, was not pleased. She fears that I will 'spoil' him on our sojourn. Could be." Another time, he and Carol drove 150 miles to Chowchilla, central California, "looking for this old geezer who makes creatures from abandoned farm equipment. My bride just can't believe we drove 300 miles in the rain for this stuff; the best part of her day was the Taco Bell lunch." Yet another time, he told me that a friend, Lynn Farrar, retired from the Southern Pacific Railroad and the most knowledgeable railroad authority in the state, "swears there are two tunnels between Auburn and Newcastle on the 1910 line. Today, after a flea market, I'm gonna walk those four miles myself. Leaving my flashlights at home. Do you like persimmons? If so, which type? The pointy ones, or the round, apple-shaped ones? We have access to forty acres of persimmons. I'd ship you a box if you like them."

I began to look forward to those yellow-envelope letters, having no idea what he would say next. One note was just one paragraph long:

In the Sacramento Cemetery is the son of Alexander Hamilton, William Stephen Hamilton, died of cholera in Sacto in 1850. The bronze bust on his headstone is a pic of his dad. Also there is Edwin Bryant Crocker and Mark Hopkins [of the Central Pacific

Railroad]. And Lewis Keseberg, the fellow that ate Mrs. Donner. Remember John Sutter, who founded Sacramento and on whose mountain land began the 1849 Gold Rush? His son, John Sutter, Jr. (1826–1897) is buried here, removed from Acapulco, Mex. to Sacto in 1964. Guess who moved him? HIS DAUGHTER! The granddaughter of John Sutter Sr. died in Sacto in 1970, 144 years after the birth of her grand dad. John Sr. (1803–1880) is buried in Litiz Moravian Church Cemetery, Lancaster Co., Penna. Odd world, eh?

What really seemed to spark his imagination once we started corresponding was my planned transcontinental driving trip with the family. Knowing people all over the West who shared our interests in trails and rails, he began pelting me with addresses, phone numbers, and anecdotes about people in Utah, Nevada, and California "who can enrich your journey." And within a short period of time the phrase "perhaps I could be of assistance" evolved in our letter exchanges to "should you have an interest, I'd meet you somewhere in Nevada," to a full-blown plan to rendezvous in Ogden.

It sounded great to us. "Carol says she's had a bellyful of desert dust, out-of-the-way motels, and kicking through sagebrush, so she's staying home," he said. But two of his Ogden friends would be joining us for a 200-mile stretch across Utah and borderline Nevada. One, Chuck Sweet, was, Chris said, "an avid railroader, but not to the extreme of not being able to discuss other issues. He has walked the U.P. and C.P. grades in Wyoming, Utah, and Nevada." He was an indefatigable collector of railroadiana, "so much stuff that he can barely walk around in his apartment—rail, spikes, locks, lanterns, emigrant tokens, wood tokens, documents, handbills, books—a lot of stuff." The other was Bob Chugg, a seasonal National Park Service ranger at the Golden Spike National Historic Site at Promontory Summit, north of Ogden. "Bob was in Vietnam," Chris wrote. "His platoon was all killed 'cept him, and he was shot in the neck. Needs a cane, etc. An awful nice fellow. His roots in Utah go back to the first pioneers. Loves all things historical."

Chris and I coordinated the Nevada-California itinerary I'd sketched, to which he had added considerably, and after we embarked from Vermont in June, discussions continued every couple of days by cell phone. I remember receiving one call from Chris as we drove beneath Castle Rock at Green River, and another while standing in the bedroom of Buffalo Bill's house at North Platte. Bob and Chuck, at that latter moment during the Fourth of July weekend, had driven out from Promontory across the wastes of northwestern Utah into the Pequop Mountains of Nevada, "a dry run in every sense of the phrase," as Chris said, to refresh their memories of terrain, railroad grades, and ghost towns. On July 20, Chris got up hours before dawn to drive his

Chevy pickup from Newcastle, California, to Ogden in time for dinner. We were all to meet at the Timbermine steakhouse, at the mouth of Ogden Canyon, at 6:00 P.M. After my family had checked into our motel, we found our way across town, awed by the abrupt and hulking Wasatch Mountains standing as a solid wall, particularly Mount Ogden, just east of town and towering at 9,575 feet, and Mount Ben Lomond, 9,764 feet, rising above the northeastern outskirts. The big breaks of Weber Valley and Ogden Canyon likewise beckoned the eyes.

The Timbermine steakhouse was crowded for a Thursday evening—the place sprawled beyond a large, high-ceilinged waiting lounge stuffed with antiques and curiosities, into several bustling dining rooms, all done in mining and Western motifs. The place was as wonderfully jumbled as the larger community museums we'd seen, whimsically curated and everything suffused with a grilled surf-and-turf aroma. We looked around for our guides-to-be, not knowing what to expect, although we had seen a snapshot of Chris standing proudly in his driveway next to his yellow '70 Volkswagen Beetle, the top of which came up about to his waist: Chris was, we'd been told, six feet eight inches tall. Abruptly, we saw him, parting the waves of diners who did their best to get out of his way as he elbowed forward and then emphatically pumped the hands of each of us four, bending down and beaming at Mimi and Davey, and exclaiming something like, "Gosh darn, it's good to meet the Bain-Duffy clan, all covered with road dust from eleven states—you must be *exhausted!*" We were, and a little overwhelmed, but in Chris's wake there were his two smiling Ogden friends, Chuck Sweet and Bob Chugg. Our names came up, our table was announced, and we forged ahead into this new phase of travel consciousness.

Anecdotes collided in the air over our heaping table—talk of ghost towns, railroad accidents, and pioneer mishaps, mingled with yarns from our cross-country drive and from our guides' previous adventures out in the desert. At one moment I sat back looking proudly at my bright-faced children, who were holding their own in this "adult" conversation, and at my lovely and vivacious wife, who, like the kids, had so avidly adapted to the long days of transport and historical sites, and devoured every tale I told them about the bygone times out on the old iron road. I looked across at these strangers who were about to escort us out into the God-Knows-Where wilderness, a barren corridor several hundred miles long and far away from civilization: Chris—gray-haired and pink-faced—spearing away at a huge steak and roaring at a manic string of puns from Chuck—medium height, little brushy mustache and bottle-bottom glasses—now reading the dessert menu with it held an inch away from his eyes; and Bob—quiet and direct of gaze—powerfully built and, strongly dependent on his cane, back from the men's room after a painfully slow and utterly majestic walk across the dining room. Back East, a month ago, my mother and mother-in-law thought I was crazy to trust three strangers with myself and my family out in the desert. Looking at my family, and looking at these three men, I was filled with confidence. And utterly happy.

The next morning after breakfast our family found the way to Ogden Union Station, at Wall and 25th. In the peak of its heyday during the 1940s it was one of the busiest in the West, with 119 passenger trains received daily on seventeen tracks. Today with the nearest passenger rail service some forty miles south in Salt Lake City, the elegant, buff-colored stone building with its red-tile roof, two imposing columned entryways, and a large fountain outside is maintained by the city of Ogden. It houses several splendid little museums and galleries, and a full-service restaurant, the Union Grill. Built in 1924, the station replaced a Gothic-looking turreted predecessor destroyed by fire, which in turn had been erected in 1888 on the site of the original wooden depot from 1869. One traveler who paused to take in Ogden in those earlier years was an Englishman, James Bonwick, who commented in his book, *The Mormons and the Silver Mines* (1872), on the liveliness of the railroad hub. "At train time," he said,

> it was an interesting sight to watch the make-up of the railroad passengers, amongst whom one would see blue-coated, brass-buttoned officers and soldiers of the United States Army; mining men; prospectors; longhaired, buckskin-dressed mountaineers and trappers; red blanketed Indians from the Indian country north, west, and south; Chinamen of the old primitive time wearing the bamboo, top-like hat. Added to these were well-dressed, well-to-do travelers from the eastern cities going to California and quite aristocratic-looking English, French, Dutch, and Germans traveling by way of San Francisco and the Pacific to China, Japan, New Zealand, or Australia.

Even then, the station district was sublimely colorful, with hotels, cheap rooming houses, saloons, gambling halls, and cathouses lining 25th Street for four or five blocks. After the turn of the century, the stretch increased in its boisterousness, noisy and brightly lit and crowded shoulder to shoulder all night long; 25th became known as "Two Bit Street" for the price (twenty-five cents) of pleasure there. Now almost squeaky clean though old-fashioned in aspect, historic 25th Street has some of the best restaurants in town, and is thickly settled by antiques emporiums. At the head of the street Union Station draws many strolling patrons.

We stepped inside the main ticket room, large and high enough to contain a church with massive darkwood timbers and murals depicting the building of the first transcontinental railroad. In the time we had available it was hard to see everything at Union Station with its four museums and art gallery, but we found an excellent guide in Bob Geier. The bearded and bearlike executive director was a transplant to the West from a Long Island town near my wife's hometown; from the moment we all met he became a part of the extended family. The Utah State Railroad Museum was the highlight for us, so we covered it most carefully, saving for later the Natural History Museum, the John M. Browning Firearms Museum (named for the famous gun company's founder, an Ogden native), and the Browning-Kimball Classic Car Collection (ten automobiles are displayed, including an American-made 1928 Rolls-Royce Phantom I, once owned by Marlene Dietrich, and a very rare 1909 Pierce-Arrow).

Inside the railroad exhibit we found the Transcontinental Wall—a relief map of the West scored by white fiber optics to mark the route of the Union Pacific and Central Pacific railroads. Red lights marked the meeting point at Promontory Summit and also Ogden (giving the opportunity for jokes about Ogden's onetime red-light district down the street from the station). The sprawling, colorful geographic relief map helped illustrate what the locating engineers faced back in the 1860s, and wall placards and photographs depicted scenes and provided biographical information about a number of the personalities of the era.

The Transcontinental Wall was at that point a work in progress—it was completed in 2001, Ogden's sesquicentennial—and Bob Geier was as proud of it as he was of the exhibit on the Lucin Cutoff. Built in 1904 by 3,000 laborers for the Southern Pacific Railroad at a cost of $8 million, the cutoff crossed the northern end of the lake—5 miles of fill projected inward from the eastern and western shores, connected by an 11¾ mile trestle, which was at the time the longest trestle in the world. It required timber from 38,000 trees, including 25,000 wood pilings, the longest of which was 90 feet; at one point a splice piling was needed—it was 120 feet long. Hundreds of railroad cars of rock fill were dumped there.

The Lucin Cutoff was laid between Ogden and the sleepy station town of Lucin, and reduced the distance westward around the lake, eliminating forty-four steep and curving miles through the Promontory Mountains from transcontinental passenger and freight business. It represented a dramatic savings for the Southern Pacific, and became a popular tourist site. Especially interesting were the station and crew houses sitting on deep rock fill at midlake, and passengers often talked of the optical illusions afforded by nearby islands seemingly floating on air. At station gift shops they could buy curios manufactured by local artisans—miniature houses, barns, windmills, and other figurines—which they would hang in the lake. Within a few days, the WPA guide reported, they were "covered by a half inch of sparkling salt crystals." Later, the Lucin Cutoff trestlework was replaced by rock fill, forming a dike across the northern end of the lake.

Elsewhere in the Ogden station exhibit was material on the Central Pacific Chinese, on railroad technology, and on rolling stock. Also there was the Wattis-Dumke Model Railroad Museum, the centerpiece of which was a diorama depicting the terrain between the Wasatch and the California Sierra, across which twelve scale-model trains ran through canyons and across deserts, over both the Lucin Cutoff and the Promontory Mountains line. Up there, I pointed out to the children, was our destination for the afternoon— the Golden Spike National Historic Site, where, thanks to our new Ogden friends, we were to be given a special tour. We said goodbye to director Bob Geier, knowing we'd see him for dinner at the Union Grill that evening. In the next two years, as it would turn out, I would return to Ogden Union Station three times, to deliver talks and speeches, becoming more and more certain that it is a sublime historical attraction, operated by a wonderful community of professional staff and volunteers.

The state of Utah presented a problem to the "good roads" promoters of the Lincoln Highway Association and others in the early years of the century. Routes from Wyoming and through the Wasatch Range could follow the old wagon roads, though with much blasting necessary, but whether Ogden should be included in a route passing through Salt Lake City was actively debated, as was the path across the Salt Lake Desert. Would it tend westward across flood-prone salt flats to favor the old California Trail's Humboldt Valley route across Nevada (today's Interstate 80), or strike southwestward toward the Nevada town of Ely and thence along the old Pony Express route (today's U.S. 6/50)? Lincoln Highway promoters favored the straightaway Pony Express route, which stubbornly contended with five mountain passes with punishing grades and 7,000-foot altitudes, with deep valleys in between. These difficulties made the choice a hard sell, and in Salt Lake City for a long time it seemed as if no one was interested in buying anything. State officials were disinclined to cooperate with outsiders, which held up the process for years. Choice of what became the federally designated route—linking with the Humboldt Valley—was delayed until 1923, and from that year the Utah link of the Lincoln Highway—bypassing Ogden in favor of Salt Lake City, and following the Pony Express into Nevada—began to disappear. There was never any serious discussion of the original 1869 railroaders' route around the north end of Great Salt Lake, hardly surprising given the operative political climate in the early 1900s.

When, in 1868 and early 1869, the race-maddened railroad companies commissioned their parallel grades up the valley north of Ogden into the Promontory Mountains, the lines passed solid little family farms and orchards green with irrigation water from the Wasatch. The pleasant farm town—called Brigham City by hopeful Mormon boosters—was also on the route along with the last-gasp Hell on Wheels town, Corinne. This wild place, founded near the Bear River lake outlet by anti-Mormon speculators who dreamed of capturing prominence from Ogden and Salt Lake City, grew in two weeks to more than 300 frame structures, shacks, lean-tos, tents, and combinations thereof; at least 19 saloons and 2 dance halls opened for business, catering to the rough graders imported late in 1868 and to the traders, teamsters, drifters, and ne'er-do-wells who naturally flocked in. After Corinne lost a furious political struggle with Ogden over which would become the railway junction of the Central Pacific and Union Pacific, the town thrived for a few years more catering to wagon freighters and other rough types. In 1870 town leaders unsuccessfully petitioned the U.S. Congress to move the capital of Utah to Corinne. There were even attempts to secede northern Utah and join it to Idaho. These efforts failed, and when the historians writing the state guide for the WPA turned to address the roads connecting Ogden with Brigham City and moving beyond the thirty miles to the Idaho border, they whimsically wrote that "Corinne dreams in the sun, like an old man remembering his youth. The once roaring, fighting, hilarious rakehell town has little to show for its riotous past. A handful of houses, a few weather-stained business buildings, a church, and a school, are all that remain of a city of more than 2,000 people."

Sixty years later when we drove through, Corinne continued to dream, possibly even thankful that it did not disappear with barely a trace like others along the railroad route. The highway leaves Ogden and fairly rubs between the base of picturesque Mount Ben Lomond and the wetlands and flats of Bear River Bay, a vast refuge of migratory birds. One of my fondest hopes while planning the itinerary along the old iron road was to pause here at Bear River and see it with our friend Terry Tempest Williams, a poet and naturalist with whom I had worked at the annual Bread Loaf Writers' Conference in Vermont, and who had visited Middlebury College several times to lecture and meet with students. She had lived for years in Salt Lake City with her husband, Brooke, but had recently moved to the Colorado River Valley in southeastern Utah, and was temporarily unreachable. Terry's memoir, *Refuge: An Unnatural History of Family and Place*, had been published in 1991. It was a striking meditation on place and loss, chronicling the endangered habitat of egrets, herons, and owls, at Bear River, a place of great meaning for the author; that narrative was set in parallel with the slow death (from breast cancer) of Terry's mother, who had been exposed to fallout from atomic bomb tests in the 1950s. So had her two grandmothers and two aunts, and all had suffered from breast cancer. Terry, too, had been exposed as a child, and at that writing had endured several alarms and thought of the illness as an inevitability—she was already, she said, a member of "the clan of one-breasted women." Her book was of inestimable meaning to us, not only because it was beautifully written and exquisitely structured between descriptive journeys into the Bear River refuge and her own family odyssey, but also because Mary had become a member of that clan herself in 1997. In so many ways Terry felt like a family member—her name was my own sister's name; also, Mary and I had named our daughter Mimi after Mary's grandmother, and Terry's grandmother was also named Mimi. To have been able to turn west from Brigham City into the bird refuge to rendezvous with Terry, as I had done so many times in the company of her book, would have been a delight.

But it was not to be, and in fact the mood was quite different from a meditative hike seeking snowy egrets, dark-eyed juncos, and avocets. Instead, in quite a jaunty and expectant mood, we were driving in a two-car formation following the parallel railroad grades dutifully scraped by Mormon farmers at their elders' behest for the Union and Central companies. Chuck Sweet was riding shotgun with us, while Chris Graves followed in his truck; Bob Chugg had gone up to Promontory earlier to work. We left Interstate 15/84 at Brigham City, turned onto State 83, and drove through sleepy Corinne, having Chuck point out the burg's last tavern, now a favorite haunt of park rangers and volunteers. The Promontory Mountains were rising steadily just west where, across marshy, indistinct Bear River Bay, we saw the rocky brown, twenty-five-mile-long peninsula whose tip was called Promontory Point. The mountains continued northwest where, as Chuck told us, a break in the peaks opened the way toward Promontory Summit, where the last spike had been driven.

A vast column of smoke from a brushfire rose from somewhere in the mountains ahead. We crossed a number of little creeks and passed ponds, signs for a public shooting grounds, and a ranch where men on horseback, wearing cowboy hats, were hustling horses into a paddock. Then, as we approached an industrial area—it was the headquarters of Thiokol, makers of rocket engines for the space shuttle—we turned off State 83 onto Golden Spike Drive, a two-lane blacktop. "Safety Our #1 Priority," read a big Thiokol sign, and I marveled at this juxtaposition of the Iron Horse with rocket engines in the same region of dry Utah hills. Later, rangers at the Golden Spike site told me that while out on the Union Pacific grade they could sometimes hear from over at the site a clanging steam engine bell in one ear and a Thiokol rocket engine being revved through a test in the other.

Rocky protrusions began appearing, and then cliffs so worn through with caves that they looked moth-eaten. Then we were skirting the bottom of a slope and looking up toward the parallel railroad grades running only a few feet from each other along a sidehill and steadily gaining altitude. A ravine opened, interrupting the grade, and we could see the site where the Union Pacific's Big Trestle had stood—it had been 85 feet high and 400 feet long, subject of admiring stereograph portraits, and so quickly and carelessly built that when it was seen swaying and groaning in the breeze in April 1869 by the Central Pacific's contractor and founding partner, Charles Crocker, he balked at permitting his company's trains to run over it.

Just beyond the trestle site was the seventy-foot-high "Big Fill," the Central Pacific's solution to the riddle of the ravine and the shaky original trestle, a gargantuan railroad embankment dumped into the deep gap with donkey carts little bigger than wheelbarrows. Guiding our eyes, Chuck knew the terrain so well that he could prime us, driving at 40 mph, to look at specific places up on the hillside where we could see the parallel rail cuts deeply blasted into rough, solid gray rock, which were visible only for a few seconds from the road. The car was climbing heavily now, straining to reach the level of the railroad, and then suddenly we crested, crossed the original grades, which veered off to the south for one more curving rise, and we emerged on the floor of Promontory Summit valley. "Another half mile," said Chuck, "and we rejoin the grade, literally—it folds back to us and we're driving on the actual railbed of 1869." No sooner had he said this than it was a fact. Across the summit valley floor, the enclosing mountains rose to define the northern horizon, which was now completely enveloped in dense white smoke: the brushfire, easily a mile wide judging from the smoke plume, was burning up the far side of the mountain. Ahead, we could see several low brown buildings, and after crossing a single track, we arrived at the Golden Spike National Historic Site.

Every time I go there it is like a pilgrimage. I have seen it in aromatic spring, in high, blasted summer, in bleak autumn, and in frosted winter, and it always wells me up with reverence and anticipation, like some historic sites can do. They are all hallowed ground. Promontory is hallowed. The soil was not sanctified in a few hours or days by the blood of combatants, as in a battlefield, nor by a few minutes of a historic speech, as on the Washington Mall or in halls in Philadelphia or Boston. One does not meditate on some endeavor done by a single individual or a small team, no matter how earth-shaking, such as those created by Eli Whitney, Thomas Edison, Alexander Graham Bell, Guglielmo Marconi, Philo Farnsworth, or the young gawky electronics engineers and software designers in California's garages. Instead, one stands there at the culmination of years and years of dreams and toil, perspiration and peril, and—I'd be the last to leave them out—ambition and greed, of so many human beings.

How many? Of the struggling laborers, the Chinese and the Irish and the other immigrants and the war veterans, the graders and tunnelers and track-layers and woodcutters and water haulers, the tie and timber cutters and bridge builders, the masons? Of the surveyors and other engineers, the drivers, firemen, brakemen, the loaders and handlers and movers? What numbers does one estimate in those Central Pacific and Union Pacific years in the 1860s? Twenty, twenty-five thousand, thirty, thirty-five? More? And I can't forget the anxious, toiling farmers of Zion, worn out by fighting grass-hoppers and drought and hopeful of earning some coin while blasting and scraping the double grades across the dry northern width of this territory. I won't leave out the managers and executives and interested politicians. And what numbers of deaths? How can we ever know? And what of the actors in the side dramas, the settlers who plunked down at towns that were only a name on a map, a station, and a water tank; the women and men who tailed the tracklayers to make whatever they could of that wild postwar, no-holds-barred era; the Native Americans who cooperated and were betrayed or the Native Americans who resisted but only delayed the inevitable?

All contributed to the moment when a silver maul lightly tapped a golden spike—all added to the urgent, symbolic, volcanic energy of that moment on May 10, 1869.

Behind them all were thousands upon thousands more—the explorers and wagon and handcart pioneers who were born too early to take advantage of that epochal enterprise but nevertheless stepped off into the unknown to discover something greater about the world and about themselves, who died by the thousands or suffered mightily and survived long enough to greet the Iron Horse as it edged across Nebraska and California, Nevada and Wyoming, and Utah where the two strands connected and the nation went wild.

Here we were at this place, my family there for the first time—not an espe-cially promising or welcoming spot. The *New York Herald* reporter Albert Richardson wrote about that summit valley when he rode through on one of the first transcontinental trains in May 1869. "Promontory is neither city nor solitude, neither camp nor settlement. It is bivouac without comfort, it is delay without rest. It is sun that scorches, and alkali dust that blinds. It is vile whiskey, vile cigars, petty gambling, and stale newspapers at twenty-five cents apiece. It would drive a morbid man to suicide. It is thirty tents upon the Great Sahara, sans trees, sans water, sans comfort, sans everything."

But there, by agreement, the Central and the Union companies met. When-ever I go, I think about that moment—the tracklayers and teamsters and subcontractors; the engineers and executives; the men of the 21st Infantry, the Tenth Ward Band, the dignitaries from Ogden, Salt Lake City, Carson City, Sacramento, and San Francisco; the sharp-eyed pickpockets, saloon keepers, and souvenir sellers. The two bright engines. The telegraph wires strung eastward and westward toward home or "Old Home," as emigrants called it. Follow those wires hundreds or thousands of miles across deserts, mountains, plains and prairies, to the places where vast throngs filled city and town streets looking up at bell towers, where uniformed gunners were stationed next to celebratory cannons, where hands paused waiting to pull bell ropes on assembled fire engines, or yank the cords of steam whistles on locomotives in yards and ships moored in harbors and on rivers. Think of the collective indrawn breath held across the nation waiting for the Reverend Doctor John Todd of Massachusetts to get done praying, and for the speeches shouted out into the open Utah air and sunlight.

All the expectations of a nation focused on the people there perspiring into their starched collars or soft work clothes and wedding the rails symbolically for the whole country, and, they were convinced, for the world. And all the energy that was about to be convulsively released, from Presidio to Plymouth Rock.

One cannot resist the colorful rhetoric of the day. I turn again and again to the three high points of speechifying at Promontory Summit, just before and just after the wedding of the rails, the Big Event. Ponderous Leland Stanford, Central Pacific president and one-term California governer, at base a failed lawyer turned storekeeper who liked the limelight of politics and the perquisites of power, who try as he might could not elevate to the occasion: "This line of rails, connecting the Atlantic and Pacific, and affording to commerce a new transit, will prove, we trust, the speedy forerunner of increased facilities." No sooner had these eloquent words fallen on celebrants' ears than he said more, and said it all: "I will add that we hope to do ultimately what is now impossible on long lines: transport coarse, heavy and cheap products, for all distance at living rates to trade." The shortest book in California in the 1800s would be titled *Leland Stanford's Greatest Speeches*.

But we have Grenville Dodge, chief engineer of the Union Pacific, who told the crowds at Promontory: "The great Benton prophesied that some day a granite statue of Columbus would be erected on the highest peak of the Rocky Mountains, pointing westward, denoting the great route across the Continent. You have made the prophecy today a fact. This is the way to India."

But my favorite is that of James Campbell, rolling stock supervisor of the Central Pacific. "The work is finished," he said.

> Little you realize what you have done. You have this day changed
> the path of commerce and finance of the whole world. Where we
> now stand, but a few months since could be seen nothing but the

path of the red man or the track of the wild deer. Now a thousand wheels revolve and will bear on their axles the wealth of half the world, drawn by the Iron Horse, darkening the landscape with his smoky breath and startling the wild Indian with his piercing scream. Philosophers would dream away a lifetime contemplating this scene, but the officers of the Pacific Railroad would look and exclaim, "We are a great people and can accomplish great things."

James Campbell gave the last speech buoyed by champagne bubbles in the C.P. construction superintendent's car, and it is an eloquent one. We can study his words, and those in the other speeches, the journalism, the diaries and memoirs; we can peer at the indelible images caught by photographers Andrew Russell, Alfred Hart, Charles Savage, and S. J. Sedgwick, hoping to inch ourselves even a half step closer to that wild jubilation that struck every citizen in this country at a most personal and meaningful level. We can look out at the soil, the rocks, the mountains, most certainly the fills and cuts that point unerringly across the waste and uncertainty toward, in those days, civilization, the comfort of numbers, the rule of law, the reassurance of faraway loved ones. We can let our eyes play over the artifacts displayed in that national shrine or stored there in archaeological collection drawers—the shovels, the insulators, the multicolor medicine and whiskey and opium bottles; the wooden dominoes, buttonhooks, skeleton keys, and coal buckets. The china doll. And we feel the human connection. And we are moved.

Inside the visitors center, Bob Chugg was stationed behind the counter in his National Park Service uniform. He introduced me to Bruce Powell, then the director of the Golden Spike National Historic Site, and to rangers Rick Wilson and Bob Hanover, who toured us through the elegant little museum and displayed the rarities kept behind the scenes, waiting for an upturn in the economy and enlargement of the facilities. Among the artifacts were many from the C.P.'s Chinese tracklayers—ginger tea and soy sauce containers, an opium can, Chinese coins, and some 65 of the 361 pieces for the 4,000-year-old game of Go. There were, representing the Irish workers, whiskey and beer bottles, spittoons, coal buckets, spikes, tie plates, and machinery. In the storeroom I saw a beautiful ambrotype mounted in royal blue velvet with a gilded, hinged case: it was a portrait of Charles Crocker taken by a Sacramento photographer and a one-of-a-kind rarity never before published.

Outside in the sun, Chris Graves and Chuck Sweet were studying Andrew Russell's photographs of the Last Spike Ceremony. Thirty yards away, gray-bearded John Ott, a Park Service volunteer, stood at trackside next to the warm flank of the Central Pacific's Jupiter locomotive (Union Pacific's No. 119

was back in the shop, having its steam pipes adjusted). Ott pointed to the resplendent engine, its bright red, royal blue, and gold paint, and its blinding brass, looking almost otherworldly against the dim colors of desert and mountains. "Our two engines are replicas," he told some thirty tourists standing nearby. "The originals were ignominiously scrapped between 1903 and 1906. They were meticulously re-created in 1975 by O'Connor Engineering Labs of California, who had no original plans to work from. The National Park Service had commissioned a draftsman to render diagrams, but he couldn't stand the deadline pressure and had a nervous breakdown and burned all of his fifty drawings. O'Connor started from scratch. They had to rely on the old Russell, Hart, and Savage photographs from 1869 for many details, exhaustively measuring, building casts, machining tolerances, and finally when the whole complicated work was done, which took four years, deciding from the sepia photographs and observer accounts how the engines and tenders should be painted. They delivered these two beauties by May 10, 1979, the 110th anniversary."

His spoken narrative to the assembled tourists vaulted us back to May 1869, as the succession of trains pulled up to disgorge their loads of dignitaries, and a wagon train of entrepreneurs unloaded cases of whiskey and kegs of beer and shoved them into the shade of canvas tents. A brass band shouldered their instruments in the bright sun and began to play contemporary airs, bluecoated soldiers from the 21st Infantry lined up at parade rest, telegraphers wired the last golden spike and a silver-plated maul to send the signal of the final blow, and photographers captured the scene. "The Union Pacific's Thomas Durant, whose train had been held up in Piedmont, Wyoming, by strikers demanding their back pay, stayed inside his palace car as long as he could at Promontory," John Ott said. "He was prone to migraines. The only known remedies were opium or whiskey. Emerging from the dimness of his car into this sunlight must have been a stab through the eyeballs."

As Ott continued to his attentive audience of thirty, I reflected on that original ceremony as documented by participants and observers. News account estimates of the 1869 crowd size varied from 500 to 3,000. In contrast, a century later on that empty, trackless summit valley (the abandoned tracks had been removed in 1942 and were reused at various military installations), with its raw new visitors center and parking lot, some 28,000 attended the reenactment on May 10, 1969, when local actors in period costume revisited the orotund speeches beside two steam engines borrowed from the Virginia & Truckee Railroad in Nevada. Thereafter, every anniversary the big event was restaged by the National Park Service and the Golden Spike Association of Box Elder County; the arrival of the gleaming locomotive replicas in 1979 completed a grand scheme. They are scrupulously cared for, and every day during the summer tourist season are driven out, clanking, chuffing, hissing, and ringing, to the obvious delight of onlookers. Reenactors frequently stage ceremonies during the summer. During one of my visits, on May 10, 2002, when I was invited to deliver the keynote speech, I got to sit on the reviewing stand with impersonators of Leland Stanford, Thomas Durant, and Grenville Dodge, with a Taiwanese dignitary from San Francisco and representatives of the Hibernian Society of Utah, and with the charming and dignified president emeritus of the Golden Spike Society, Delone Bradford-Glover, listening to the oompah band and looking between Jupiter and No. 119 at a healthy and enthusiastic crowd of 3,000.

But even that paled before an honor during our family journey in late July 2000, when my children and I were helped aboard Jupiter by fireman Keith Corbridge and engineer Rick Tomanto and conveyed up and down the track, with Davey yanking the bell cord, while Mary caught the adventure on videotape. Feeling the heat from the firebox on one's knees and hearing the clank and hiss and the piercing locomotive whistle, seeing the summit valley and the Promontory Mountains from that angle, and the tracks stretching east and west until they disappeared from view, brought up ancient connections and caused a cascade of images in my mind, frozen in time more than 130 years before, and I felt an unmistakable twinge, a palpable tug from the West.

Later, I stood outside the visitors center with director Bruce Powell, ranger Bob Chugg, and volunteer John Ott. "I've been out here five years," said Ott, "and I've seen these trains going back and forth hundreds of times, and it never fails to move me." He gestured out toward the mountains. "This is one of the few national historic sites where the physical setting is unchanged. No four-wheel drives, no motorcycle scars on the mountainsides. . . . It's beautiful."

I pointed to the northern horizon, ominous with dark smoke from the far side. "We could see that in Ogden," I said. "Doesn't it concern you?"

"It started yesterday with a lightning strike," said Bob. "We've had one-one-hundredth of an inch of rain in three months, since May."

"And yes, we're pretty concerned," said Bruce. "We're monitoring the radio bands, listening to weather reports. We had a bad fire in the valley in 1996—had about seventy-five volunteer firefighters camping out around here. They fought it for several days."

"Then," continued Bob, "it seemed to be going out, like they often do, but then suddenly it really popped, moving 100 yards a minute."

"We had to evacuate the historic site," said Bruce, "and were scared for the buildings and contents and of course the locomotives. We pride ourselves on how natural it looks up here, how unspoiled, but then Mother Nature steps in and decides to rearrange things. So we keep our fingers crossed."

That evening we gathered in the high foothills beneath Ben Lomond's peak at Bob Chugg's house for dinner, an informal meal spread on plates and containers across a kitchen island, with a wide balcony deck beckoning us outside. The view was extraordinary. With the abrupt wall of the Wasatch nearly at our back, we looked down across grassy slopes and saw the tidy roads and houses of North Ogden, and then Ogden itself, ablaze with lights and life in the dimming evening, and beyond the city the great blue lake, with its craggy islands and scrim of mountains south and west.

One could look across this vista and see terrain crossed by trapper and guide Jim Bridger, who is credited with being the first white to describe Great Salt Lake; and by the fur trappers who for years used the meadows beneath present-day Ogden for their annual rendezvous; and by the pathfollower, John C. Frémont, who, unaware that a number of Americans had already seen the lake, came in 1843 to the confluence of the Weber and Ogden rivers with the Salt Lake and proclaimed himself the discoverer of all he saw. We could look down toward where the Connecticut-born fur trapper Miles Goodyear had built a cabin for his Indian wife and two children, and a stockade, in 1844 or 1845, becoming the first permanent settler.

We could turn our gaze from the opening of Weber Canyon and follow the pathway of the Pacific railroad through Ogden and up the valley floor toward the Promontories.

And I could see the approximate place where Bernard DeVoto had been born (in 1897) and raised. As a boy he had scrambled all over these foothills with his chums and penetrated the many Wasatch canyons to camp, hunt, and fish, until he escaped the puritanical, anti-intellectual village for an education at Harvard and an active literary life. I had walked reverently in De-Voto's footsteps many times, whether in Boston, Cambridge, New York, or on the green Vermont meadows at the Bread Loaf Inn where he had taught, laughed, quarreled, and caroused with gathered writers like Robert Frost, Wallace Stegner, Catherine Drinker Bowen, and A. B. Guthrie Jr. I had followed DeVoto's lead through an infatuation with Mark Twain in particular and Western history in general; whether anyone else agreed, I considered my book on the first transcontinental railroad and three transformative decades of American history to be an extension of what he had done in his *Year of Decision: 1846*, using one of its last paragraphs as the opening epigraph of *Empire Express*.

To see DeVoto's hometown grown to a city, the overlooking foothills of play and make-believe now a development where my friend lived and entertained his historical-minded, railroad-obsessed comrades, completed some kind of circuit. And though I did not know it at the time, I would revisit this house with its commanding balcony view several times each ensuing year as research, speeches, and book signings brought me back to Utah, each time finding Bob Chugg's house full of lively people and invigorating talk, often about history. It was just the sort of gathering Bernard DeVoto would have relished.

Early the next morning, a Friday, the sky to the north of Promontory Summit was clearer but the entire south face of the rimming mountains was charred, narrowly sparing a ranch at its base. The Golden Spike site was only a couple miles' remove across brush and fields of winter wheat. The rangers and volunteers knew enough not to stop monitoring radio waves, though, as there was still enough fuel on the mountains to keep a wildfire smoldering.

We had driven in a three-vehicle caravan up from Ogden, linked by CB radios. Under the fliptop lid of Chris Graves's pickup were coolers brimming with bucket upon bucket of motel ice that was cooling down an ample supply of water, soda, and beer. There was also a heavy-duty automotive hydraulic floor jack: our mission that day was to trace the original 1869 grade of the Central Pacific Railroad for 100 miles across the northern Utah desert from Promontory Summit, and every map and guidebook had cautioned that there were "no services" for the entire length. "Driving on old railroad beds is fun," said Chris. "But there are a lot of old spikes out there still, working their way out of the soil with every winter cycle and waiting for a nice, soft rubber tire to come along." "That would puncture your aspirations," said Chuck Sweet. We obligingly groaned. "And fixing a tire out in the desert sun really flattens you out." Chuck had unrelentingly, exhaustively filled the previous day with puns, low comedy, and trivia between savvy nuggets of history and touching comments about human nature in vanished eras. In such succession the puns seemed at times to weary Bob and Chris, who adopted a long-suffering attitude, but they awakened something devilish in me and I felt myself responding in kind.

"We'd better tread lightly around this guy," I cautioned my kids.

More groans. "Now we've got two of them," Mimi exclaimed. "This is going to be a long day." She paused and grinned slyly. "Tiresome!"

With Chuck riding in Bob's pickup, and me in Chris's, and Mary driving with the children, we set off from the battered "Last Spike" concrete obelisk, which stood a few yards from the visitors center and which for many decades had been the only object marking Promontory Summit. We drove parallel to the re-created track for a half mile or so until it ended, and then were driving on the grade. Chuck's voice came over the radio, indicating an old Central Pacific ramp to a "borrow pit," where fill was removed to be used on the line, and to the first of many small trestle sites. The mountains opened up to the north. Then, to the southwest, we saw the purplish vista of Great Salt Lake. A large sign appeared on the south of the roadbed and we pulled over to examine it—

a famous sight memorialized in historic railroad photographs: "10 Miles of Track, Laid in One Day. April 28th 1869," erected proudly by the Central Pacific Railroad to preen at winning a famous corporate wager.

General Jack Casement, commanding the brigades of Union Pacific track-layers, was a thoroughly modern brigadier general, not only in terms of keeping his army well organized, productive, and well fed (and entertained), but also in telegraphing press dispatches at each new accomplishment. Correspondents were always welcome in his tent, and his tracklayers avidly put on displays for the visitors, as much for the rewards of bonus money and beer as for the satisfaction of setting a new record. His counterparts in the Central Pacific, particularly the burly former Sacramento shopkeeper Charles Crocker, now head of construction, winced at successive dispatches of new U.P. track-laying records. The issue had long since become a personal one, rooted in Western disdain for the effete East and most definitely in class issues. Charley Crocker was born in industrial Troy, New York, the son of a bankrupt liquor wholesaler who began anew at farming in Indiana, and he was used to hard physical labor on the farm and in a blacksmith's shop; Charley had crossed the West along the Platte Valley and California Trail during the 1849 Gold Rush, and prospered selling supplies to miners in the Sierra. Working up to the status of dry goods merchant in Sacramento, and serving as one of the founders of the California Republican Party, Crocker's by-your-bootstraps philosophy had little patience for the New York moguls in their Pullman palace cars such as Dr. Durant and Colonel Silas Seymour, nor for the retired cavalry officers in the U.P. who clung to their old military ranks. "We have no men on our line with high sounding titles—Generals colonels &c. &c. as we hear of on U.P.," he wrote to an associate from the Nevada trackside in 1868, "but I think we can build R.R. as well & as fast as they."

The Union Pacific people had crowed over putting down four miles in a day, and Crocker's men had gone better with six; then Casement had spiked eight. This was the last straw. Crocker complained that it was an abnormally long day, extended into deep-night darkness by scores of lanterns at track-side. "Now we must take off our coats," he told his construction supervisor, James Harvey Strobridge, a tough labor boss of Irish heritage, "but we must not beat them until we get so close together that there is not enough room for them to turn around and outdo me." In the last week of April 1869 the Central Pacific was well into Utah territory and crawling up the west slope of the Promontory Mountains, while the Union Pacific was edging up the east slope toward the summit. Crocker told Strobrige they would lay ten miles in a day. "How are you going to do it?" the supervisor wondered. "The men will all be in each other's way." Crocker laughed. "You don't suppose we are going to put two or three thousand men on that track and let them do just as they please? I have been thinking over this for two weeks and I have it all planned."

At dawn on April 28, Crocker's army was impatient to begin. The Union Pacific's Grenville Dodge, Jack Casement, and other brass were invited to watch, grumbling a little in the early light that Crocker had cheated a little by embedding its railroad ties beforehand. But what ensued over the daylight hours was a miracle for the era. One by one, platform cars dumped their iron, two miles of material in each trainload, and teams of Irishmen fairly ran the 500-pound rails and hardware forward; straighteners led the Chinese gangs, shoving the rails into place and keeping them to gauge while spikers walked down the ties, each man driving one particular spike and not stopping for another, moving on to the next rail; levelers and fillers followed, raising ties where needed, shoveling dirt beneath, tamping, and moving on—"no man stops," Charley Crocker directed, "nor allows another man to pass him"— and no man, Irish or Chinese, did more than one task, each a cog in that large, dusty, sweating machine advancing up the incline toward the summit. At midday Crocker sent the last train forward, "two train-loads with two engines upon it," he commented proudly, "and it was the heaviest train that ever went over the road," and "it went over the very six miles we had laid in the morning, and went safely, and if the track had not been good, it could not have gotten over." He offered to relieve the tracklayers with a reserve team— but they were adamant about finishing out the day themselves. All would get four days' pay for the feat.

A detachment of soldiers was nearby, and came out to watch the work. "Mr. Crocker," the commander exclaimed later that day, "I never saw such organization as that; it was just like an army marching over the ground and leaving a track built behind them." He had walked his horse alongside the men and measured their progress as "just about as fast as a horse could walk . . . a good day's march for an army." The afternoon's pace even included the necessary bending of rails for curves, as always done by brute strength with sledges. Other Central Pacific workers kept the telegraph wire unreeling during the day, stringing it up on freshly planted poles, and when they were all finished the news was flashed to Sacramento. "We got our forces together and laid ten miles 185 feet in one day," Crocker would chortle later, "and that did not leave them room enough to beat us on: they could not have done it anyhow."

"That $10,000 check that Dr. Durant had to write Charley Crocker to pay off the wager," said Chris Graves, squinting in the sunlight at the weather-faded sign, "must have been one of the hardest things he ever did." The place of victory had later acquired a freight siding and another signpost with its name, Rozel, all of which was now only memory. Chuck walked us a few yards to show us remnants of an old telegraph pole—they can be found the length of the Central Pacific byway in Utah, and in that dry country many were actually the ones planted by the railroaders of 1869. Not far away in the dust was a manhole cover placed over an AT&T fiber optics cable trench.

"They pretty much follow the Pacific Railroad, from Nebraska all the way to the California Sierra and over the top," Chris commented. "And why not? A perfect right-of-way across the West! I'd love to know what those telephone ditch diggers turned up, going within a few feet of the original tracks, where thousands of workers had walked and worked, and where the old woodburners steamed by."

Rolling down the dusty grade again, noting that the two parallel grades hacked, scraped, and blasted by the rival companies were merely yards apart, we stopped below a place where a short, dry-stone chimney stood leaning in a scooped-out burrow, all that was left of a habitation site. It was one of some 500 habitation sites found by archaeologists in the Promontory Mountains and approaches, ranging from single huts to encampments of entire crews, some being Mormon contractors, others Chinese, others Irish. A few days before the 1869 ceremony, a member of Charley Crocker's entourage from California looked out of the elegant directors car as it was pulled by Jupiter through Utah hills, desert, and salt flats toward the Promontories. On sidings he saw "the novel sight of a town on wheels, houses built on cars to be moved up as the work progressed." Then there were acres of tents and the smoky cook fires of the Central Pacific's Chinese workers. Days later, the adventurous New York newspaperman Albert Richardson rolled through on a westbound long-distance train, one of the first to make the trip, and saw on the same stretch of track—where now we were driving—"hosts of Chinamen, shortening curves and ballasting the track. Nearly four thousand are still employed in perfecting the road. They are all young, and their faces look singularly quick and intelligent. A few wear basket hats; but all have substituted boots for their wooden shoes and adopted pantaloons and blouses. . . . They are tractable, patient, and thorough; they do not get drunk, nor stir up fights and riots."

I thought about those acres of canvas tents staked out along the railroad grade, and cook fires redolent with peanut oil, garlic, cuttlefish, and pork. Now, replacing so much human activity, the vast terrain supported little but creosote bushes, though sunflowers grew in profusion along the grade as we passed. I saw bleached ribs out in the brush. Ten antelope appeared and scattered at our approach. We rounded the base of the North Promontory Mountains to see the source of the column of smoke visible at the Golden Spike Site as well as fifty miles away in Ogden—amid a blackened char thousands of acres across, an angry red glow smoldered still.

With billowing road dust choking us as badly as if it had been smoke, we allowed our three cars to become separated enough to let the dust settle before the next vehicle rolled through. At this point, Bob's truck was leading with Chuck riding shotgun; our Dodge was second, carrying Mary, Mimi, and me; Chris's pickup was last, carrying Davey as a passenger, way back on the other side of our dust cloud. I was looking out across the enormous emptiness when suddenly our CB radio crackled to life. "David," Chris said in a commanding voice, "you'd better get back here—pronto!" I immediately went into anxious parent mode, assuming some kind of accident, stomped on the brake, did a three-point-turn on the desert road, and roared off back toward Chris's truck, a mile or so back and now, I could see anxiously, stopped in the middle of the road next to a strange pickup. Bob's truck was not far behind us.

Reaching them, I leaped out of the car, seeing that my son looked all right and that the strangers with Chris were smiling. "David," Chris called. "Get two copies of your book out of the back! These fellows would like to buy signed copies!" Relieved almost to a limpid state, I shook hands through a cloud of adrenaline haze and heard Chris's explanation. He had come to a crossroads, encountered this truck, and figured that the only people crazy enough to be following the byway through this blasted terrain would be railroad buffs. He stopped to chat about what they'd seen out here, asked them if they were familiar with the new book on the Pacific railroad, *Empire Express*, and when the men answered in the affirmative, Chris told them "I have the author right here just ahead in my group, and he'd be glad to sell you first editions, no extra charge for the inscription!" I could hardly believe the situation, selling books to strangers out in this wasteland, but the two grinning railroad buffs had pulled out checkbooks, and the deal was soon done, and two satisfied customers headed down the old grade in the direction of Promontory Summit.

"Chris, I can't believe you," I exclaimed, deciding not to say anything about my parental alarm. "Who else but you could sell books to perfect strangers—in the middle of the desert!" And he did it again and again, in Utah, Nevada, and California over the next several weeks, affording me some welcome gas money and some wonderful, unexpected interchanges with readers; probably the most memorable sales site (after the west slope of Utah's Promontory Mountains) was outside the western portal of the Central Pacific's Summit Tunnel, at the height of the California Sierra. I sold *three* copies there, to people who stowed them in their backpacks and disappeared around a bend of the summit pathway, while Chris grinned and grinned.

We drew closer to the lake, seeing white salt flats and marveling at the eerie purple water, which Bob explained over the radio was due to a "ferocious algae bloom." Large rocks stood offshore like beached whales, instantly recognizable as Monument Point from a famous 1869 stereograph taken by California artist Alfred A. Hart, whose camera had captured scenes during the building of the Pacific railroad. Hart's eye was, if anything, even more artistic than that of the Union Pacific's Andrew Jackson Russell, and his Monument Point view was beautiful as well as historically poignant. On May 8, 1869, the Jupiter with its load of West Coast dignitaries had paused for a tourist stop at Monument Point. Hart hurriedly climbed the hillside with his heavy and ponderous camera, having sighted a marvelous opportunity—a wagon train was slowly approaching from the East. As the lead team plodded past Jupiter, Hart exposed a negative: five white-topped covered wagons; the bright, precise Jupiter and its palace cars, with white smoke rising on a breezeless day; Monument Point's rocks lapped by a warm salty sea; the big sky. As we walked along the water's edge and put our hands out onto the warm sides of those monumental rocks, Mary stared with her painter's eye at the scene, resolving to return here one day to paint it.

"If the brine flies will let you," Chuck commented, swatting the air around his ears. Looking at Mary, though, he could see that not even swarms of insects would deter her, so compelling was the composition of desert, beach, rocks, water, and sky.

At Rozel and the ten-mile victory marker, at the crossroads encounter with railroad buffs, and at Monument Point, Bob Chugg had stayed comfortable behind the wheel while the rest of us walked around; with his combat disability, walking was enormously difficult. When we pulled up at the vanished ghost town of Kelton, though, Bob struggled out of his truck. In the 1870s Kelton had its highest population of 200, while it served as a section station and connection to rangeland and mineral holdings in northwestern Utah. Now the only things left standing were tombstones surrounded by rickety fences. Bob leaned on his open door reflecting on the bleary scene, and it gave me time to study him, handsome and powerful, broken, and overcoming it by sheer force of will. He was born in March 1948 in Ogden, a fifth-generation descendant of Latter-day Saints pioneers. The family moved to Washington State and he attended the University of Puget Sound for a year and a half before dropping out in favor of the marines. In Vietnam Bob was a radioman and patrol leader to a company that had just suffered 75 percent casualties. On December 7, 1968, the 27th anniversary of Pearl Harbor, he suffered a bullet wound across the spine that left him totally paralyzed and unable to speak above a whisper.

Eventually, miraculously, over great time, he was able to walk with just a cane. Spending more than nine months in hospitals before despairing of the care, he left on a weekend pass and didn't return; he went home. I have never seen him bitter or heard him voice any kind of opinion that he, and thousands like him, were betrayed. But he has internalized his pain and his many disappointments into a kind of mute nobility. After more college work and two failed marriages (he obtained custody of his two young daughters a few years later and raised them), for a while he operated a bicycle shop, volunteered "a lot with the schools, my church, and the community," and was tremendously attracted to the National Park Service, working at the Great Basin National Park before moving to the Golden Spike National Historical Site. "I've loved history as long as I can remember," he once told me. "It's probably in my genes—one of my great-great-grandfathers founded the Utah Historical Society and the Utah Genealogical Society, and some of my family lines can be traced back to the late 900s."

For Bob, his work at the Golden Spike Site, and his readings and explorations into regional history, have helped give his life meaning. I have watched him in action behind the docent counter at Promontory as he interacted with historical heritage tourists; he uses his power and knowledge well. At windswept, empty Kelton, next to a sad little cemetery, he pulled his heavy cane out from behind his pickup seat and led the way into the enclosure, a sprig of flowers in his free hand. There were ten or eleven little family plots, most of them fenced with weatherbeaten wood. At one there was a Mormon pioneers concrete marker for the Forbes family, and there he laid his tribute. "Look at these dates," said Bob quietly. "See the human drama? The Forbes family: Henry Clay, May 21, 1835, to January 1, 1878. Ester Seddon, his wife, February 2, 1844, to March 2, 1878. And Baby Henry, March 1, 1878, to March 2, 1878." He leaned forward into a long silence. "Henry, dead at forty-two on New Year's Day. Ester, thirty-four, gives birth to his child on March first, it's a difficult delivery, and both mother and son are dead a day later. And friends or family raise these stones." He shook his head. We stood there for a long time. Later, as Bob made his way past another family plot, he broke his silence. "Whenever I go past anything old and deserted, like a town or a farm," he mused, "I think about these quiet little tragedies out in this great nowhere. The ones that didn't die of disease or accident or violence, I always wonder: what made them give up after all their hard work? Was it a loss of jobs, like when the railroad folded up here or the mines played out, or death, or illness, or just what? I always feel sad and sentimental—and nostalgic, in a strange way—about these kinds of places." He shrugged with his own kind of eloquence, stowed his cane behind his seat, and hoisted himself into the truck. We pressed on.

Low hills stood across the empty barren flats. At a crawling pace we followed ruts representing the old railroad right-of-way through pebbly dirt and abundant weeds; dry creekbeds would appear across our path, the ruts would lead us down off the embankment and across the dry creek and thus we would avoid an ancient culvert or small trestle, the wood still holding its 100-year-old form but hardly able to bear any weight. At intervals Chuck's voice would come over the CB radio, pointing out town sites, foundations, refuse piles, vanished sidings, and bumper spurs, and a guidebook gave us the names of these vanished places: Zias, Peplin, Ombey, Romola, Matlin (where railroad ties were still embedded in a "wye" siding), Red Dome, Terrace. At the last we pulled up near the cemetery, and again Bob got out to pay respects; Chuck came over to our car. He had first followed the right-of-way across northwest Utah twenty years before and had extensively studied its history. "Terrace was once a big deal," he told me. "It was the maintenance and repair center for the whole Salt Lake division of the Central Pacific Railroad that stretched between Ogden and Wells, Nevada. Look around. Once there was a train switch-yard, a sixteen-stall roundhouse, machine shops, coal bins, water tanks—they gave up trying to find well water and pumped all their water through a twelve-mile wooden aqueduct. There were hotels, restaurants, saloons (of course), stores, even a big communal bathhouse that had a lending library! There were more than a thousand people living here, maybe a couple thousand at the peak if you count the large Chinese community, which of course in those days didn't count. The Chinese ran laundries and mended clothes and sold vegetables, and in 1880 the only woman here was a twenty-eight-year-old Chinese prostitute."

Now Terrace was an untended cemetery in the weeds, under the noonday sun. "There are acres of broken glass out here," Chuck said, and indeed there were—purple, green, and milky glass, reflecting the sunlight, and rusty barrel hoops and oyster tins, and pottery shards, lying all across the surface of the earth. Every step dislodged them from resting places. Archaeologists have found medicine bottles, opium tins, rice bowls, telegraph insulators, Chinese coins, and bottles originally containing whiskey, ginger beer, Cathedral Peppersauce, and Budweiser beer (the last drunk sometime between 1878 and 1883). Except for these officially sanctioned finds and retrievals, removing items from where they lie on the ground or beneath the surface is illegal, as Chuck and Chris cautioned us, a violation of a federal law, the 1906 Antiquities Act. Anyone caught with metal detectors along historic trails or sites can be arrested, though of course in this empty quarter travelers were on the honor system. As if speaking to that issue of abundant artifacts at Terrace, we found a sign mounted on posts, which cautioned visitors to "please protect the historical resources so that others may enjoy them. Do not dig, or collect artifacts. Take only pictures, or soon there will be nothing left." Inside the cemetery fence on the ground lay another metal sign, riddled with bullet holes. "Anyone found molesting these graves," it warned, "will be prosecuted to the fullest extent of the law. Warren Hyde, Sheriff, Box Elder County." From the looks of the sign, Sheriff Hyde's vested authority was long gone. I paused before a narrow marble tombstone to copy down the inscription of one Henry Gray, "born State, NY, Died July 3, 1899, aged 76 years. Kindred to begin, O Mystery, why, Death is but life, Weep not nor sigh."

Mary came over then, took my hand, and guided me back to our car and showed me one good reason to sigh: a rear tire of our Durango had found an artifact of its own, a sharp fragment of an old railroad spike, and it was flat. I did sigh and considered weeping, as it was 101 degrees, with the nearest shade miles and miles away. But Chris Graves came to the rescue with a can of beer dripping with ice water from his cooler, and with his hydraulic mechanic's jack, which he effortlessly hauled out of his truck. Soon I had stowed the flat with the spare tire in its place, and we were rumbling westward. In a short while we were edging around the Muddy Range of the Grouse Creek Mountains, where, at the portal of the deepest rock excavation on this 100-mile stretch west of Promontory Summit, a narrow, gracefully curving man-made canyon was blasted by the Central Pacific's Chinese in 1869. The temperature climbed to 107, and Chris showed my excited son fragments of railroad spikes on the ground and dozens of ancient railroad ties tossed down a hillside embankment. The vanished station towns or freight sidings with their curious, appealing names appeared every few miles—Watercress, Walden, Bovine, Medea, Umbria, and finally, Lucin, the hamlet on the edge of the Great Salt Lake Desert, with the Pilot Range rising less than ten miles westward, signifying the border between Utah and Nevada. Lucin was the place where in 1903 the Southern Pacific Railroad anchored its great 100-mile cutoff across the salt flats and the lake to Ogden, efficiently eliminating the original route through the Promontories and across the north end of the lake, and, perhaps mercifully, choking off much of what little life was left there.

At Lucin the Central Pacific byway ends, and within a few miles one can rejoin a hard-surfaced road, Route 30, and find one's way back to civilization. We headed across the state line along Thousand Springs Creek in a pass of the Pilot Range, edging between Rhynolite Butte and China Jim Mountain toward Wells, Nevada, for motel rooms and their welcome air-conditioning, showers, and dinner. Eerily, as soon as we crossed the state line the brushy country turned entirely green. We were still following the original Central Pacific engineers' route, and by now the temperature had dropped to 90. Southeast, standing 10,716 feet, was Pilot Peak, the live-saving landmark beckoning wagon pioneers like the Donner-Reed Party off their harrowing Utah desert transit toward Nevada springwater and life. At Oasis, suddenly Interstate 80 was again roaring beside us. We climbed Pequop Pass through a jungle of nettles and weeds, dipped across an intervening valley, surmounted Moor Summit, and glided downhill toward the raffish neon truckstop and slot machine haven of Wells. On the slow, bumpy, and dusty journey from Ogden that day we had gone back a century and more, and were glad to glide back to the present, for the night, so effortlessly. Tomorrow Chuck and Bob were going to take us up into the high and dry mountain desert country of eastern Nevada, tracing more original mileage of the Pacific railroad, but for the time being we would rehydrate, relax, and gird ourselves for another hot day.

In early 1869, the frantic sense of urgency felt by the railroad executives of the Central Pacific in Sacramento and Washington, D.C., almost tingled the hands of their fellow founding partner, Charles Crocker, as he read their hurry-up telegrams in northeastern Nevada and urged his construction gangs to quicken the pace. He desperately hoped to increase it to twenty miles per week (he could not know, of course, that in four months they would be capable of the ten-miles-in-a-day stunt). But there in Nevada the men were hampered by frigid weather, a lack of dry wood for locomotive power, and disease. There was a smallpox epidemic among the tracklayers in early January, as the work edged up the Humboldt Valley past the Ruby and the East Humboldt mountains. There were no doctors available; as was common in isolated settlements everywhere, when workers sickened they were summarily quarantined in "pest cars" until they were carried out for burial or until they sweated through the disease and stumbled out into the wan winter daylight, weak and disfigured by sores and scars. "Nearly all died who went into the pest cars," bloodlessly commented the Central Pacific treasurer Mark Hopkins from his comfortable office in Sacramento, "and those who did not die increased the panic among the men more than those who died and 'told no tales.'" Charles Crocker was a more humane man though always a tough taskmaster, and he was grateful to the workers willing to stay on for their $35 per month. The track reached the snow-whitened meadows at Humboldt Wells—nowadays shortened to simply Wells, on February 8, 1869, and the men planted the C.P. milepost 532, the distance from Sacramento. But almost immediately work was paralyzed when, all the way back on the high western slope of the Sierra Nevada in California, ten feet of snow fell in two days and blocked every bit of traffic for a week. "For God's sake, push the work on," came the plaintive wire from Collis Huntington, their partner far away in the East where he tried to manage loan officers and politicians for this final push. "If I was there I would not take off my clothes or change my shirt until the rails were laid to Ogden City."

Early the next morning after our dusty arrival in Wells, we assembled in a diner for breakfast. Chuck Sweet was puckishly wearing a red T-shirt with the legend "Roadkill Café," to the delight of my children, who read the shirt's menu offerings out loud at the table until shushed. "We'll see enough of

those today," I said. This launched Chris Graves into a merry story about finding the fresh remains of a kit fox near a telephone booth outside Reno,
Nevada, and having it tanned for display at home (later he thoughtfully sent
the pelt to Davey for his birthday, and it is on his wall to this day). Then the
eggs and flapjacks arrived and we attacked them. Chuck and Bob Chugg had
been there in Wells over the Fourth of July weekend, refreshing their memories about the stretch of grade we were going to trace today. This began a few
miles east of Wells at Moor Summit and led westward across high and dry Independence Valley and Toano Valley, threading between the two parallel
mountain reefs of the Pequops and the Toanos.

We would begin the day with less than ten miles on I-80, which now effortlessly followed the horrific wagon trail of the Hastings Cutoff from Salt
Lake (and its shining city) across the Great Salt Lake Desert, past enormous
air force ranges and the famed Bonneville Salt Flats of racing car fame, and
into washboard Nevada. There was an interstate exit at Moor Summit that
gave off on an unmarked and undistinguished dirt road, which in turn led to
the original railroad grade. "Wagon trails and railroad trails just a hundred
yards from the interstate," Chuck said, grinning, his glasses flashing in the
fluorescent light, "and millions speed by, not knowing how many suffered
right there in sight of the breakdown lanes." Outside the diner, the Ruby
Mountains rose to the southwest. A man walking on the weedy side of the
county road tugged against his leashed black and white dog, at the same time
trying to sweep the ground with a metal detector in his free hand.

Soon we were at Moor Summit (elevation six thousand feet), bumping off
pavement and through weeds and nettles to the vanished town site of Moor,
a woodcutting camp for the railroad. Nothing was left but a nineteenth-
century dump cascading off into stunted cedars. Not far away ruts appeared
in the brush and led assertively off eastward, the grade scraped down by Chinese and Irish laborers in 1868, laid with rails in February 1869, and used
until a new and improved grade was created seventy-five yards away to the
north by the C.P.'s successor, the Southern Pacific, in 1904. Now of course,
modern-era corporate mergers had consolidated the Union Pacific with its
ancient enemy, and the regular freights passing us on the modern steel rails
were painted Union Pacific yellow. We stopped our three vehicles at an old siding. Most of us walked a little into the sagebrush and rabbit grass, and in no
time Davey found a reddish hunk of iron like a big, thick ruler with holes
drilled through. "A tie plate," pronounced Chuck with his expert ironmon-
ger's eyes, "which spliced the rails together. Definitely pre-1904, and probably
used here on the first line in 1869." He hefted it for a minute, turning it over
and over, before sending it back toward its place on the ground. "See how intact these are," he said. "Very neutral soil, no salt or alkali. When you hear
the *click-click* on a train, you're rolling over these splicings of separate rails."

We went a little farther in the vehicles, some of us playing musical chairs from truck to truck in different stretches for variation. The foliage opened up and the Pequop Mountains stood ahead defining the horizon. The Central Pacific tracklayers had looked up from their work at this bleak view in late February and early March 1869, nothing but sagebrush and rabbit grass and mountains lightly furred with piñon and juniper. There had been little snow here, but the frigid weather had frozen all creeks and streams that in warmer weather trickled weakly down from the mountains. Ahead, for more than forty miles on the trackless grade the creeks were so alkaline as to be useless; early on in Nevada Territory they had learned that alkaline water was not only undrinkable but it foamed up the boilers of steam engines, broke them down, and would have in short order corroded the machinery were the locomotives not immediately flushed out with fresh river water carried across Nevada from the California Sierra.

We stopped again at one of those dried creekbeds. "This is a great trestle site," Chuck exclaimed, pointing to mummified wood fragments of the old Central Pacific and to another obvious railroad grade built parallel to the C.P. "Grenville Dodge's subcontractors of the Union Pacific were dispatched here to Nevada in 1868 to begin grading eastward toward Utah, as if they could claim every inch of territory between here and the Wyoming line, which violated the U.S. Railroad Act. I guess they gave Nevada up about right here and went back to Utah, and didn't bother addressing this creek defile." A few minutes later, I had walked into a stand of prickly pear and thought I heard a train bell—but it was Davey, Chris, and Chuck, finding more tie plates and tossing them back onto the ground, where they rang. I turned back toward the others, my foot brushing the intact and bleached skull of a coyote, its teeth with long canines still solidly depending from their sockets.

Farther east, we drove out onto a high plateau against the northernmost slope of the Pequops, and stopped again, with the peak of Rocky Point Mountain, 8,275 feet, jutting nearby. "Now we're at Holborn, a freight siding," said Chuck, hopping out onto the sand. "Great nineteenth-century dump here!" A few yards away an acre of debris poked out among the weeds—barrel hoops, broken patent medicine and whiskey bottles, unfathomable fragments. Looking down, as eight-year-old boys invariably do, peering for collectibles, Davey spied a gridlike pattern in the light sandy soil and kicked it, calling excitedly to Chuck and Chris. "Excellent!" exclaimed Chuck. "Do you know what you found?" He picked up a wafflelike chunk of iron. "This," he pronounced solemnly to Davey, whose eyes were like saucers with archaeological excitement, "*this* is a fragment of the firebox door of an 1860s or 1870s steam locomotive! There were a number of accidents up in this region, and this may have been left after a derailment or crash or boiler explosion." My son was beside himself with pleasure.

Chuck with his perennial good cheer and enthusiasm could wring drama out of the humblest piece of junk out in the weeds, and I was as delighted as my boy. A few minutes later, he was calling us over to see a pile of old rusty tin cans. "Look here," said Chuck, picking one up. "This is an oyster tin—you find them all over the leavings of nineteenth-century America. People ate oysters until they were coming out of their ears. Look at the top of the can, at the smaller lumpy circle on the lid, and see how they sealed it closed." He paused for effect. "They used, by the way, lead solder to do that, a wonderful heavy-metal addition to your daily diet."

As we stood there in the weeds, squeezing history, stories, and life out of rubble and debris, it occurred to me that Chuck was simultaneously an enormous distance from his former profession and yet right next door. As a cytotechnologist at an Ogden hospital, he had peered countless times into his microscope at slide samples prepared from human tissue, examining for cancer cells or infections and thus looking at a patient's history and future, the basic ingredients, for better or worse, of individual human narrative. Chuck was a Colorado native (born 1949, two months after I was) raised in Arizona and Southern California, educated in the California state college system, and trained in cytotechnology at a hospital in Grand Junction, Colorado. Following this he had a challenging and rewarding twenty-year career at St. Benedict's Hospital in Ogden, but his own failing health forced him out of the profession. Chuck had been born with a mild-to-moderate form of cerebral palsy, resulting in left-side weakness, and he had suffered from asthma as a child. When, in 1995, his eyesight began to seriously deteriorate and the stories told in human cells could no longer be augured, he was left without a job; now he worked as a handyman and custodian for the Baptist church to which he belonged, and it was in Western history that he searched for his stories.

Always a voracious reader and collector of artifacts, historical postcards, books, and railroadiana, he had begun volunteering at local historical sites, including an acquisition project at Ogden's Fort Buenaventura and even more time spent docenting at the Golden Spike National Historic Site. Chuck was a fireman on Jupiter and No. 119 at the site, volunteering his weekend time to stoke the engines' boilers—hot but satisfying labor for the summertime tourists—and he was working his way toward an engineer's rank, which showed the poetry of history: he was related to Cyrus Sweet, the fireman on the original Jupiter at Promontory in May 1869.

And now we were standing next to vanished iron rails across a high valley in Nevada, over which the Jupiter had drawn moguls, governors, and other dignitaries on the night of May 7, 1869, on the way to the Utah "wedding of the rails." Chuck tossed the ancient oyster tin back into the brush and ushered us back to the trucks. "Onward!" he exclaimed.

At Independence Wells, with no wells in evidence, we rumbled by a lean-to and a cabin constructed of railroad ties, its ramshackle corral as uninhabited as the cabin. We passed through a grade cut and then the road turned into a path so narrow that sagebrush scratched the vehicle sides. Some minutes later we struck off on a gravel road toward Pequop Summit, with the Toano Mountains now in sight, and bested a deeply rutted, steep, and tooth-rattling defile, sending enormous clouds of dust into the air. "Our back windshield," said Mary, driving, "is becoming its own geological formation." We stopped again—I marveled that Chuck and Bob had committed this almost feature-less desert to heart and knew where they were taking us, even with the aid of a high-resolution terrain map scribbled with detailed notes. On the far side of the Southern Pacific tracks, after slipping carefully through barbed wire fences, we came to another old Central Pacific fill and cut. Everyone but Bob, who had remained in his truck, looked around for a while. Chuck wandered off down the empty grade and disappeared. After a half hour I grew concerned: it was well over 100 degrees in this open terrain, and Chuck's left-side weakness was growing more pronounced as the day wore on; he was limping a little and moving slower. I was just about to call everyone together for a search party when he reappeared, proudly carrying—just so we could get a look at one—a big, rusted black powder can once used by the Central Pacific graders in 1868 as they blasted out rocky ledges in preparation for the tracklayers. His delight was infectious.

Some miles and several stops later we drove out onto a hot, dusty hillside, having reached the long-gone hamlet of Toano, which had splayed across the bottom of the hill with the Central Pacific skirting the top. No buildings remained but the burial ground awaited us, and all, including Bob, got out of the vehicles to pay our respects beside wrought iron fences surrounding two children's graves: they were of marble and featured carved figurines of lambs. *"In memory of Mary, daughter of David and Olive Morgan,"* read Bob, squinting in the bright afternoon sun. *"Born December 9, 1878, died February 14, 1880.* Fourteen months old—maybe walking and pulling things down off the table when she sickened and died. *In memory of John David, son of John and Margaret Lewis,"* he went on, turning to the baby's companion out here in the sagebrush, *"born June 15, 1891, died December 21, 1901.* Ten years old, and died just days before Christmas." He shook his head. "Who were these parents? What were they hoping for out here, and where did they go in their sorrow? We'll never know."

We moved on, stopping at the town site of Cobre, altitude 6,921 feet. It had been a Southern Pacific railroad junction whose heyday was during the first decade of the twentieth century when it was also northern terminal of the Nevada Northern Railroad, bringing silver ingots up from mines in the Goshute Valley. We stood looking down across the flats past a desolate ranch toward the braided paths of Interstate 80 and the old California wagon trail, six or eight miles to the south. Where we stood the original Central Pacific grade turned northwest through the Toano Range, heading toward the Utah line and the white-hot emptiness beyond, and as we had traced that yesterday, it was about time to call a halt to our desert expedition. We drove the few miles down to Oasis, a negligible town next to the interstate that would carry Bob and Chuck back east toward Salt Lake City and home to Ogden. In a light-dimmed, deserted restaurant I talked a counterman into reluctantly reopening and extending lunch into mid-afternoon for the benefit of two dusty and hungry children and five wilted adults.

There wasn't much energy left for talk as we ate, but we were filled with images and experiences and wordlessly happy now for quiet companionship. I snapped a photo of our three intrepid and congenial guides who had given so much time and energy on our behalf as we traced the old iron road far from civilization. After bear hugs and fond thanks, Bob and Chuck struggled into Bob's begrimed pickup and pointed it toward home, and I waved to two of the most generous and goodhearted men I've had the pleasure of knowing. My kids and Mary clambered into the Dodge, and Chris Graves got into his truck, and we turned west toward Wells, old Humboldt Wells, for some motel pool time, dinner, and HBO for Mimi and Davey.

The next day we would begin a new part of the journey, with Chris leading the way as we followed the Donners and the other wagon pioneers, as well as Charley Crocker's legions of tracklayers, down the historic Humboldt Valley across the state of Nevada toward rambunctious Reno. We had a long day ahead of us, and crowded days beyond, but we had the infinite comfort of plentiful stories to keep us company.

PART V

13

Following the Humboldt

A t Humboldt Wells in the 1840s wagon-training emigrants discovered dozens of springs, which they likened to wells, across a verdant meadow. During the traveling season it was always crowded with wagons and camps, the livestock turned out to fatten and strengthen in preparation for the long trudge down the Humboldt River, which rose nearby. The distance across Nevada from Utah to the California line was some 400 miles, but as a pathway the Humboldt lengthened it considerably over three major bends between the mountains. From its headwaters to its feeble end at the vast evaporative plain called the Humboldt Sink, it was 350 miles as the buzzard flies, but countless twists and turns stretched it out to about 600. As the WPA guide to Nevada reported, wrathful early travelers declared "it was the crookedest river in the world." Emigrants were often forced to follow these snake curves because the only grass for their oxen grew along the river.

The farther west it flowed the less drinkable was Humboldt water, becoming increasingly alkaline until it was poisonous. Narrow and sluggish, the river may have afforded emigrants a pathway toward their dreams but they mostly found it to be a nightmare; they resented it bitterly. One forty-niner, Reuben Shaw, complained,"The Humboldt is not good for man nor beast. With the exception of a short distance from its source, it has the least perceptible current. There is not a fish nor any other living thing to be found in its waters, and there is not timber enough in three hundred miles of its desolate valley to make a snuff-box, or sufficient vegetation along its bank to shade a rabbit, while its waters contain the alkali to make soap for a nation, and, after winding its sluggish way through a desert within a desert, it sinks, disappears, and leaves inquisitive man to ask how, why, when and where."

West and southwestward from the Wells, the Humboldt curves between the East Humboldt Range of the Ruby Mountains and the Snake Mountains, bending down across bleak flats indented with numerous dry or barely moist creekbeds whose watery blue color on maps give a misleading impression. True to experience, engineers followed the pioneers, who in turn were only tracing game trails and the footprints of aboriginal hunters: the Central Pacific Railroad stretched across Nevada along covered wagon wheel ruts, joined

in the twentieth century by dirt and gravel automobile roads that, paved, became first U.S. Highway 40 and then Interstate 80. Far overhead in the wide
and deep blue sky are the jet trails of transcontinental air traffic. Many is the
time I have flown across the West, 30,000 feet above those wagon ruts and
railways, peering down when it is clear to locate myself by a bend in the
Humboldt, or by a position relative to Great Salt Lake and the Wasatch
Canyons, or by triangulation of Wyoming's Green River, mountain ranges,
and river valleys. The first time I drove the Humboldt Valley I was in the hippie
van in August 1973, and it was a reddened evening and inky night, with parallel mountain ridges silhouetted against the stars. Now, with my family and
our genial, encyclopedia-minded guide, Chris Graves, I was on the far side of
deep study of the Humboldt for *Empire Express*, whether by car, by map,
pioneers' accounts, engineers' diaries, railroaders' letters, and nineteenth-
century travel journalists. But now the Humboldt would release even more
stories, thanks to our guide. For much of the width of the state our two vehicles drove in tandem, dwarfed by screaming tractor trailers, land yachts, and
excursion buses taking Utahans to the slots and card tables of Reno. I sat as a
passenger in Chris's truck, his easy listening tapes rolling on at low level, his
CB radio squawking periodically, and his radar detector randomly burping.
We sent citizens band bulletins of historical nuggets and information about
the terrain over to Mary, Mimi, and Davey in the Dodge, and regrouped at rest
stops and historic pullovers.

"Ten or fifteen miles across those flats," Chris was saying, gesturing off to
the north as we sped westward, "there's a ghost town called Metropolis. Superman did not grow up there. Nothing there now but ruins—an archway, a
pile of red bricks, a couple of cemeteries, and a lot of weeds and broken junk.
I was last through here about ten years ago. Over on the original main railroad line out of Wells, you still had the original telegraph poles. But the
ranchers take 'em down and use 'em for fence posts. And they shoot the glass
insulators for target practice. I guess they sell or melt down the copper wire. I
would love to have one of those telegraph poles. But then there's the old Antiquities Act of 1906. I ran afoul of that bit of work in 1968 while I was chasing the California Trail. My bride and I walked it in Nevada and California
over a period of years, then advertised that for a fee we'd take others, too.
Relics guaranteed. And that was our mistake. The U.S. attorney in Carson
City was not nice. We got a long lecture about leaving historical artifacts
where they lie, so they can rot or get covered over by a state or federal highway project. And there the enterprise died. But not the urge to see the wagon
ruts and the faint imprint of railroad ties out there in the middle of nowhere,
and to hear the echoes of those brave people passing by."

Chris, born in 1941, had been following trails and chasing ghosts his
whole life, starting from his childhood in Southern California, where his

mother was a primary school teacher who took her son on long summertime expeditions to trace Western history. He loved history courses in school and at Pasadena City College, where he obtained an associate's degree.

Afterward he found himself working the night shift at a factory, "no social life, twenty years old, so I went to an employment agency and said, 'I want to wear a tie and a white shirt.'" He was hired by a finance agency in Pasadena and later transferred to Las Vegas. He was six feet eight inches and was assigned to what was deemed "peaceful repossession" of cars when loans went bad. "We slipped into garages at night, hot-wired the cars, and drove them to a secure area. We'd do five or ten cars a night—$350 a month legally stealing cars, sometimes with police backup just down the street, and that's what you did for a living." Being a car repossessor in Las Vegas was not as dangerous as one might think. "The mob boys always bought their cars with cash," he told me, "so you never encountered them. Mostly the people were down on their luck or just getting by, although I have to admit that I repossessed some celebrities' cars, people who were working the big cabaret shows in Vegas as singers or comedians." He would not divulge any famous names, instead telling an anecdote about "stealing a '58 yellow and white Ford Fairlane off a prostitute who'd stopped making her loan payments. I locked it up in a lot. She came that night and stole it back, and utterly disappeared. Only time I lost a car going the wrong way!"

He tired of the life after only a year. "Besides," he said, "I was ready to get married and settle down, and Las Vegas doesn't inspire much confidence in that area." Back in Los Angeles, he was hired by Security First National Bank, eventually rising into managerial echelons.

He met Carol Lee Miller from Long Beach on a blind date on November 22, 1963. Having one's first date the night of the assassination of John F. Kennedy did not cast a pall on the relationship; they were married in January 1964, ultimately producing three daughters. On weekends and vacations, Chris dragged them all off on jaunts to see the territory, re-creating the wanderlust he'd experienced as a child with his mother.

In 1989, Chris left Security Pacific National Bank, as it was in the throes of the financial challenges of the late 1980s, and joined First Interstate Bank as a first vice president. It was a long way from hot-wiring cars for the loan department, and his executive years at both banks still give him satisfaction. But another kind of life beckoned, and after Chris retired in 1995, he was able to make history and heritage his primary occupation. As a member of the Auburn chapter of the Native Sons of the Golden West and the Sacramento Corral of Westerners, he participates in the civic groups' active historical presentations and public service programs. For a long time he was a member of the Oregon-California Trails Association. He spends "entirely too much time" at the California Room of the California State Library, where they have all the newspapers in California from 1846 to the present on microfilm. On occasion he guides people along the old trails and rail rights-of-way in California and Nevada, more for fun and diversion than anything else, although in the years since I first met him, I encouraged Chris to make his expertise useful to producers of documentary and feature films scouting for locations, which he has now done for PBS's *The American Experience* and Bill Moyers, for Hollywood, and the BBC. "All in the hope that they'll tell history's stories right," he said. "But I don't hold my breath." Lately he has used his research skills to unite adoptees with their birth parents, or vice versa, using California's public records available over the Internet. He has collected many success stories of reunification, always approaching the people carefully and never making the final connection between them unless all parties are willing. His file of thank-you letters warms the heart. He has a mania for Volkswagens, which is hilarious, given his size: he must bend himself like a carpenter's ruler to squeeze into his lemon yellow 1970 Bug. His 1966 red and white, twenty-one-window VW microbus is a little better fit, although when standing alongside it he's still taller by a foot. He collects Western artifacts, California art, and Americana, and with the help of a brook that runs across his property, he and Carol garden in a place that is a little Eden.

"Carol's joining us in Reno," he said, as we fairly flew down the interstate across sagebrush flats, between rounded, dull green and burnt-brown mountains of eastern Nevada. "I'm a motel camper—I love to stomp through the brush, but I gotta be clean when bedtime comes, and a warm meal is a must. Carol paid her dues all those years, following me to hell and hereafter, walking the emigrant trails and poking around in ghost towns, mining camps, railroad rights-of-way. Now she never goes to a place more than once. She gets dirty once, then says, 'Been there—done that.'" Carol developed a small wool products business—she sold items at arts fairs—and she tended a small flock and spun wool. I was looking forward to meeting her for the first time—we had sheep in common, since Mary and I had kept a small flock on our farm in Shoreham for a while, but I was also interested in what kind of person had shared a life with such an interesting and preternaturally curious man as my friend beside me at the wheel.

About ten miles west of Humboldt Wells, the historic California Trail from Fort Hall (in present-day Idaho) joined the Humboldt Valley from the northeast. Among the throngs of emigrants and gold seekers was one Alonzo Delano, one of my favorite chroniclers, who reached the Humboldt on July 30, 1849, weakened from several days of fever and chills but trying to move beyond it by walking instead of riding inside his canvas-topped wagon. "It is indeed hard to be sick in a wagon," he commented, "while traveling under a burning sun, with the feelings of those around you so blunted by weariness that they will not take the trouble to administer to your comfort." Delano was from Illinois, and had gone West on doctor's orders, hoping to improve flagging health and establish himself as a merchant. He had a flair for writing and once situated on the Coast wrote under the name "Old Block" for several California newspapers. His *Life on the Plains*, published in 1854, went through two editions, and is one of the best emigrant narratives of the era; I had used his first edition at the New York Public Library in preparing to write *Empire Express* and found myself so absorbed it held up research for several days while I finished the book. For Delano, following the California Trail down the Humboldt left him (like most wagon train pioneers) all choked up. "One of the most disagreeable things in traveling through this country," he complained, "is the smothering clouds of dust. The soil is parched by the sun, and the earth is reduced to an impalpable powder by the long trains of wagons while the sage bushes prevent the making of new tracks. Generally we had a strong wind blowing from the west, and there was no getting rid of the dust. We literally had to eat, drink, and breathe it." Another diarist who remarked on this was Reuben Shaw, in 1850: they breathed in "fine white dust, more like flour. Our men were a perfect fright, being literally covered with it. Our poor animals staggered along through the blinding dust, coughing at every step!" Some wore veils, others handkerchiefs or goggles, all blinded and hacking with coughs.

Speeding into a headwind across the Dennis Flats, we passed by the Halleck interchange, and twenty miles beyond Halleck, we pulled into the town of Elko, a spruce and prosperous place bolstered by a gold-mining boom in the area for the past fifteen years, which had doubled the population to its present 20,000. It bustled even on an early Sunday morning—moderately busy traffic past clean, modern buildings, shopping areas, medium- and higher-end chain motels and restaurants, and car dealerships full of gleaming new cars and customers. There was neon everywhere amid all the newer illuminated signs. Elko, with its altitude of 5,063 feet, had been born nearly 150 years before as a freighting point from the California Trail up into the gold rush camps of the Eureka and Hamilton districts. The Central Pacific Railroad reached Elko, 469 miles from Sacramento, on February 8, 1869. The predominantly Chinese tracklayers had a cold time of it; the valley had the dubious distinction of being one of the coldest spots in the nation. Frigid winter winds blasted down from Idaho, concentrating and intensifying before walloping the 11,000-foot Ruby Range and coming to a frozen standstill over their heads.

While the laborers struggled across those frosty sagebrush flats, the C.P. supervisors worried about supplies of fresh water and fuel and the high cost of obtaining them; they required thirty carloads of wood daily. "With the

exception of a few cords of stunted pine and juniper trees," complained
Lewis M. Clement, the Central Pacific's locating engineer, "all the fuel was
hauled from the Sierra Nevada Mountains. Not a coal bed on the line of the
Central was then known, and the only one yet discovered is a poor quality of
brown lignite. . . . There was not a tree that would make a board on over 500
miles of the route, no satisfactory quality of building stone. The country offered nothing." This was seconded by the Vermont-born James Harvey Strobridge, construction superintendent. "Supplies cost enormously," he recalled
about their Nevada labors, and offered a telling illustration.

> I sent a wagon load of tools from Wadsworth to Promontory, and
> the expenses for the team and trip were fifty-four hundred dollars.
> I found a stack of hay on the river near Mill City [in the Humboldt
> Valley], for which the owner asked sixty dollars a ton. He said I
> must buy it as there was no other hay to be had. The stack was still
> standing in his field when we moved camp and it may be there
> now for all I know. Another settler had a stack of rough stuff, wil-
> lows, wiregrass, tules and weeds, cut in a slough. I asked him what
> he expected to do with it. Not knowing that he had a prospective
> buyer, the man answered, "Oh! I am going to take it up to the rail-
> road camp. If hay is high I will sell it for hay. If wood is high I'll sell
> it for wood.

Despite the rising heat, we stretched our legs on a self-guided walking tour
of historic Court Street, seeing a Knights of Pythias temple in ornate red
brick and terra-cotta; Nevada's first county high school, opened in 1895 and
now a county office building; and several expansive Gilded Age houses. But
our main focus was on Idaho Street: the Museum of the Northeast Nevada
Historical Society, housed next to a green park in a modern building with a
brightly painted stagecoach on display out front. The museum didn't open
until noon, but we had a long day of driving ahead of us and couldn't linger
that long. Fortunately, Chris had contacted the museum's director, Jan Peterson, and charmed her into opening the museum just for us.

Jan was a kindred spirit. Her father had marked the California Trail after
having spent his youth being conveyed to historical spots along Nevada's by-
ways by his father and grandfather. Embodying a long stretch of history and
tradition, Jan presides over a national-level award-winning facility and is active in a number of regional historical associations. She is also on the board of
a new $14 million California Trail Interpretive Museum planned by the Bu-
reau of Land Management and set to be built near the Hunter exit of Interstate 80, where the Hastings Cutoff came into the Humboldt Valley. In 2002

they would hire a director and break ground for the museum, aiming to complete it in time for summer 2004.

Meanwhile, for this trip we had her modern, open-plan facility in Elko to explore. There was the new Wanamaker wildlife wing, with its impressive collection of preserved animals from four continents exhibited in natural habitat scenes; and its exhibits on gold mining, ranching, local Native American culture, and prehistoric life. There was an extensive firearms collection, military artifacts and photographs from Forts Halleck and Ruby, branding irons, Shoshone tribal baskets, mastodon fossil bones, and one of the largest petroglyph replicas (taken from local mountain cavern walls) in existence. There was a fine display on the Chinese of the Tuscarora mining camp, men who had moved from building the first transcontinental railroad to find riches in placer claims abandoned as no longer profitable by whites. One interesting display was the battered saloon bar from Halleck; Jan told me that the museum obtained it by agreeing to pay annual rent "in the form of a bottle of Beefeater Gin, served right here over the bar." I even found myself admiring the cabinetry of the display cases. Jan told me the museum had obtained the professional help of inmates of the local minimum security prison—"many of the third-offense DUI offenders happen to have carpentry and electrical skills," she noted.

But the case that held my closest interest was that devoted to Bing Crosby, certainly a big surprise to find the Hollywood crooner out here in the Nevada desert. It seemed that ever since Crosby appeared in the dude ranch picture *Rhythm on the Range*, which was filmed in the Sierras in 1936 and co-starred Frances Farmer and Martha Raye, he had wanted to try ranch life, even if he would be the kind of cowpoke as in the Johnny Mercer tune he sang for the picture, "I'm an Old Cowhand"—"I'm a cowboy who never saw a cow, never roped a steer 'cause I don't know how, and I sure ain't fixin' to start in now, Yippie-eye-oh-kay-yay." In 1943 he bought a 3,500-acre ranch on the Humboldt River east of Elko, and in 1947 sold it in favor of a 25,000-acre spread, high rangeland at an elevation of 6,400 feet and some sixty miles north of Elko. He (or, more accurately, his ranch managers) ran 3,500 head of cattle there. His three sons spent part of every summer working as hired hands there, receiving cowboys' wages, and Crosby and his wife, Dixie, entertained many Hollywood friends at the ranch, including Clark Gable, Gary Cooper, and Randolph Scott.

Crosby never flaunted his wealth or celebrity, and this was true of his time spent at Elko, although he was always known for his generosity. "He would slip into the back pew of the Catholic church," Jan said, "and just as unobtrusively slip out at the end of mass, but there would be a $50 or $100 bill in the collection plate that morning." Once, in 1948 during a furious blizzard, the

Crosbys arrived at a banquet honoring the crooner to find 350 patrons lined up in the driving snow to buy tickets. He bought out the house and ushered everyone inside, out of the weather. He raised funds for local hospitals and other charities, became an Elko tourism booster, and arranged for Elko to host world premieres for some of his films, bringing out the leading actors and crew for the ceremonies. Crosby was named honorary mayor of Elko in 1948 and held the post for many years. He was also made an honorary tribal member of the Western Shoshone-Paiute tribe, the first white to be inducted; his Indian name was *Sond-Hoo-Vi-A-Gund* ("Man of Many Songs"). In 1951 he was given a tuxedo tailored of blue denim. Reverently hung and lit, it reposes in a display case at the museum. "He must have been a tiny man," said Jan Peterson, "or at least his tuxedo was tiny. I certainly can't fit into it and I'm not that big. But he fit in beautifully in Elko—never wore his toupee, and in those days in Elko, no matter how important you were, you still had a ring-down telephone on a party line." She pointed to a label reproduced along the text caption: "See this label? The original is inside the coat. It says that Bing Crosby must be admitted to any and all hotels. This grew out of a much told story in which Bing was up in British Columbia on a hunting trip, and at the end, when he tried to check into a hotel, the clerk wouldn't admit him—he was dirty and unshaven and probably dressed in jeans and a hunting coat, and, being truly bald, he was unrecognizable to the poor clerk."

We stood there in the museum admiring the lines of the little blue tuxedo. "When his wife, Dixie, died in 1952," Jan said, "he visited Elko less and less. The place must have had a lot of memories—they were here constantly for five years, and their boys experienced so much as ranch hands, being treated as normal guys. Bing remarried in 1957—that was Kathy, of course—but I guess this was Dixie's place, because he never brought Kathy here. Finally he began selling off the acreage in the late 1950s. And moved on."

"I may not know much about a white-faced steer," Crosby once told a reporter, "but I know people. That is my business. And I must say I'm in love with the folks in Nevada."

When initially following the California Trail along the Humboldt Valley, emigrants encountered the Bannock tribe of eastern Nevada, who were largely found to be friendly and supportive. This was not true of the Northern Paiute, who lived a barely subsistent life along the Humboldt across central Nevada, and who made the emigrants' lives miserable. Not as numerous or as aggressive as the Plains Indians, they huddled in bands of five to ten families in the wintertime, but in summer were forced to scatter into tiny groups to forage for small game and pine nuts, a major portion of their diet. It has been said

that each Paiute required some 50 to 100 square miles for survival. Explorers
and the emigrants commented on their extreme poverty, desperation, and
"backwardness." John C. Frémont noted in 1844 that these Indians were
"humanity . . . in its lowest form and most elementary state." The constant
emigrant traffic of the 1840s and 1850s was, of course, devastating to the
ecosystem of the Humboldt, with dire consequences upon the Paiute (com-
monly they were called "Diggers" by the whites for their practice of looking
for edible roots and burrowing insects). They picked off stragglers behind the
wagon trains, stole cattle, foodstuffs, and personal belongings. "Scarcely a
night passed without their making a raid upon some camp," recalled Alonzo
Delano of his passage in 1849.

Elezer Stillman Ingalls poignantly described in 1850 how, along with the
cruelties of terrain, the desperate preyed on the desperate, sometimes reduc-
ing the smaller wagon parties to their most elemental. "A man with his wife
came into camp last night on foot," he recorded, "packing what little property
they had left on a single ox, the sole remaining animal of their team; but I was
informed of a worse case than this by some packers, who said they passed a
man and his wife about 11 miles back who were on foot, toiling through the
hot sand, the man carrying the blankets and other necessaries, and his wife
carrying their only child in her arms, having lost all their team."

In early 1869, the Central Pacific's official photographer Alfred Hart followed the newly constructed track eastward along the Humboldt into Ten-
Mile Canyon—later called Palisade Canyon. There, red basaltic walls towered
more than 800 feet over the narrow, boulder-choked riverway. Wagon parties
of twenty and twenty-five years before had despaired of squeezing through
unless the river was especially low, electing to follow a higher, straighter route
over Emigrant Pass, 6,121 feet in altitude, the divide between the Tuscarora
Mountains to the north and the Cortez Mountains to the south. The railroaders, though, could not afford such an expensive climb from the level of the
Humboldt, and so sent 3,000 Chinese with 400 horses and carts some 300
miles in advance of the track to carve a path through the canyon. All supplies
for men and horses were hauled by wagon teams from the base of the California Sierra. Somehow, the photographer lugged his heavy camera equipment
to the heights of the northern canyon walls. He brought with him a local
Paiute or Bannock man, attired in animal skins and leggings and a headband
of rawhide or cloth around his long, nearly shoulder-length hair, and posed
him at the edge of the precipice, looking down on the narrow, winding river
canyon and, far below, the straight, no-nonsense railroad track that had conquered the canyon at tremendous expense and some loss of human life. It was
a magnificent view and one of the most striking photographs taken along
the entire length of the transcontinental railroad. It is a powerful sight: that

Native American gazing at the twin bands of iron that were sealing the fate of one civilization as they were ushering in another. It became Alfred Hart's most enduring image.

I had always wanted to see Palisade Canyon for that reason, and another. There was also a notable Hart photograph taken around January 1869 at Palisade, showing the first construction train emerging proudly from the north end of the canyon, smoke pouring from its wide funnel stack and blowing in a winter wind away and into the canyon. It, too, captured a particular moment, with a few railroad superintendents or engineers posing at trackside about to be enveloped in a burst of steam, while others wave from atop a tool car.

And so we left the interstate again, followed U.S. 40 and the path of the wagon trains straight up and over Emigrant Pass. South of the old Southern Pacific shop town of Carlin, we followed the rails along the two-lane blacktop of Route 278. Cowboys were rounding up a herd of cows along the roadside. We took a twisting dirt road up and around a mountainside—we could then see the wide green valley of the Humboldt abruptly narrowing into the red canyon—and then we drove across a new concrete bridge to the town site of Palisade, at its peak populated by some 300 souls. A few modern ranch buildings stood nearby, but in the undergrowth there was an old hip-roofed structure constructed of railroad ties, and an even narrower road climbed and switched back past hovels, foundation stones, and ruins standing above and below it, to a hilltop cemetery, where we stopped.

Below us at the mouth of the canyon, on the far side of the blue river, portals of two railroad tunnels gaped. With the familiar terrain, and with the 1860s imagery from Alfred Hart's photographs suggesting that a gleaming Iron Horse might burst out of the darkness at any moment, I reflected on the air of wonderment pioneers experienced as the railroad sliced across the barren miles. "Think of that, ye Sage Brushers," wrote a Nevada journalist in a breathtaking single sentence in 1869, "who used to occasionally span the Humboldt Deserts on the hurricane deck of a mustang, famished for water, eaten up with alkali dust, chilled by Washoe zephyrs or petrified with blasts like simoons, think of getting into a close, comfortable carriage and speeding away, at a pace which would kill a Norfolk in an hour, on and on, in a single night passing all that dreary stretch of country this side of the Humboldt, that country paved with the bones of animals, that died from exhaustion in the great exodus to California, that country that still lingers in the memory of former emigrants as the dark and fearful ground. Verily the Railroad is a benefactor, a beneficent engine of glory and strength."

A few miles west from where we stood, near the siding town of Dunphy, back in the 1930s' WPA days, the streamliners of the Southern Pacific normally reached their highest cruising speed of the whole distance between

Chicago and San Francisco, some 85 to 90 miles an hour. While we poked around in the Palisade cemetery, reading the short stories engraved therein, an eastbound Southern Pacific freight train thundered out of the left-hand tunnel. It curved around into the canyon and disappeared, leaving only echoes and a tingling in the feet.

Four or five toilsome months on the California Trail wore down even the strongest and most resourceful emigrants, with the Nevada portion an exhausting affliction: no wonder that quarreling broke out so often, sometimes with dire consequences. How can one not understand it, with Paiutes turning their oxen into pincushions, with wolves howling nearby every night, with the constant choking dust and the barely drinkable water? Women "went a little mad," as many trail diaries noted, while normally placid husbands lost their tempers and beat their wives, or exploded into violence with other men over minor issues. People made choices in the heat of anger or the confusion of desperation that they would never have considered back in the conventional East. There was a common saying then, that "alkali works into the nerves," which, if it were true anywhere between the Missouri and the Promised Land, it would have been so in Nevada along the Humboldt. But today, looking back, one can see it only as a traumatic stress disorder, suffered on one level or another by thousands, punctuated sometimes by heedless violence, enacted in countless forgotten places.

About ten miles west of Emigrant Pass, down at the river crossing known as Gravelly Ford, two such dramas played out, one relatively obscure, the other famous. In one story, an emigrant was taking his wagon and team across when somehow the wagon box became detached from the undercarriage and began to float away in the current, his wife clutching her babies and screaming for help. Whether flustered or simply imbecilic, the husband kept his attention on saving his animals, and it was left to onlookers to rush into the stream to save the woman and small children from drowning. The rescuers then seized the husband and would have either hanged him or drowned him in the Humboldt, were it not for the desperate pleadings of his wife.

The other altercation near Gravelly Ford took place in 1846, among the Donner-Reed Party. Bad choices and rotten luck had dogged them since Fort Bridger in Wyoming, when they had taken the unproven Hastings Cutoff across the Wasatch Range and the punishing southern route around Great Salt Lake and the white Utah desert. Pilot Peak's springs saved them, they achieved Humboldt Wells, but then they were lured south down the appealing-looking Clover Valley—which impelled them to circumnavigate the Ruby Mountains, a needless, time-consuming 200-mile detour.

They reached the Humboldt on September 26 with ragged tempers and a growing forboding. For the next week as they followed the river west, they were harassed by Indians and lost several valuable livestock. James Frazier Reed, the organizer of the emigrant party that history has identified with his associates, the Donner brothers, recorded these losses in his diary. Reading the entries, and knowing the fate of the Donner-Reed Party, one cannot help but like the forty-five-year-old Illinois merchant, who had emigrated as a young child from County Armagh, Ireland, had engaged in mining at Galena, Illinois (Ulysses S. Grant's hometown), and served as a private alongside Abraham Lincoln in the Black Hawk War; Reed had dabbled in railroad build-ing during the Midwest's interior improvements romance, which soured when the state of Illinois defaulted on its payments, but he had succeeded again at the mercantile business and then at farming. When he took his family West, there were some hard feelings among several in the wagon train—envy, really—at his relative wealth at being able to hire laborers and transport his family in an impressively large and comfortable wagon (which, as it turned out, he was forced to abandon on the Utah desert).

On October 5, near Gravelly Ford, they reached the bottom of a steep and sandy hill and were obliged to double-team to get their wagons over. However, one man named John Snyder, a hired man, was in a foul mood and refused to link his team with any other, whipping his oxen cruelly and filling the air with profanity as the poor beasts struggled, tripped, and strained to make headway in the soft soil. James Frazier Reed stepped in to halt the beating, of-fering his own team in aid, but Snyder abusively yelled at him to mind his own business. Words escalated, and suddenly Snyder leaped upon Reed and began raining heavy blows on Reed's head with the stock handle of his bull-whip, cutting deep gashes and blinding Reed with his own blood. Reed's wife, Margaret, ran over to get between and stop the fighters, but then Snyder be-gan hitting her on the head and shoulders. In an instant, Reed had un-sheathed his hunting knife and stabbed Snyder's breast, breaking two ribs and puncturing a lung.

Reed, appalled at what he had done, threw away his knife and shrugged off the ministrations of his wife and daughters to stanch his bleeding wounds, ascending the hill to where the dying Snyder had been carried by others. For the entire journey from Independence, Reed and Snyder had been on warm terms, aiding each other when cattle were lost or burdens got too great. Now, apologizing profusely, Reed tried to make Snyder comfortable, and inevitably, in about fifteen minutes, Snyder died.

The teamster's companions were for immediately lynching Reed, though his claim of self-defense, defense of his wife, and genuine remorse should have swayed them; they tilted a wagon tongue in the air against an ox yoke and would have hanged him but for calmer heads. More than 200 miles from the western base of the California Sierra, it was decided that Reed should be banished from the wagon train. His wife and daughters could remain—and would be cared for, they promised. With no alternative, Reed acquiesced to the wilderness justice and rode off alone. His level head might have saved the emigrants from disaster in the Sierra mountains, far ahead.

"We went down to Gravelly Ford, oh, must have been ten or fifteen years ago," commented Chris Graves, partly to me in the passenger seat and partly to my family over the CB radio. We were barreling across the flats west of Emigrant Pass on Route 80, just a few miles north of that historic crossing of the Humboldt. "Some cowboy took us there. It's an unremarkable stretch of dirt, and a sandy hillside. When we were there the river was maybe one or two feet deep and at most some thirty yards across. Somebody had put up two white wooden crosses on the hillside. One was unmarked. The other simply read, 'Pioneers.'" Chris gestured off along the sagebrush. "Down that valley, which was a nearly permanent campground for the Paiute, is the old station town of Beowawe, which the Central Pacific reached in November 1868. Just beyond, the C.P.'s construction engineer, James Evans, discovered some nifty hot springs and geysers, very much like Old Faithful, with beaches of pink opal and boiling mudpots nearby. They're gone now—tapped for energy by some enterprising son of a gun." He chuckled ironically. "Beowawe was also the site of a fabulous ranch owned by none other than Dean Witter, founder of the investment firm. Somewhere I've got a picture of Witter's wife when she was a little old lady, standing alongside her driveway. She'd had it lined with 100 ox yokes she'd found out along the covered wagon trail. Guess that would have been in violation of the 1906 Antiquities Act, but I doubt anyone would have been stupid enough to prosecute Mrs. Dean Witter. None of the yokes are still there, of course—who knows where they've gone? And now the ranch is owned by Germans."

We were passing through a narrows between the Sheep Creek Range to the north and the Shoshone Range to the south, the latter still supporting immense mining enterprises. Signs for the town of Battle Mountain appeared, the eponymous peak loomed to our left, and we signaled for an exit. Battle Mountain (4,511 altitude), named to commemorate a skirmish between emigrants and Paiute raiders in 1861, was the site of the first recorded Women's Suffrage Convention, in 1870; Nevada would grant suffrage to women in 1915, five years before the U.S. constitutional amendment. We pulled up in front of the Owl Hotel and Café for lunch. Stepping out into the open sun was a punishment. Inside, grimacing people smoked and played the slots, and immobile barflies huddled over drinks in the dimness. Occasionally one gave forth with a racking cough. "This is like an Edward Hopper painting," whispered Mary, as we waited for sandwiches and iced tea. Chris smiled and shook his head. "You better believe it." We ordered from an unpromising menu. "Emigrants hugged the north slope of Battle Mountain," Chris then said. "It was less muddy. That mount has been yielding riches since 1866—silver from the Little Giant Mine and others, copper from the Copper King Mine, gold from the Black Rock, and even some lead, believe it or not. Still giving out millions every year. I came here as a child in the 1940s with my mother during one of her spells of what she always called 'itchy feet.' We'd just get into the car and drive, and a couple of days later, we'd be halfway across the West. There was only one tree in town here in Battle Mountain. It stood behind an Orange Julius stand. We would set out under the tree and gasp from the heat."

In similar conditions we followed the Humboldt as it took another snake bend to the northwest, past bare, brown mountains jabbed by numerous mine entrances above cascades of lighter colored tailings, crossing intervening flats. Past Pumpernickel Valley, planted between the high Sonomas and the Osgood Mountains, perched the little town of Golconda (4,392 altitude), site of hot springs used by the wagon trains for rare personal hygiene and after the railroad by resort goers seeking therapeutic relief from assorted maladies. Flanking the railroad tracks were streets named for Central Pacific moguls Leland Stanford and the brothers Charles and Ed Crocker, now the address for trailers and tumbling shacks. A few miles west, at Bliss, the Humboldt changed its direction again, turning southwest, flowing turgidly past the town of Winnemucca, settled in 1850 and named—as was the sheer-faced, overbrooding mountain—for a Paiute leader friendly to whites. The Central Pacific reached the place, then called French Ford, on October 1, 1868.

A number of scholars believe that Winnemucca was the site of a bank robbery on September 19, 1900, by Butch Cassidy's Wild Bunch gang. On that day three armed strangers entered Winnemucca's First National Bank on Bridge Street, announced a holdup, and cleaned out the vault of gold coins, which they poured into ore sacks. They galloped out of town, ducking and returning bullets, eluded shots fired from a pursuing train, outran a posse, changed mounts at a prearranged cache, and disappeared.

Because the Wild Bunch gang had just pulled off a train robbery on the Union Pacific line weeks before near Tipton, between Rawlins and Rock Springs, Wyoming, some historians have said it would have been impossible for them to hightail it all the way to Winnemucca. But supporters say that the Wild Bunch—Cassidy, Will Carver, and Harry Longabaugh (the Sundance Kid)—could easily have boarded a Denver & Rio Grande train less than 100 miles' ride from Tipton, changed to the Southern Pacific line at Ogden, and disembarked at Winnemucca with generous time to spare. That Cassidy may have been planning something at Winnemucca was suggested by fragments of discarded letters torn into tiny pieces and found at the outlaws' camp: one letter was on stationery of Rock Springs lawyer Douglas Preston, whose most illustrious client was Butch Cassidy. Local people claimed to have talked with friendly strangers at the camp who matched the description of Wild Bunch members. The leader rode a white horse, which a boy at a nearby ranch said he admired, and the stranger said that someday the youngster would own it. As the outlaws fled the robbery on fresh mounts, the leader directed a by-stander to give the white horse to "the kid at CS Ranch," which was done. As an old man, the recipient of that white horse confessed to a writer that he had worshipped Butch Cassidy for some seven decades. "For a man," he wrote, "when he was crowded by a posse, to remember his promise to a kid—makes you think he could not have been all bad."

Past Winnemucca we now traced the great bend of the Humboldt. Seeing that the river had dug itself into a canyon, we flanked East Range, still rich in gold mining, and drove past the brown Eugene Mountains, known for their tungsten deposits. We drew closer to the mineshaft-honeycombed Humboldt Range; there, below the crossroads hamlet of Mill City, was a large area on the lower flank of the mountains down to the interstate, even jumping to the highway median strip, that had been burned down by a wildfire. The ground still smoked in an upper valley.

We were even with the twenty-mile-long Rye Patch Reservoir. In the 1930s the federal government built a dam impounding the Humboldt for irrigation projects; now it was enjoyed by bathers and Jet Skiers, the comparison of which to toiling, dust-choked and thirsty emigrants was just as startling as the imagery of hurtling tractor trailers and land yachts contrasted with rattling, canvas-topped prairie schooners. We got out to walk across the Rye Patch Dam, the bright blue reflective water and sudden humidity in the air coming almost as a physical shock in that otherwise dry and brown terrain. About twenty miles southward, toward the town of Lovelock, lay the place along the West Humboldt Range where—to nineteenth-century explorers and emigrants—the Humboldt River disappeared after spreading into a grassy meadow and a tule swamp, evaporation and percolation simultaneously taking away the water. It was famed as the Humboldt Sink; the equally well-known Carson Sink sat a few miles away across some hills, dead-ending the Carson River. Nearby springs afforded the first good water in many miles, though not everyone found them. Some had to content themselves with alkaline river water. Here in 1828 the fur trapper Peter Skene Ogden, thought to be the first white in the area, had caught beavers along the Carson and commented on swarms of pelicans flapping above the sink of the Humboldt. Forty-niner Reuben Shaw wrote in his diary that he had imagined the Humboldt Sink to be "a great rent in the earth, into which the waters of the river plunged with a terrible roar." Instead, he and his companions found "a mud lake ten miles long and four or five miles wide, a veritable sea of slime, a 'slough of despond,' an ocean of ooze, a bottomless bed of alkali poison, which emited a nauseous odor and presented the appearance of utter desolation."

Samuel Clemens had passed by during one of the prospecting frenzies in Humboldt County. His party camped for two days near the Humboldt Sink. "We tried to use the strong alkaline water of the Sink," he reported puckishly in his memoir, *Roughing It*, "but it would not answer. It was like drinkng lye, and not weak lye, either. It left a taste in the mouth, bitter and every way execrable, and a burning in the stomach that was very uncomfortable. We put molasses in it, but that helped it very little; we added a pickle, yet the alkali was the prominent taste, and so it was unfit for drinking. The coffee we made of this water was the meanest compound man has yet invented."

For emigrants there at Humboldt Meadows the barely drinkable water had to be gathered in casks, along which would be piled cut grass that must have tasted bitter indeed to cattle, but at least did not kill them. They soaked their wagon wheels so the wood would swell and lock tight against the iron rims; they walked around and around their wagons, making sure all was right. What lay ahead on the California Trail was the dreaded, furnace-hot wasteland called the Forty-Mile Desert. It was a twenty-four-hour crossing, usually begun in late afternoon to take advantage of more merciful nighttime, across deep sand that could stall a wagon and quickly exhaust a team. One diarist wrote: "Even the wagons seem to know that we are off today for the great adventure in sand, volcanic ash, alkali, furnace heat, and the stench of putrid flesh— We crossed along the edge of an immense baked plain with the fetid stinking slough for a guide, although the wreckage along the way almost paved our route. . . . It must have been here that one emigrant said he counted a dead animal every 106 feet." In 1850 Franklin Langworthy made the crossing and he compiled a list of what he saw as "peculiarly abundant." His inventory speaks eloquently and horrifically about the profound suffering and desperation that spurred the abandonment of possessions carefully and sometimes lovingly carried all the way from Old Home. Strewn across the desert were "log chains, wagons, wagon irons, iron bound water casks, cooking implements, all kinds of dishes and hollow ware, cooking stoves and utensils, boots and shoes, clothing of all kinds, even life preservers, trunks and boxes, tin bakers, books, guns, pistols, gunlocks, gun barrels, edged tools, planes, augers, and chisels, mill and cross cut saws, good geese feathers in heaps, or blowing over the desert, feather beds, canvas tents, and wagon covers."

We drove past Lovelock. It seemed as if the rushing interstaters were aware that we were on the edge of that terrible transit of old, and surged ahead at higher speed. About five miles across that joyless waste, my wife's voice came over the CB radio in Chris's pickup.

"Guys," Mary said, "is the Forty-Mile Desert exactly forty miles, and are there any rest stops? Because I just looked at the gas gauge, and the car computer says I have thirty-two miles of gas left."

We turned around and went back to Lovelock.

"Water barrel hoops are still plentiful in the Forty-Mile Desert," Chris announced as I pumped gas for Mary, checked the oil, and inspected the tires. "Rusted to nearly nothing. Every once in a while you can see a concentration of nails—all that's left of a piece of furniture or even a wagon. You can see dark hummocks out on the sand, the shape of an ox that has lain down and breathed his last, and those darkish shapes are exactly that, returned to dust, or something like it. Sometimes there's a piece of backbone the wind has uncovered. And you see a collection of rocks in a rectangle and you know you've found an unmarked grave—those emigrants had to search a great distance to find even one rock out in that wasteland."

Off we went again, at 70 or 75 miles per hour. At an exit sign for White Plains, earlier called Jessup, Chris gestured toward the railroad grade, close by between the highway and the Trinity Range. "Chinese pottery covers the ground there, in smithereens condition," he said. "Left over from the 1868 construction camps. And somewhere along by these hills, Jacob Donner got so discouraged he cached his gold and his books in a cave here. People have been looking for it ever since."

Halfway across the desert flats, Chris piloted us down an exit ramp that put us onto a gravel road, which seemingly came from nowhere and headed for nowhere. We went southwest, parallel to the interstate, dipped through an underpass, and stopped at the sudden and surprising sight of high chain link fences with pale buildings beyond—it looked like a top secret government installation. "This," announced Chris, as we stood out in the furnace heat, clutching the hot fence for a second and staring through at bubbling pools, "this is the Emigrant Springs, what some folks called 'the Spring of False Hope,' and what was later called Brady's Hot Springs, and what is now called Nightingale Springs."

I felt a rising tide of excitement. In the 1840s as the emigrants struggled across the desert, their oxen would sniff the western winds and sense the moisture, and they would begin pawing the sand and bellowing, and sometimes break loose and stumble into a run, and plunge into the scalding water, sometimes escaping with terrible burns and sometimes just collapsing and dying in the pools. In 1844, just before the rush of emigration began, mountain man James Clyman stopped here; he had a little water spaniel who had been his constant companion since he had left Milwaukee. "Not Knowing that it was Boiling hot," Clyman wrote, "he deliberately walked in to the caldron to slake his thirst and cool his limbs when to his sad disappointment and my sorrow he scalded himself allmost instantly to death." Another traveler,

Harriet S. Ward, wrote in 1853 that she saw the bones of a man "who, being deranged, threw himself into it, and immediately perished. A woman and a child were also killed by falling into it last summer."

Signs along the road near us warned of SCALDING WATER so we wouldn't be tempted to climb the high fence and leap into the 400 degree water. Hissing steam pipes ran behind the chain link fence. "People in the wagon trains would put scalding water out to cool for the next group," said Chris as we stared into the bubbling froth and the clouds of blowing steam. "Such a touching thing. When James Frazier Reed arrived here after his banishment from the Donner wagon train, he had a cup of 'fine Tea' here, as he noted on his map."

A big red truck rumbled beneath the interstate and slowed at the gate: it was hauling two open-topped trailers that we could see were full of white garlic bulbs. "Gilroy Foods," explained Chris. "They haul these all the way from their fields in California's Central Valley—up over the Sierra Nevada mountains and down to this spot. They harness the heat from the springs to dehydrate the bulbs." He pointed to cooling towers above the processing plant. "When the wind's right it makes you think of all the best Chinese food you ever ate." As it did when we left.

We pushed off once more and regained the interstate, squeezing into late afternoon traffic and rushing between the Truckee Range, northward, and the Hot Springs Mountains, with air conditioner blasting and water beading on cold soda cans from the cooler. "The amount of suffering on the latter part of the route was almost incalculable," recalled James D. Lyon in 1849, who, fearful of his wagon train's snail-like progress, went ahead on a packhorse. "No one except those who saw or experienced it, can have any idea of its extent—sights, the thoughts of which, would make the blood chill in any human breast. After I left the [wagon] train, I saw men sitting or lying by the roadside, sick with fevers or crippled by scurvy, begging of the passerby to lend them some assistance, but no one could do it. The winter was so near, that it was sure death literally, and the teams were all giving out, so that the thought of hauling them in the wagons was absurd. Nothing could be done, consequently they were left to a slow lingering death in the wilderness."

In 1868, as the Central Pacific Railroad inched its way across the Forty-Mile Desert, the task of locating water for its locomotive boilers, its construction workers, and their beasts of burden seemed insurmountable. Test drillings all along the route found a surprisingly adequate-seeming water table not far beneath the surface, but the water was poisonously alkaline and murderously corrosive to engine boilers. The geothermal springs were more trouble than they were worth, besides being not nearly productive enough for the railroad's great thirst. The nearby mountains disclosed no obvious springs, so they were forced to transport dependable snowmelt water that had tumbled down the

California Sierra in the Truckee River, which they did in large caulked wooden tank cars. In his photographic documentation of the railroad work, Alfred Hart produced a moody landscape of a long train being pulled across the desert, the silhouetted tanks shaped like huge tom-toms.

Ahead of us was a low plateau. When parties rounded it they could see in the distance the path of the Truckee River and they knew they would live. "You can still walk those ruts," said Chris. "You can find places where the rocks are lined up along the wagon ruts, where pioneers would lift up a rock and set it aside to make it easier for the next party coming through." We neared the river and the town of Wadsworth (altitude 4,077 feet), where the Truckee River bends abruptly northward to skirt between the Truckee and Pah Rah ranges, where what has not been taken by reclamation projects will finally empty into blue and crystalline Pyramid Lake, twenty miles away. At the bend in early 1844 John C. Frémont, heading down from his expedition to Oregon, found a seasonal Paiute village there, and he named the willow-fringed, "clear and pure" river Salmon Trout after the plentiful and large-sized fish speared by the Paiutes in weirs they constructed along the river. Later in the year, though, the pioneering Townsend-Stevens-Murphy wagon party renamed the river Truckee, after the Paiute who had guided them across the Forty-Mile Desert.

Truckee the man was the father of Winnemucca, who late in life had been a conciliator but who, as a younger man, stood for no nonsense from the white interlopers. In 1860, two Paiute women were abducted by some whites on the Carson River, who kept them imprisoned in a trading post. A tribal party learned where their women were held, attacked the post, and killed five of the abductors. Some days later an angry force of more than 100 miners swept over from Virginia City and down the Truckee looking for retribution. In a river gorge leading to a wooded draw the posse was ambushed by Winnemucca's warriors and mostly cut to pieces, although survivors made it back to the mining camps. A larger force, bristling with Regular Army troopers, met the Paiutes in another battle, but the Indians soon vanished into the mountains above Winnemucca Lake. Infrequently afterward there would be a raid on a stage or wagon train, but for the most part there was peace. This continued, in contrast to the pervasive Indian problem experienced by the Union Pacific in Nebraska and Wyoming, when the Central Pacific Railroad entered the area in late spring and early summer of 1868. Construction bosses Charles Crocker and Ed Strobridge instituted a policy by which every Paiute would receive a railroad pass, which all but ended any mischief along their line.

It was in May 1868, however, when the railroaders experienced a new kind of racial problem. Chinese tracklayers began quietly filtering out of the work camps and heading back to civilization. Boss Strobridge got to the bottom of it and reported the story to general manager E. B. Crocker in Sacramento. "Worthless white men have been stuffing them with stories," Crocker explained, "that east of Truckee the whole country was filled with Indians 10 ft. high who eat chinamen, & with big snakes 100 feet long who swallowed men whole." To counteract this Strobridge organized a junket for fifteen or twenty men—"& we sent good men along with them to show them 100 or 200 miles of the country where the road was to be built. . . . The fact is there are no Indians on the line until we reach Winnemucca, & then they are harmless like the Piutes on the Truckee, & the Chinese despise them. We have a ticklish people to deal with—but manage them right & they are the best laborers in the world. The white men we get on our work here are the most worthless men I ever saw."

Now we were following the Truckee River canyon between the Pah Rah and Virginia ranges, the river still edged with majestic cottonwoods. For many of the wagon train emigrants they were the first sight of trees since leaving the Platte River in eastern Wyoming and western Nebraska, and they were as good on the eye to them as they were to us. We rounded a bend and were confronted with the awe-inspiring sight of the first line of the Sierra Nevada range, slate blue and granite gray and steeply soaring toward the clouds. When the Donner Party arrived there in 1846 they could see snow on the mountains. They walked up the Truckee alongside the few wagons left to them, in the rain, and met a white man and two Indians who told Margaret Reed that they had encountered her banished husband in the Sierra—"he looked very bad," their daughter, Virginia, wrote to a cousin much later, "he had not ate but 3 times in 7 days and the three last days without any thing his horse was not abel to carrie him." Reed hoped to secure provisions and return over the mountains to save his family, they said. The men advised the Donners to take it slow and not worry about the conditions in the heights: plenty, they said, had made the transit even in midwinter. They were insane.

The ranges on either side of our highway were heavily creased, and we found ourselves speeding through deep cuts and canyons in low brown hills. We emerged onto flats suddenly crammed with industry—it was the city of Sparks, an engulfed suburb of Reno—and then we shot into a strait of truck traffic and veering family cars. There were tall buildings covered with lights in the late afternoon light, signs too numerous and confusing to read, and the mountains rising all the higher ahead. Chris grabbed his cell phone and called the hotel where his wife was waiting for us; Carol had driven over the mountains that afternoon. We signaled for an exit, Mary following faithfully through the crush of traffic, and Chris reached Carol and asked for directions. "I'm up on the twenty-fourth floor," she told him. "I can see you driving up the exit ramp."

High-rise Reno sprawls over ground once verdant with extensive meadows that rewarded the stressed and starving ox teams and pack horses of the emigrant hordes. James Frazier Reed noted on his hand-drawn map on October 13, 1846, that the valley was so lush with grass it could support 20,000 head of cattle. Sometime later when the Donner Party reached Truckee Meadows, it made what has been termed "its final fatal mistake" of lingering one day too long before beginning to seek a path across the Sierra Nevada. Later guidebooks would caution emigrants that if they encountered snow in the foothills they should assume it was impassable above; they should break their wagon trains into small parties, scatter across the lowland, and winter over. When the Donners entered the foothills on October 31 the snow was already three feet deep. They met another party who said they had attempted to cross and failed. "Well we thought we would try it," remembered Virginia Reed.

Back down on the Truckee Meadows, to which they should have retreated, in future years emigrants heeded the Donners' mistakes. "Dont let this letter dishaten anybody," Virginia Reed would write her cousin after surviving her horrific winter ordeal in the Sierra, "never take no cutofs and hury along as fast as you can." News of the Donner Party tragedy would depress emigrant traffic in 1847 and 1848, but reports of gold discovered along California's American River unleashed a torrent of eager-eyed pilgrims in 1849. Many left their impressions behind—we would summon them, as well as those in the ill-fated Donner Party, when we crossed the California line and addressed the Sierra ourselves—but standing on metropolitan concrete and asphalt laid down over the old Truckee Meadows I kept returning to the words of James Wilkins, a forty-niner, who lost his faith out on a barren trail and could take no comfort from the soft green grass or the sight of hundreds of teams placidly, gratefully munching and strengthening for the big climb ahead. "I wish California had sunk into the ocean before I had ever heard of it," despaired poor Wilkins. "Here I am alone, having crossed the desert, it is true, and got to some good water, but have nothing to eat all day, my companions scattered, our wagons left behind. That desert has played hell with us."

The first settler here was C. W. Fuller, who in 1859 began with a dugout cabin in the south bank of the Truckee and eventually erected a toll bridge across the river, rebuilding it after a washout. In 1863 he sold the property to M. C. Lake, who extended his business with a trading post and small ranch. Railroad graders inched down the Truckee River in 1867, not to be followed by the Central Pacific's tracklayers for a year, so obdurate was the mountain summit and weather. Charles Crocker intended to build a station town on the meadows—the company had, after all, struggled for five years to conquer the Sierra without being able to establish (and sell) a single large town along its route, while the Union Pacific boasted of countless future metropoli across Nebraska and Wyoming. Crocker settled with entrepreneur Lake and intended to name his railroad town Argenta, anticipating traffic with the burgeoning Nevada mines to the south and east. But some of Crocker's men begged him to name it after their old Mexican War and Civil War commander, Jesse Lee Reno, killed at South Mountain.

Reno it would be then, and it was formally born on May 9, 1868, in a real estate auction. Bidders appeared well before auction day and camped on the ground, buying up all the available blankets against the still cool nights, and eating all the food. The first lot brought the railroad $600, and about 200 more lots were taken that one day. In a month Reno would boast 100 houses. In June 1868, railroad chief engineer Samuel S. Montague wrote to an associate. "Reno is a lively place," he said, "and one month from the day the lots were sold could boast about 200 buildings, stores and dwellings. The day I passed through there on my return, an opera company was to entertain the good folks. On Sunday last I saw a circus tent in process of erection—a circus company performed there last week to an audience of 1,000 people."

We were booked into a hotel called Circus Circus, as a matter of fact. Thanks to Nevada's legalized gambling, enacted in 1931, Reno had secured itself its brightest future after years of successive booms and growth, and, as Chris Graves had declared to us much earlier, "We can take advantage of other people's greed, and the casinos' hunger to attract customers and keep their rooms filled every day, not just on weekends, by staying in a casino hotel—thirty-five bucks a night, Sunday through Thursday, the rooms are luxurious, the food is not only good but cheap, and no one's forcing you down onto the gambling floor!" Back in the 1930s Reno began to swell with hotels and casinos built by entrepreneurs like the onetime California carny Bill Harrah, who reinvented himself in Reno with a chain of bingo parlors that evolved into casinos. At the same time (this was in the era, many decades long, when most states had draconian, unrealistic divorce laws), Nevada's easy divorce law required only a six-week residency, and this helped create another boom. A few years later, Reno's growth surged as amorous Californians began crossing the border to evade their state's three-day-notice marriage law and the blood test requirement. Presumably, many of those impatient marital customers later found it necessary to return to Reno to sunder their wedlock, again requiring solace while resting elbows on green baize or grasping handles of one-armed bandits. But the biggest boom was saved for the 1970s, when the city exploded into an orgy of construction with new casinos and hotels changing the cityscape and the culture forever.

Upstairs in the quiet hotel corridor of the twenty-fourth floor, we met Chris's wife, Carol, after they had warmly and lingeringly embraced. She was tall enough not to be dwarfed by her towering husband, with laughing eyes and a bright smile, funny, quick, peppery to keep up with her declamatory husband, and very warm. She charmed the children by greeting them first, and in a few words pulling relaxed smiles from them, and then embraced Mary "to congratulate you on putting up with this bozo" (pointing to Chris) "and with *this* guy" (me), "who drags his family across every desert and mountain between here and Vermont!" Mary would have nothing of it: "I've loved every minute of it," she declared. "I love to drive, and I haven't had to cook for nearly two months!"

Our room was next door. There was an appropriate hiatus for showers—after mine, I glued myself to the plate glass window, looking back eastward, trying to mentally remove the urban sprawl from the flats on either side of the tamed, canalized river, looking past Sparks to the Virginia Range and the Truckee corridor, stupefied at the visual overload of the big city after hundreds upon hundreds of miles of virtual emptiness. At our height, to the people of the wagon trains we would have been as if in the clouds drifting down over the Sierra.

Regrouped, we all headed downstairs for dinner. We had no intention of lingering in the casinos. In my grandmother Rose Donahue Haward's household, gambling was a touchy subject because of "the shame of Kansas City." Boss Tom Pendergast in the 1930s had opened the town to gaming and gin mills, making it one of the most corrupt in the nation. The discomfort with gambling had unerringly passed through my mother and lodged in my heart. "Playing cards are the devil's picture book," I always liked to joke, echoing the kind of thing Rose used to say—but I don't even buy lottery tickets. That night in Reno we had tickets for a more wholesome kind of entertainment at the Eldorado—the flashy theatrical extravaganza *Spirit of the Dance*, its Irish folk dancing possibly bearing at least a tiny resemblance to something Rose's emigrant father, Peter Donahue, might have recognized back in County Galway in the 1840s, albeit with quieter music and no laser beams. But for the experience, we picked our way across the low-ceilinged gaming floor, four adults and two children, overwhelmed by the deafening clang of bells and electronic burps and beeps, the flashing and clashing lights, the smoke, the thousands of hustling bodies, the histrionic overload making us shrink and shirk and wince.

"That's it!" cried Mary, clutching the children and forging through the crowd toward our restaurant. "Take me back to the desert!"

14

Silver State

Twenty days after jumping off from St. Joseph, Missouri, on the morning of Wednesday, August 14, 1861, the Overland Stage creaked to a halt in front of the Carson City station in a cloud of acrid dust. The Nevada territorial capital had been officially designated for barely a month, and consisted of four or five blocks of tightly packed low wooden buildings connected by rattling board sidewalks, beyond which was a scattering of houses, shops, lean-tos, warehouses, cabins, and tents in no particular order. Two of the passengers who descended into this scene were Orion Clemens, newly appointed secretary to the territorial governor, and his brother, Samuel Clemens, former Mississippi River steamboat pilot and deserter after three weeks' service in a Missouri militia regiment. Stale Eastern newspapers in the stagecoach mail bundles would have reported on the Confederate victory at Bull Run on July 21, but broadsheets and mail alike were the worse for wear, having been sat and lounged on by the Clemenses and fellow passengers all the way across the plains, mountains, and deserts, with much newsprint probably transferring to their clothes and hands. Little matter; Carson City was a world away from Washington and Richmond, the territory of Nevada itself was only twenty-two weeks old, and the Comstock silver rush was in full and glorious uproar.

"We were not glad, but sorry," recalled Clemens a few years later in *Roughing It*. "It had been a fine pleasure trip; we had fed fat on wonders every day; we were now well accustomed to stage life, and very fond of it; so the idea of coming to a stand-still and settling down to a humdrum existence in a village was not agreeable, but on the contrary depressing."

All was barren and sterile: no trees, no vegetation but sagebrush and greasewood, enveloped in bitter, choking alkali dust, the road into town heavily littered with the bones of horses and oxen, the town itself dribbled onto the bottom of a bowl of bare brown hills, with the wild Sierra Nevada ridgeline towering nearly a mile overhead just a few miles to the west.

"I said we are situated in a flat, sandy desert," Sam wrote to his mother.

And surrounded on all sides by such mountains, that when you gaze at them awhile,—and begin to conceive of their grandeur—

and next to feel their vastness expanding your soul—you ultimately find yourself growing and swelling and spreading into a giant—I say when this point is reached, you look disdainfully down
upon the insignificant village of Carson, and in that instant you
are seized with a burning desire to stretch forth your hand, put the
city in your pocket, and walk off with it.

By contrast, thirty-six days out of Orwell, Vermont, we had that morning
followed U.S. Route 395 south some thirty miles from Reno along the Washoe
Valley, for a few miles slipping between blue Washoe Lake and the imposing
greenish brown Carson Range of the Sierra, passing old and new mine enterprises and signs for resorts and "fine living" developments. We climbed
the low pass of Lakeview Summit and dipped down into Eagle Valley where
Carson City (population 40,000, altitude 4,600 feet) sprawled out before us—its high-end suburban developments in the north, the historic
nineteenth-century district, the government complexes with the bright and
shining silver dome of the capitol, the casinos and hotels, and the urban grid
of shopping centers, motels, sandwich-and-fries franchisers sprouting south
and east. In two vehicles and still accompanied by our trusty Old West guide,
Chris Graves, we would, over ensuing days, travel a great circle embracing almost a quarter of Nevada's landmass, veering slightly into southeastern California territory in order to accomplish it.

Imagine that great circle as a necklace draped over stretches of the Sierra
Nevada, the Toiyabe, the White and Palmetto mountains; the forbidding Esmeralda alkali flats; valleys of the Big Smoky, the Reese River, Edwards Creek,
feeble Dixie Wash, and Walker River; the circle rising to completion across the
Desert Mountains, the lower Carson Valley, the Virginia Mountains, and finally the westward-beckoning Truckee River. On that necklace hang ancient
nuggets of silver and gold, symbolic of the going-going-gone mining boom
camps, Carson, Virginia, Bodie, Goldfield, Tonopah, Ophir, Park Canyon, and
Austin. On "The Loneliest Road in America," U.S. 50, we would finally rejoin
the paths of the Pony Express and the Lincoln Highway. And we would follow
the footsteps of people like Kit Carson, John C. Frémont, and our most voluble
pilot, young Samuel Clemens, who in Nevada would define himself and establish his life's true course, and (in many ways) literature's.

A brochure in our motel advertised a driving tour; signs would indicate
that we pause in front of certain sights and tune the car radio to a certain frequency, giving us a ninety-second broadcast of history. After unpacking we
set off. On North Carson Street we passed the state capitol, a square stone pile
with numerous annexes and a silver dome; the main building had been built
of stone from the state prison quarry. Down the street was the Old Federal
Courthouse, an ornate 1891 redbrick and sandstone edifice with a noble

clock tower, recently renovated for the state tourism offices, and the Roberts House, a Gothic Revival cottage with decorative bargeboard on the eaves and an ornate columned porch, almost the duplicate of our next-door neighbor's house in our Victorian village in Vermont. (Noting this, the children allowed that they were beginning to feel the slightest tinge of homesickness.) We paused at the Governor's Mansion on North Mountain Street, with its columned portico and broad verandahs; the Bliss Mansion, on West Robinson, some fifteen rooms with seven marble fireplaces, and with porches sprawling across each wing; and the Ormsby Home, at Third and Minnesota, built by Major William M. Ormsby, an unprepossessing frame house raised by the Carson City booster who died very badly at the hands of Paiute warriors. Ormsby had commanded the troops who pursued the vengeful Indians under Winnemucca in 1860, up on the Truckee River near Pyramid Lake. The volunteers intended to punish the "red devils" for daring to object to the kidnapping and serial rape of their women by white miners; instead, the militia was cut to ribbons.

Though it was only one of several houses in which Orion and Sam Clemens resided in Carson City, we had to pause in front of the peak-roofed and gabled frame house at Division and Spear streets. From our children's perspective, the Clemens House, with an inviting little front porch linking catty-corner wings, had its twin back home in Orwell, Vermont, "the old Allen place that burned down a few years ago!"—but it took little effort to look at the privately owned house and imagine a man with dark, curly hair and droopy mustache lounging beneath a cloud of cigar smoke on that porch.

Orion Clemens was busy in the weeks after their arrival, helping to prepare the makeshift village for state governance and on the side trying to revive his theretofore lackluster law practice. Governor James W. Nye—"an old and seasoned politician from New York," commented Sam many years later in his autobiography, "politician, not statesman," who had winning personal qualities and a shrewdness Sam admired—resided in a one-story, two-room cottage. The rest of the government, including chief and associate justices, rented rooms around "with less splendor," he would write in *Roughing It*, "and had their offices in their bedrooms." The first territorial legislature was elected on the first Saturday of September 1861 (the population of Nevada at that point was between 12,000 and 15,000, scattered across a great deal of mountainous and barren terrain). It would convene on the first of October. Having no capitol building, lawmakers met in the primitive Hot Springs Hotel near the Carson River. Because by federal law the state senate had to have separate quarters from the assembly, Orion hung a canvas curtain across the middle of the hotel's public room, but the federal government declined to pay the expense for the canvas, subtracting the cost, $3.40, from Orion's $1,800 annual salary.

While Orion was thus occupied, Sam lit out to inspect the territory. With an acquaintance he rode over the Sierra's Carson Range to Lake Bigler, later redubbed Lake Tahoe, "a noble sheet of blue water lifted six thousand three hundred feet above the level of the sea," he wrote, "and walled in by a rim of snow-clad mountain peaks that towered aloft full three thousand feet higher still. . . . I thought it must surely be the fairest picture the whole earth affords." The pair staked out a claim on public land for a timber franchise, built a rude cabin and a primitive fence, and spent some weeks fishing, exploring, and mostly lazing about, until Sam's careless campfire building ignited a wildfire that destroyed not only their claim but most of the abundant pine-forested mountainside besides. Clemens turned hopefully to mining.

Bulletins of new discoveries pulled him in all directions. First he headed southeast into the Esmeralda district in the Toiyabe Mountains and bought shares in the Black Warrior Mine in Aurora, being encouraged by acquaintances to invest in a number of other claims. With several companions in the late autumn, he rode 175 miles northeast of Carson toward the eastern flank of the Humboldt Range, to the Unionville mining camp in the Buena Vista district, buying more shares, spending two months of the winter at hard labor, yielding little. Everyone seemed to be offering shares—so many feet of ledge, at so many dollars per foot—that for many in Nevada the boom was only a bubble, with the only money to be made in passing dubious shares on to the next sucker. Investors kept receiving invoices from those actually digging and blasting, to help pay expenses. Nevertheless as he kept dipping into savings Clemens kept predicting in letters home to his family that he would soon strike it rich. In the spring he was back down at Aurora, cautioning Orion in a letter on April 13, 1862. "Don't buy anything while I am here—but save up some money for me," he said. "Don't send any money home. I shall have your next quarter's salary spent before you get it, I think. I mean to make or break here within the next 2 or 3 months." A week later: "No, don't buy any ground, anywhere. The pick and shovel are the only claims I have any confidence in now. My back is sore and my hands blistered with handling them to-day. But something must come, you know."

Nothing did, and in early summer 1862, Clemens hit bottom. "I abandoned mining and went into milling," he wrote in *Roughing It*. "That is to say, I went to work as a common laborer in a quartz mill, at ten dollars a week and board." This palled quickly. "I caught a violent cold . . . which lasted two weeks," he wrote Orion on July 9, "and I came near getting salivated, working in the quicksilver and chemicals. I hardly think I shall try the experiment again. It is a confining business, and [I] will not be confined, for love nor money."

It was time for a career change.

Sam Clemens picked up a pen. It was not an unfamiliar tool; after his arrival in Nevada he had sent four descriptive letters signed "SLC" back east to Iowa and the *Keokuk Gate City*, fulfilling a promise to the editor of Orion's hometown newspaper to tell about his Overland Stage experiences and to send dispatches from the mining districts. The missives were, the *Gate City* readers were assured, "peculiar and interesting, and probably quite satisfactory." They were all that, having been rewritten from letters Sam had sent his mother, and being destined for later reworking into *Roughing It*. He was not paid, but reaction to the letters was positive enough that, desperate for cash in the spring of 1862, he reworked them and sent them off to the editor of the *Virginia City Territorial Enterprise*, the largest paper in Nevada, with high hopes. By early May they were appearing—they were signed only "Josh"— but paid a pittance, and on July 23 Sam was growing so desperate that he asked Orion to canvass the *Carson City Silver Age*, the *Sacramento Union*, and other papers to publish a series of letters for $5 or $10 a week. "The fact is," he said, "I must have something to do, and that *shortly*, too . . . my board must be paid." On July 30, he wrote that the editor of the *Enterprise* had offered him the job of local reporter at $25 per week. While mulling that over in ensuing weeks he lost himself in prospecting in the White Mountains on the California-Nevada border, but by late September he had removed to Virginia City, that fabulous and wild mining camp sitting astride the hillsides and valleys of the Comstock Lode, to wield his pen with a vengeance, an activity he would not interrupt for the rest of his long life.

We stood in the bowels of Carson City's Old Mint on North Carson Street, built of sandstone with narrow, arched Italianate windows, bracketed cornices, portico, and belvedere, where gold and silver were minted until 1893 when production fell and the building was converted to a state assay office. Its current incarnation is as the Nevada State Museum. We had paced through an eerie walkthrough diorama of a Nevada ghost town, inspected extensive cases of hard currency, paused before the reconstructed skeleton of an Imperial Mammoth found in the Black Rock Desert, regarded dioramas about Native Americans in the Great Basin, and lastly peered at the museum's weapons gallery with its wealth of stories. One exhibit was simply a large pocketknife with a five-inch blade that had been found between two skeletons in Death Valley—a threadbare story inviting reams of speculation. Another display was a Colt .22 officer's model target revolver owned by the flamboyant frontiersman and U.S. senator Key Pittman. In June 1933 the senator attended a sixty-six-nation economic conference in London, and he argued energetically on behalf of silver. He frequently got drunk during the trip and amused himself by shooting out London streetlights with his six-gun.

The last exhibit was deep in the basement level: a claustrophobic wiggle through low and narrow and dimly lit mine tunnels past life-size dioramas, an extremely effective reconstruction that brought to mind a letter Sam Clemens wrote to his family in early 1863. He said if his sister, Pamela, were to visit a mine, she would find herself in "great, dark, timbered chambers, with a lot of shapeless devils flitting about in the distance, with dim candles flickering in the gloom, and then she could look far above her head, to the top of the shaft, and see a faint little square of daylight, apparently no bigger than one of the spots on a chessboard."

I gasped with relief when we found our way out the exit door into bright, late afternoon sunlight.

The next morning, while Mary and the children relaxed around the motel pool, Chris and I drove down in his pickup along the steep and green eastern edge of the Sierra, traveling ten or fifteen miles to the elysian little town of Genoa (4,750 feet in altitude), flanked by flood-irrigated hayfields and pastures often shaded on their margins by cottonwoods. Genoa had been Nevada's first permanent white habitation. Originally settled by a handful of Utah Mormons around 1849, which gave it the name of Mormon Station, it acquired a solid trading post with a log stockade in 1851. Renamed Genoa in an 1855 survey, the post was briefly abandoned when, in 1857, Brigham Young called all Latter-day Saints back to Utah to garrison his kingdom against the approaching U.S. Army. Gentiles gleefully swooped in and snapped up the garden spot at distress sale prices. "The recalled Mormons left a curse on Genoa as they departed, so angry as they were about being forced to sell so low," Chris said, as we stood on the picturesque, tree-shaded main street in front of the old brick Genoa courthouse, later a school and now the town museum. "But I'm hanged if I can sense anything like a curse on this place. This is as close to paradise as most people are going to get. Many's the time I've been here," he sighed, his view taking in the Sierra slope, the wooden and brick houses in town, the green farm fields, "and I often dream about this place. Can't tell you how many times I've been tempted to sell our place in California and buy something here, although prices have been rising steeply for some years. But Carol always says, 'What would we do here?' Me, I'd take the cap off a bottle of cold beer, put my feet up on a porch railing, and just eyeball it all."

On a hill we found the Genoa Cemetery, where reposed the bones of an extraordinary local character, John "Snowshoe" Thompson, a native of

Telemark, Norway. "A man made immortal," read a nearby plaque, "for his unbelievable treks through the most severe storms of the Sierra to bring the mail to pioneers of western Nevada. He was never paid for this hazardous service." His life path had taken him into the story of the first transcontinental railroad when he helped keep winter-bound tunnel crews in touch with the lower, populated regions of the Sierra. Thompson's preferred mode of conveyance across the otherwise impassable Sierra snows were then called "Norwegian snowshoes," long and narrow strips of wood upon which he would strap his boots ("Canadian snowshoes" were paddle-shaped wood forms across which a crosshatch web of buckskin or fiber was stretched). With the addition of poles, Thompson glided quickly over the slopes above Lake Tahoe, stringing together isolated communities with mail and news, a century before those self-same mountains would support many thousands of similarly equipped skiers, drawing brightly colored moving lines across white mountainsides in that cold, extraordinarily clear blue-sky atmosphere.

A little way south of town we passed an improbability in this high and dry country, a golf course, with piped Sierra snow-melted water spraying the greens and making rainbows; a woman looked for her ball in the sagebrush margin. Then we came to a narrow crease in the Sierra wall, down which a creek wildly tumbled. It was called Jack's Valley, and the road wound steeply upward through precipitous grades, curving and switching back, leaving and then rejoining the creek, with dull sagebrush and some red-leafed bushes clinging to the slopes. John C. Frémont had urged his party this way in 1844; the Pony Express riders in 1860–61 exhausted the mounts obtained in Genoa by the time they reached the next station on the southern Lake Tahoe shore, a mere thirteen miles west. In 1859 Horace Greeley, the nation's most celebrated newspaper editor, would, in a stagecoach leaving Carson City, importune his driver, the crusty Hank Monk, to get them quickly to Placerville on the far side of the mountains where he was to lecture. In *Roughing It*, the writer formerly known as Sam Clemens picked up the story in an immensely lighthearted way:

> Hank Monk cracked his whip and started off at an awful pace. The coach bounced up and down in such a terrific way that it jolted the buttons all off of Horace's coat, and finally shot his head clean through the roof of the stage, and then he yelled at Hank Monk and begged him to go easier—said he warn't in as much of a hurry as he was awhile ago. But Hank Monk said, "Keep your seat, Horace, and I'll get you there on time"—and you bet you he did, too, what was left of him!

At one clifftop curve, we pulled off the road to get out and stare eastward across Carson Valley toward the 9,000-foot ridge of the Pine Nut Mountains, fifteen miles away. "Look at these goose bumps," said Chris, showing his fore-arm. "God put his foot down in Genoa." Several hundred yards uphill, Jack's Valley Road was barred to further progress, auto-gated with an electronic passkey kiosk. We had managed to ignore the lobbyists' and politicians' hide-away estates tucked deep into canyon margins or planted on barely visible, screened cliffs, but now, as Chris turned the pickup around in a driveway and headed back down the mountainside, there was no necessity to state the obvious. We motored back to Carson City in something like silence.

All of us lunched together in downtown Carson City with another of Chris's far-flung friends, Wendell Huffman, animated and erudite, whom Chris introduced as "one of the great authorities on early Nevada and California railroads." In only a few minutes' conversation, this was readily apparent. Wendell was finishing a book on the subject. Currently serving as librarian at the Carson City Public Library, he had written his master's thesis on the eminent nineteenth-century naval engineer and architect James Gilliss; he also contributed to historical quarterlies and consulted with the Nevada State Railroad Museum in town.

After lunch he took us on a tour. Located on South Carson Street on a thirteen-acre site, the museum was an imposing glass-fronted exhibition building with two enormous tunnel-sized portals on its pebbly concrete southern facade and with a large annex in the rear for reconstruction and maintenance of rolling stock, the largest collection in the country. It included seven steam locomotives and more than fifty passenger and freight cars. On weekends the museum operated vintage train rides on a one-mile loop of track next to a restored Southern Pacific depot and wooden water tower. Wendell introduced us to manager Richard Reitnauer, and to my children's delight they were allowed to clamber aboard a handle-pump car and with some straining to get it moving on the track. Then we went inside to inspect the superb collection, much of it built around the remnants of the old Virginia & Truckee Railroad. In its heyday the V&T connected Virginia and Carson cities with Reno, where immense riches of the Comstock Lode were transferred to the Central/Southern Pacific Railroad. The V&T had been built between 1869 and 1872, and became known as the only line "to carry its entire weight of locomotives, track, and equipment in silver," not to forget all the gold. Literature about the railroad disclosed that investors received dividends as high as $100,000 a month. It operated until 1950, although from the 1930s it had suffered from declining freight and passenger revenues—enough to cease Virginia City runs in 1938; those tracks were taken up during the patriotic scrap drives around World War Two. Local railroad enthusiasts resurrected the V&T in the 1950s on a two-and-one-half-mile portion of the right-of-way out of Virginia City, running vintage steam locomotives pulling open platform cars (with benches for tourists) and cabooses; the tantalizingly short run took passengers past ruins of some of the richest bonanza mines of the Comstock. The operators had plans for extending it an additional seventeen miles to the state capital, as soon as a public-private partnership had raised an additional $15 million to meet grants from state and federal agencies.

Inside the museum, Wendell took us past the huge, brightly painted old engines. "The Virginia & Truckee had flush times and bought excellent equipment," he said, gesturing toward a Baldwin 1875 coalburner, V&T #20, and to a Baldwin 1913, V&T #27, both built in Philadelphia, and to V&T #18, built by the Central Pacific Railroad shops in Sacramento in 1873. "But then of course it went into a deep decline. The desert is a harsh place, but in the scheme of things it was fairly friendly to this equipment. And many of the cars are in such good condition because in the 1930s they went to Paramount Studios in Hollywood, where they were spruced up for the movies and employed pretty regularly." Along that line of history we also saw a Cooke 1861, V&T #8, which had last been owned by Desilu Productions and had been in many movies such as *The Gambler*.

The main part of the museum contained a wealth of smaller memorabilia, but I was particularly curious about the mechanics' shops in the rear. There, in the large, echoing space we saw locomotives, baggage cars, flatcars, cabooses, and boilers, and a real oddity, a pointy-ended McKeen car with round portholes, not squared windows, stripped down to bare metal and lacking an engine, which the engineers were going to rebuild from the ground up. "These are all works in progress," explained Wendell, "and you have to have a little faith that one day they'll be back together, cleaned up and bright and running." With the amount of peeling paint, rust, corrosion, splayed wood and wood rot, it did take a certain amount of faith to imagine them whole again, but then we had the examples of the brilliant finished work in the public area of the museum. Before we left, Wendell took us over to a deteriorated Virginia & Truckee coach. "We think," said Wendell, "that this was the original officers car of the Central Pacific, and that it was present at the Golden Spike Ceremony in 1869. "Imagine that!" said Chris in a reverent whisper. "You could say this car was built to float on champagne! This was pulled by old Jupiter over the Sierra from Sacramento with Leland Stanford presiding over the wine, oysters, and quail for his high-falutin friends, all the way across Nevada and Northern Utah to Promontory." Much restoration was obviously ahead for the battered hulk on wheels. "Good luck, old buddy," he said, and patted its splintery side.

In the fall of 1863, when Samuel Clemens took up the pen, Virginia City was the center of the continent's greatest mining activity, a town sprawling on the steep, barren, rocky flank of Sun Mountain (now Mount Davidson) above a vein of silver ore from fifty to eighty feet thick. Gone were the days when a man with a pan, or with carpentry skills to build a sluice box, could reap the alluvial riches in the gravel of tumbling Sierra streams. The rush started in late 1859 with word of the Nevada Comstock strike—worth perhaps $4,000 or $5,000 per ton in silver and $3,000 in gold. For most prospectors their freelance days were over as they became miners toiling deep underground for $4 to $6 per day, not that being "wage slaves" dimmed their hopes for a big strike. The first arrivals found no town, certainly no city, and high frigid winds spirited away their tents and blew down their shacks. Those who survived the winter were, like as not, forced to scrape holes in the ground for shelter, and they warmed themselves with wan sagebrush fires. Because there was little food they drank whiskey and hoped that the heat of greed would keep them alive. But by the apex year of 1863, with its population of 15,000 to 20,000, the town was solidly built of wood and brick and ran up the mountainside in stair steps—"the entire town had a slant to it like a roof," recalled Clemens in *Roughing It*. The town was visible from 50 miles away and from its top-most street one could see the Humboldt Range, 150 miles away to the northeast, the white sandy wastes of the Forty-Mile Desert to the north, and the towering Sierra Nevada marching north to south. "The scene when the sun is low has few rivals in America," pronounced the WPA writers in 1940.

"Virginia had grown to be the 'livest' town, for its age and population, that America had ever produced," Clemens wrote.

> The sidewalks swarmed with people—to such an extent, indeed, that it was generally no easy matter to stem the human tide. The streets themselves were just as crowded with quartz wagons, freight teams and other vehicles. The procession was endless. So great was the pack, that buggies frequently had to wait half an hour for an opportunity to cross the principal street. Joy sat on every countenance, and there was a glad, almost fierce, intensity in every eye, that told of the money-getting schemes that were seething in every brain and the high hope that held sway in every heart. Money was as plenty as dust; every individual considered himself wealthy, and a melancholy countenance was nowhere to be seen. There were military companies, fire companies, brass bands, banks, hotels, theatres, "hurdy-gurdy houses," wide-open gambling palaces, political pow-wows, civic processions, street

fights, murders, inquests, riots, a whiskey mill every fifteen steps, a
Board of Aldermen, a Mayor, a City Surveyor, a City Engineer, a
Chief of the Fire Department, with First, Second and Third Assistants, a Chief of Police, City Marshal and a large police force, two
boards of Mining Brokers, a dozen breweries and half a dozen jails
and station-houses in full operation, and some talk of building a
church. The "flush times" were in magnificent flower!

"Elsewhere in the West," said the peerless historian Bernard DeVoto in his
Mark Twain's America (1932), written while he was editor of the Mark Twain
papers, "the miner labored in inaccessible gulches and, for a bust, made infre-
quent pilgrimages to the big town—Denver, San Francisco, Helena. But here
hard rock and the big town met in one continuous bust. The West consum-
mated itself." Feeling immensely at home, Clemens covered the city beat with
alacrity, and when a story lacked a meaty quotation or when a day lacked a
story, he made them up. His salary was increased to $40 a week, but he
hardly had to draw it; as a reporter for the most prominent Nevada newspaper in the territory's most prominent boomtown, he was forever rained on
by favors, whether of free drinks and food or of the hard cash variety, or in
speculative shares of mine enterprises.

Crawling along C Street like those buggies of 1863, we nosed our vehicles
past pedestrian-jammed wooden sidewalks and heavily patronized antique
and gift shops, restaurants, and picturesque saloons. Roads from Carson City,
sixteen miles to the southwest, have been vastly improved since the Comstock
days, and Virginia City was enjoying yet another boom. It had been a long
time in coming. Making note of the town's long slide from the mid-1860s
well into the twentieth century, WPA writers went through in the late 1930s
noticing that "most of the houses have long been unpainted and the elaborately turned wooden balustrades along the high retaining walls are beginning to sag. The few that still have most of their decorative urns and finials
are marked exceptions." C Street was a faded and cobwebby collection of curio shops and wanly colorful last-gasp saloons. "Each year there are a few less
buildings," noted the WPA, "for annually in the spring the undermined earth
sags a little more at one spot or another." A decade later the peripatetic journalist John Gunther wandered into town and gave it a few lines in his massive
"taking the pulse of postwar America" epic, *Inside U.S.A.* (1947). "Virginia
City is, then, a fragrant tomb," he commented sadly. "The population was
forty thousand in its heyday; today, two thousand. Never have I seen such
deadness. Not a cat walks. The shops are mostly boarded up, the windows
black and cracked; the frame buildings are scalloped, bulging, splintered;
C Street droops like a cripple, and the sidewalks are still wooden planks; the

telephone exchange, located in a stationery shop, is operated by a blind lady who had read my books in Braille."

We parked on one of the famously steep side streets and ambled back down to C Street, soon crushed by leisure-timers with loot in their pockets. We ducked in and out of a number of stores, finding little to pique our interest except ice cream and cold drinks, until I spied the *Territorial Enterprise* building down the street. I left my family to some kind of gift shop (Chris had wandered off by himself) and, feeling the sudden urgency of a pilgrimage, I quickly slithered through traffic to gaze on its brick facade and posterlike displays. There was a women's clothing shop on the first floor, and from there, for the price of $1 admission, one could descend a flight of steep wooden stairs into a dank and deep basement, to see the remnants of the 1860s newspaper office, which had probably been consolidated from various levels of the building.

Behind ropes were press equipment and an editor's desk and wooden chair, and bound and unbound copies of the yellow and brittle newspaper. In a gilt frame on an easel was a large and meditative oil portrait of Samuel Langhorn Clemens, who in the *Territorial Enterprise* issue of February 19, 1863, stopped signing his pieces "SLC" and began using the pen name of "Mark Twain," a riverboat pilot's term for a twelve-foot depth. The amusing story was called "Ye Sentimental Law Student," and it was partly a parody of legalese and partly in the flowery language of the professional travel scribe, and it moved the writer a step away from daily journalism, as colorful and improvised and imaginative as that may have been in the Wild West, and a step further into literature. Some years after he had outgrown the *Enterprise* and Virginia City, and had begun to get a name—for his stories and sketches, his bright and ironic travel book *The Innocents Abroad*, and his standing-room-only "lectures" delivered all over the nation—he sent a fond letter of recollection back to Virginia City from the East. "To find a petrified man," he said, "or break a stranger's leg, or cave an imaginary mine, or discover some dead Indians in a Gold Hill tunnel, or massacre a family at Dutch Nick's, were feats and calamities that we never hesitated about devising when the public needed matters of thrilling interest for breakfast. The seemingly tranquil *Enterprise* office was a ghastly factory of slaughter, mutilation and general destruction in those days."

I took in the portrait, the desk, the chair, wanting to believe that Twain had once sat there, but I had the extra reassurance of the brick *Enterprise* building overhead, the terraced old frontier town, and the surrounding mountains to have a feeling of connection again with the man whose work had always so informed mine and whose life I had followed down a long road. My memories went back to the illustrated red clothbound editions of Huckleberry

Finn and Tom Sawyer my parents gave me one Christmas, to the later novels and memoirs I bought with grass-cutting earnings or checked out of the library, to the fiercely iconoclastic essays and parodies contained in later compilations like *Letters from the Earth* and *Europe and Elsewhere*, which I found in a shambling used bookstore in my hometown. Years later, when beginning to write a nonfiction book about a little known war fought between Americans and freedom-loving Filipino nationalists beginning in the year 1899, which acquired us our first colony, I began the narrative with this admission: "It was Mark Twain who sent me to the Philippines." My book was titled *Sitting in Darkness*, after Twain's angry anti-imperialist diatribe, "To the Person Sitting in Darkness." In a fashion that I hoped would have met with his approval, the narrative arced from the past to the present while weaving the two strands of the story together. The earlier thread followed two prominent and emblematic figures of the time, a Kansas-bred soldier of fortune named Frederick Funston and a rebel leader named Emilio Aguinaldo; the former captured the latter in a brilliant, audacious act of chicanery, bringing the war to a conclusion, cutting the hopes of nationalists who had looked to the United States for inspiration and protection, and knotting the two countries' destinies for many decades to come. The present-day thread followed that larger relationship to its obvious conclusion in the awkward postcolonial embrace, with various American administrations trying to accommodate the ruthless oligarchy of President Ferdinand Marcos. To tie the narrative knot, I went to the Philippines and duplicated the expeditions of Funston and Aguinaldo, and saw present-day Philippine culture and politics up close. Later there would be an epilogue covering another journey to Manila and the countryside, during the tumultuous campaign between Marcos and challenger Corazon Aquino, with the surprise ending of "People Power" overturning the dictatorship.

Mark Twain not only sent me to the Philippines, he lived with me over nearly five years for the book. He came back to me in *Empire Express*, assigned an occasional cameo walk-on, but he still helped inform the spirit of that book about the first transcontinental railroad. And of course he had popped up again and again, like a hitchhiker, all along the old iron road in the summer of 2000. I had made the pilgrimage to the "white town drowsing"— Hannibal, Missouri—eating fried catfish with hot peppers on the Mississippi side of the levee, shuffling through the tourist-clogged streets and boyhood home; someday I hoped to see his expansive house in Elmira, New York, with its shiplike prow, but earlier, I had explored his mansion in West Hartford, Connecticut, and delighted at the billiard room retreat up under the eaves in the attic, about which he had dreamed and planned when a penniless prospector in Nevada, certain he'd find the silver ledge that would send him to Easy Street. Twain was as much in those places as he was in all his volumes crowding my shelf, and I liked to think that something of his spirit ran all across my own work. In the morning we would head off for days in more hot and high-altitude and mostly ghostly mining camps, with *Roughing It* in my lap and Twain's words in my ear. A line I had been drawing for four decades had extended this whole long way to a basement on C Street in Virginia City, Nevada, and I paid homage to that glorious man for many minutes before puffing back up the basement stairs into late afternoon light to find my wandering family and rescue them from modern-day tourist traps.

We were expected for cocktails and dinner at the home of another of Chris Graves's acquaintances, Andria Daley-Taylor, who lived with her husband and children up on A Street in an elegantly restored two-story Victorian house, with exterior bric-a-brac painted in contrasting cream and maroon. The house had formerly belonged to a colorful and flamboyant pair of writers and preservationists, Lucius Beebe and Charles Clegg, who lived in Virginia City for many years. In the 1950s they helped resuscitate both the Virginia & Truckee Railroad and the moribund *Virginia City Territorial Enterprise* (they became the publishers), and together issued more than thirty books on Western Americana and railroads. The couple had moved to Nevada in 1950 after years in New York City—Beebe was a popular café society columnist for the *New York Herald*, and contributed to most national magazines (I remembered his articles in *American Heritage* when I was growing up, and one particularly lavish book about the Pullman palace cars in the Gilded Age). Besides outfitting a sumptuous old Pullman private car for their travels, Beebe and Clegg had painstakingly restored their house in Virginia City. The house had also once been owned by the theatrical manager John Piper; it was a harmonious touch that Andria Daley-Taylor, who was glamorous and theatrical in her own right, was now living there. Andria, also a historical preservationist who had earlier worked for the state of Nevada, was leading the restoration of John Piper's Opera House in Virginia City.

We went down to see the Opera House at the corner of B and Union streets with her; it was opulence fallen on hard times. Built in the 1860s and entirely of the Gilded Age, the hip-roofed Italianate structure had arched windows and a columned street-gallery facade decorated with patriotic bunting. Inside, though there was much bare wood and plaster lathe, and no fixed seats, it had a commanding proscenium stage flanked by stacked private boxes, with balconies on three sides of the spacious hall and chandeliers hanging from a grand, single-vaulted ceiling. It was startling to stand inside such a vast open space. Lillie Langtry, Edwin Booth, Maude Adams, and Buffalo Bill Cody had performed there; Charles Dickens had delivered one of his celebrated lectures, John Philip Sousa had conducted his marches, and Emma Nevada had sung the opera for which she was internationally renowned. The theater was cheerfully full of ghosts. Andria told us that her local nonprofit society had bought the place from Piper's great-granddaughter in 1997 for $475,000, and was sponsoring a young troupe of Shakespearean actors that summer; one performance was interrupted by a bat. "This place began going downhill around 1900," Andria said. "It was vacant, then was a roller-skating rink— there are a couple of organs stashed somewhere around here—and then it was a silent movie theater. It was dark—except for the odd hole that still lets in light—when we bought it, but not irretrievable. We're writing grant proposals until they're coming out of our ears." In 2002, she obtained a prestigious grant of $400,000 from the National Park Service's "Save America's Treasures" program, and the project continues to sail vigorously ahead.

Stepping back outside onto B Street, I pulled Mimi and Davey next to me on the sidewalk, we held hands in a little circle, and I got them to jump up and down with me a few times, hard. Andria and Mary saw what we were doing and laughingly came over. "Oh, ye of little faith," said Andria. Then I explained our little test to my puzzled children. "We're standing on a thin crust of rock," I said, gesturing toward the higher slopes of Mount Davidson, down the steep streets, and then sweeping off toward Gold Hill. "This whole mountainside is hollowed out, as deep as a thousand to sixteen hundred feet. You've heard the expression 'undermined'? Well, that's Virginia City." I reminded them of when we were in Rock Springs, Wyoming, looking at the big coal mining map that showed the entire city was undermined by layer upon layer of four-foot tunnels following the coal seams; we had walked back outside the museum almost gingerly, now seeing only the insubstantial crust of earth and rock beneath city streets, buildings, and buttes. "The empty spaces below Virginia City," I told them, "make the tunnel system below Rock Springs seem like the puny work of moles and voles." The children peered down at the ground beneath their feet.

Mark Twain had marveled at this miracle of engineering and blind faith when living and reporting in Virginia City. "Virginia was a busy city of streets and houses above ground," he wrote in *Roughing It*.

> Under it was another busy city, down in the bowels of the earth, where a great population of men thronged in and out among an intricate maze of tunnels and drifts, flitting hither and thither under a winking sparkle of lights, and over their heads towered a vast web of interlocking timbers that held the walls of the gutted Comstock apart. These timbers were as large as a man's body, and the framework stretched upward so far that no eye could pierce to its top through the closing gloom. It was like peering up through the clean-picked ribs and bones of some colossal skeleton. Imagine

such a framework two miles long, sixty feet wide, and higher than any church spire in America. Imagine this stately lattice-work stretching down Broadway, from the St. Nicholas to Wall Street, and a Fourth of July procession, reduced to pigmies, parading on top of it and flaunting their flags, high above the pinnacle of Trinity steeple.

To enter that awe-inspiring subterranean world, Twain stepped into an enclosed platform no bigger than a broom closet and shot like a dart down the shaft. "It is like tumbling down through an empty steeple," he wrote, "feet first." Then, armed with a single candle, he tramped through drifts and tunnels to see men blasting and digging unremarkable-looking rock that in the aggregate was yielding millions to investors. The heat in those deep galleries was almost too much to bear; miners worked stripped to the waist, never far from vats of iced drinking water, and under such working conditions the accident and fatality rate was daunting. If a drainage pump failed the mines quickly filled with hot water; mechanical aids to force breathable air deep into the recesses were notoriously quirky. It was overwhelming to Twain. "You reflect frequently that you are buried under a mountain, a thousand feet below daylight; being in the bottom of the mine you climb from 'gallery' to 'gallery,' up endless ladders that stand straight up and down; when your legs fail you at last, you lie down in a small box-car in a cramped 'incline' like a half-up-ended sewer and are dragged up to daylight feeling as if you are crawling through a coffin that has no end to it." Soaked with perspiration and shiny from dripping candle grease, he decided he need never do it again—but he did, again and again, covering new discoveries, accidents, floods, and cave-ins.

When former president Ulysses S. Grant and the first lady visited the Consolidated Virginia mine in the 1870s, they carried their own candle lamps. A short tour was all they required. Back at the top, Grant mopped his brow and had only one comment: "That's as close to hell as I ever want to get."

Walking up the incline of Union Street, I saw that Mimi was planting her feet carefully on the sidewalk, a little jumpy and annoyed when her little brother came up beside her and stamped his foot. But soon we were all back on Andria's shaded sideyard patio, forgetting about those yawning voids beneath us, with cold juices and sodas for all the children and astringent cocktails for the adults. Several of Andria's friends dropped by, and we talked about the Virginia City of the 1860s, about its long resuscitation when Lucius Beebe, Charles Clegg, and other Virginia citizens brought it back most remarkably by force of will and personality, and about its current incarnation as a tourist destination, studded with bed-and-breakfasts, museums, tours, and shops, the sidewalks populated by free-spending throngs. It was elegant, intelligent, and delightful talk and company—it would have pleased the famous sophisticates Beebe and Clegg and no doubt the original impresario John Piper. It seemed to me that Twain himself, who had prowled those meaner streets of the rambunctious and teeming mining camp in the 1860s, would have enjoyed one of Andria's gasp-inducing gin-and-tonics; even the man's famously choking cloud of cigar smoke would have been welcome (and, given the spirit of the evening, not altogether surprising) in that little convocation. Later, our large group virtually took over a Virginia City Chinese restaurant for the best I have ever found in a lifetime grazing that cuisine. We went long into the evening, and returned finally to our motel in Carson City ready for a rest.

Our next day's itinerary was a formidable one. Blinking in the morning light, wishing for more downtime, we loaded up our vehicles on autopilot and pushed off southward toward the state border with me riding shotgun in Chris's pickup and Mary following with the children in the Durango.

Our road took us due south on U.S. Highway 395 up the middle of the Carson Valley between the Sierra Nevada on our right and the Pine Nut Mountains on our left. We passed the trade and tourist towns of Minden—noting its scattering of Art Deco architecture—and Gardnerville, seeing that on the far side of town, as we passed through the Dresslerville Indian Colony, "it gets poor real fast." "These are members of the Washoe tribe," Chris explained, "who are third down in population and standard of living, after the Paiute and Shoshone. Once they lived fairly well on fish from mountain-fed streams, but when mining, ranching, and tourism came to the Carson Valley, that was it for the Washoe. A lot of them have left the reserved lands to work in various casinos in Reno and Las Vegas. The baskets their women wove prior to 1930 or so are highly prized by collectors, with prices running over $1,000 for a perfect polychrome basket." Of course nothing of that made any difference in lives in depressed Dresslerville.

A low pass took us to a high valley decorated with sage and scrub junipers; to our side a whole mountainside had been burned raw in a wildfire. The road skirted a crystalline little body of water called Topaz Lake (altitude 5,050 feet), formed eighty years before when the course of the Walker River changed and deluged an unfortunate's ranch; it would have been placid now but for its wild, rushing regatta of buzzing motorboats. We paused at a California border agriculture station checkpoint, where the sentinel was momentarily nonplussed by Mary's green Vermont license plate and suspicious about introducing New England bugs into the California ecology. Finally he decided we'd been on the road long enough to shake any pests loose and let us through.

Above Topaz Lake, then, in California, the road followed the tumbling, shallow, blue and white-frothed Walker River in Antelope Valley through forests and past cafés, barbecues, and antique motel colonies, entering a high-walled and jagged canyon in the Toiyabe National Forest. Fly fishermen in high waders stood in the sunlit currents. Boulders from on high had come to rest on the roadside, or perhaps had been pushed there after the squashed car wreckage was taken away. "In September," said Chris to Mary over the CB radio, "all the cottonwoods turn yellow and it's real purty." "It's still like spring up here," replied Mary. "Look at those meadows—lupine and Queen Anne's lace!" "And last year's snow up in the Sierra heights ahead of us," finished Chris. "It probably looked this way when old Jedediah Smith, the mountain man, crossed the Sierra here on his way to the Great Salt Lake. Captain Joseph Walker and his rip-roaring fur trappers came through going westward around 1834. And the Bidwell-Bartleson Party, first of the overland pioneers in 1841, used the Walker River as its path up to the back door of the Sierra. A lot of gold rushers rushed through later." From Sonora Junction, he said, where our road took a sharp dogleg and veered eastward, the pioneers would have ascended the Sierra over Sonora Pass, at 9,642 altitude only fifteen miles to our west, now a state road closed every winter by snows.

Higher still, following a chain of creekbeds, the road plunged between two high and blasted-looking green walls, called, unsurprisingly, Devil's Gate. "Somewhere around here in 1844," said Chris, "John C. Frémont was coming through dragging that crazy howitzer he had brought all the way from St. Louis. What a nutcase! Well, he wanted to impress the Indians and maybe the Mexicans in California, but he gave up trying to get it over the mountains and left it here. Probably acquired a fine coat of rust until some prospectors bound for Nevada—so some folks say—came through and dragged it back down the Walker River and the Carson. They say it ended up in Virginia City where they used it for patriotic occasions. At least that's what some believe." I could not resist a postscript: "Making a big noise but shooting blanks," I said. "Just like Frémont."

In the high-valley town of Bridgeport (altitude 6,473), my 1939 WPA guide to California reported that "dilapidated, partly crushed houses, long since abandoned, evidence the weight of winter snows." "We're getting to what was, last night according to the Weather Channel, the coldest spot in America," said Chris. "It was 27 degrees last night in the ghost town of Bodie, where we're headed. Under this sun it has somehow gained back nearly 70 degrees!" Twain had something to say about the local weather in *Roughing It*: "More than once . . . I have seen a perfectly blistering morning open up with the thermometer at ninety degrees at eight o'clock, and seen the snow fall fourteen inches deep and that same identical thermometer go down to forty-four degrees under shelter, before nine o'clock at night. Under favorable circumstances it snows at least once in every single month in the year."

We kept climbing canyons through piñon and pine forests, with snow visible overhead in the Sierra. Then we were turning eastward up a paved road through a side canyon, winding through tumbled boulders with the sharp smell of sage pouring into the car. A sign told us we had risen to 7,000 feet; we kept ascending along a creekbed bordered by grassy meadow bottoms and willow thickets. Finally we came up onto a high desert ringed by higher desert peaks. The road narrowed, the pavement ended, and we saw a flash of bright blue water between hills to our south—it was Mono Lake. "We're now at 8,000 feet," announced Chris over the CB. "Please make sure your seat backs are upright and your tray tables folded for landing, and make sure your seat belts are firmly fastened!" The road bumped and curved through the glaring sun, we paused at a gateway to pay a nominal fee, and we crossed into Bodie State Historical Park.

It spread across the bottom and sides of a natural amphitheater of bare brown hills that were scoured by constant high winds, beaten down and bleached out by the sun, baked in the summer and freeze-dried in the winter. This blasted ghost town—the largest in the West—in its heyday had supported some 12,000 inhabitants. The lode beneath was discovered in 1859 by two prospectors, Waterman S. Body and Black Taylor. Unfortunately for W. S. Body, while returning to his new mining claim with supplies he froze to death; the camp would be named for him, although soon its spelling was changed to "Bodie" to discourage mispronunciation. With the discovery registered as the Bunker Hill Mine in July 1861—the same year a mill was established at Bodie—the mining began, albeit quietly. Not ten miles away, across the present Nevada state line near Aurora, the Esmeralda district commanded more attention; the first gold strike in 1860 had provoked a rush that brought 10,000 hopeful miners to the district, including Sam Clemens, who, to his everlasting disappointment, despite no end of digging and speculating, never partook of even a smidgen of the estimated $30 million in bullion produced at Aurora in ten years.

Meanwhile, Bodie held only about twenty residents; the Bunker Hill Mine changed hands several times, and in 1877 was sold for $67,500 to four partners, who renamed and incorporated it as the Standard Mine. The next year came Bodie's big bonanza of gold and silver ore. The Standard itself began a run worth nearly $15 million, and Bodie's population jumped to more than 10,000 within a year. The bleak hills were quickly claimed and some $400,000 in bullion was produced per month by about thirty companies, yielding as much as $100 million during the boom years.

Under the blazing sun we stood in a dusty parking lot that held at least 100 cars and RVs. Colossal, dumbfounding pieces of rusted machinery, most dramatically a driving wheel with a twelve-foot diameter, were strewn about in the sand and sagebrush. Down a slope spread what was left of Bodie, some

170 weatherbeaten houses, shops, hotels, churches, and saloons, their tin or wood-shingle roofs pressing back against the weight of the sun's rays, maintaining a more or less upright condition despite strong persuasion from the force of gravity, and from a malevolent hot wind from the north, all in a state of "arrested decay," our brochure told us. On the hillside opposite, the remains of the Standard Mine and mill tottered, the last survivor of the crowd of enterprises at Bodie. Standard had been destroyed by fire in 1898 and rebuilt the following year as the town drifted sharply downward in vitality. Bodie had shrunk steadily; a disastrous fire in 1892 took care of many houses and buildings, and in 1932, a toddler playing with matches started the fire that destroyed most of the rest. We were gaping down on an extensive settlement in that blasted bowl, but today Bodie was less than 10 percent of its boomtown size. Standard Mine was reopened again in 1936 as the Roseclip Mine, with only a hundred or so residents left to rattle around the mostly empty town. More than fifty years ago the last residents died or slunk off, and in 1962 the state of California claimed the rarely visited ghost town and some 1,000 acres for a state park. Now at least 200,000 people venture to Bodie each year.

"Bodie was a hell of a town in the old days," said Chris, beneath his white, high-crowned cowboy hat that would have been knocked off beneath a regulation-height basketball hoop, had there been one. "At least one man dead in the streets every morning, and muggings and confidence tricks commonplace. Sixty-five saloons at the very least, and seven breweries to keep them supplied! And brothels everywhere you looked! And opium dens! And there was an old expression about outlawry here, 'the bad man from Bodie,' of which there were many. Still, there were families living here, and schools, and churches, and decent folk tried to make a go of it despite the noise and danger from the riffraff. There's an oft-told story—I think you'll like it. Once there was a little girl—she lived with her family in California, some say in Sacramento, some say San Francisco. Her parents heard the siren call to Bodie with all its riches in gold and silver, and told her that they were moving. To her this was worse than a death sentence! That night, as she was saying her prayers—or writing in her diary, as some tell it—she sadly said, 'Goodbye, God, we're going to Bodie.' This phrase got around to such an extent that everybody knew it—'Goodbye, God, we're going to Bodie!' This finally so galled the sensitive souls here in town that they began to assert that, no, everybody who told the old story had it wrong: what the little girl actually said was, 'Good, by God! We're going to Bodie!'"

With hats on and sunblock slathered, and with water bottles at the ready, I and my family picked our way down a path past wreckage and debris toward the aggregation of buildings. For his part, Chris waved and headed up the slope toward the "outcast cemetery" to pay his respects at the grave of a legendary madam of Virginia City, Carson City, and Bodie, one Rosa May, who died in this town while nursing miners through a pneumonia epidemic. Her marker had been lost over the years and the grave unknown until an amateur historian had compared antique photographs of the cemetery while standing there, triangulating himself with a view against the surrounding hills until he rediscovered the site. This led him on an obsessive, three-and-one-half-year search that took him more than 20,000 miles across the desolate regions of California and Nevada, during which he interviewed people who had known Rosa May and he had discovered a trove of her personal, handwritten letters. Chris had read the resulting book by the researcher, George Williams III, and, taken with the story, wanted to see Rosa May's resting place. He disappeared into the glare, saying he'd meet us later "downtown."

Set on the rectangular grid of the larger town that it once was, Bodie quietly baked away while curious tourists walked the streets, some with dogs on leashes, others with sunbonneted babies in backpacks, nearly everyone panting in the thin air, everyone talking in subdued tones and moving slowly in the heat. Buildings were weatherbeaten down to bare brown boards. Sad little picket fences enclosed weeds. Windows were intact, and most doors were open, and we could enter the houses and businesses before being stopped by fence wire nailed across interior doorways, through which we could see how, when the Roseclip Mine had breathed its last and gone silent, the last remaining residents had seemingly gotten up from their chairs or risen from their beds and walked out without looking back, leaving their spent old lives behind.

Dishes, tin cups, and flatware sat on kitchen tables, furred with dust; 1940s calendars yellowed on walls next to faded, turn-of-the-century maps and prints; clothes and hats waited on pegs behind half-opened closet doors; everywhere one looked were homey hand-me-downs, castoffs, bereft furnishings from vanished lives: Victorian settees, rolltop desks, leaking overstuffed chairs, mattress springs on sleigh beds or ornate iron bedsteads with graying sheets and blankets flung back as if in haste, trundle sewing machines, washbasins, Old Crow bottles, kerosene lamps, Jazz Era radios and phonographs, armoires, and gilt-edged, silver-flaking mirrors. Flowered wallpaper still adhered to some walls, brown-stained and peeling. In an austere, wooden Gothic church, heavy and baroque organs flanked an empty pulpit; in a schoolhouse, small wooden desks with empty inkwells assembled before a smudgy blackboard with the alphabet in Spencerian script; saloons waited bleakly with empty stools at dusty bars, with mousy-looking pool tables huddled in the back.

Thousands upon thousands of artifacts, together with a rich accumulation of documents, had been left behind to someday be catalogued and mined for their stories by the private support group, the Friends of Bodie, which hoped someday to open a visitors center and museum outside the town limits on the dusty approach road, so it would not affect the view and the eerie, heart-tugging feeling of connection one felt with the lives once lived here.

Chris met us back at a primitive museum and gift shop set up in the old Miner's Union Hall, where we saw a Chinese opium lamp, bottles, tins, an abacus, enameled bowls, highbutton shoes, fans, parasols, flowered hats, combs, and a corset once owned by the famous madam-with-a-heart-of-gold, Rosa May. The red light she had used to beckon customers was nearby, a delicate, wrought iron hanging lamp with red glass; her bright and friendly photographic portrait hung on a wall in an oval frame: her hair was parted in the middle and pinned in a bun; she was wearing a high-necked velvet dress with big fabric-covered buttons sewn down the arm and lace on the cuff; her arms were jauntily and confidently folded, a successful businesswoman with a round, open face and Mona Lisa smile. "She had style," said Chris. "I paid my respects up at her grave. There's a nice little board fence around it, and a decent view of the town that would not accord her burial in the proper cemetery." He looked around the museum. It was populated by murmuring tourists in shorts and hats, carrying tiny videotape cameras and fancy digital recorders. "We are 350 miles and 500 light years from Los Angeles," he exclaimed. "Good, by God! We're going to Bodie!"

Back on Highway 395, we descended and then ascended through dismal sagebrush hills, pausing at a marker and the site of Dogtown, elevation 7,000 feet, a long-vanished, short-lived mining camp that stirred local excitement in 1859 until rumors of strikes elsewhere emptied it overnight. Then, after we achieved a greater height, from 8,135 feet we dropped 1,100 ear-popping feet in less than ten minutes as the highway looped down steep but graceful curves: piercing blue and briny Mono Lake spread below, its two ash-colored volcanic cones rising from the alkaline waters, its beaches white with chemical foam.

A Los Angeles friend, Stephen Fisher, producer of a superb PBS documentary on novelist and environmentalist Wallace Stegner (and also of many nature films), had produced one on historic Mono Lake and the threats from faraway, formidably thirsty greater Los Angeles, which in the 1980s sought to "draw down" the clear-running feeder streams into diversion tunnels for the greater good, so-called, of the metropolis. Mono Lake, like Great Salt Lake, has no outlet and is a harsh chemical soup that itself supports only tiny saltwater shrimp and blackflies. However, in estuaries at the mouths of those feeder streams, the lake had been host to the second largest colony of California gulls. Numerous and diverse wildlife depended on the lake.

But once opened the urban long-distance water-guzzling lowered the lake level alarmingly, creating land bridges and endangering vast island nesting grounds. Thousands of gulls died off and other species began to disappear.

As Steve's film showed, after an improbable coalition of local Native Americans, activists, and lawyers pressed a long and heated campaign, fought ferociously by Los Angeles Water and Power and allies, the California State Supreme Court eventually ruled that Los Angeles must respect the public trust and heritage, and protect an environmental ecosystem. That particular battle was decided favorably, though predictably the front has shifted to many other endangered places. For us the sight of Mono was a familiar one because of Fisher's hauntingly beautiful nature documentary; it was thrilling to see it in its bowl of high volcanic peaks with the alkaline water's vivid blue color a treat for the eyes.

It was a "lonely tenant of the loneliest spot on earth," in Twain's words. He came to know it well, having followed rumors from the nearby Esmeralda district of a fabled, "lost" cement mine. The indomitable legend held that back in the 1850s, three brothers from Germany were on their way to the California gold fields when they stumbled upon a vein of naturally occurring cement in which lumps of gold were "set like raisins in a pudding." Supposedly, winter roared in before the brothers could do anything about their claim; two froze to death and the third was driven mad, condemned for the rest of his life to babble incoherently about the lost cement mine somewhere up near the headwaters of the Owen River. When Twain and some companions investigated the Mono region for the cement mine, they took time to explore the lake in a hired rowboat. This nearly turned to disaster when a quick storm blew up while they were out in the middle of the eighty-seven-square-mile lake. They avoided capsizing with difficulty until they were close enough to shore to touch it—and then "the sea gave the boat a twist," recalled Twain, "and over she went! The agony that alkali water inflicts on bruises, chafes and blistered hands, is unspeakable, and nothing but greasing all over will modify it—but we ate, drank and slept well, that night, notwithstanding."

We had a destination and a mission of our own at Mono Lake. Down above the shore on the highway, we stopped to examine a bronze plaque set on a concrete pedestal: it memorialized the brief life of Adaline Carson Stilts, daughter of the legendary mountain man and scout Kit Carson, upon whom John C. Frémont had depended in his Western explorations. Adaline Carson was a shadowy and tantalizing historical figure, born in 1839, about whom Carson and his biographers have said little. Her mother was a full-blooded Arapaho, with whom Carson lived in Colorado, and who died when Adaline was little. Instead of abandoning his daughter or leaving her with strangers when Frémont summoned him to another expedition, Carson took her to St.

Louis to be raised by his married sister. Adaline attended a female seminary there. By the time she was fourteen, Carson had remarried and was planning to settle at Taos, New Mexico, so he reclaimed his daughter to live with his new family. A year later, when Adaline was fifteen, Carson allowed her to marry a man named Louis Simmons, who was middle-aged, but Simmons soon either deserted or divorced her. Supposedly, according to a contemporary who knew the family, Adaline "was a wild girl and did not behave properly." At any rate, she apparently made her way to California in 1859 during the Esmeralda gold rush—or, perhaps, during the early flurries of interest at Bodie or Mono Lake—and became involved with a man remembered only for his name, George Stilts, and his connection to her. She died about 1860, aged perhaps twenty-one.

The roadside marker indicated that she was buried somewhere nearby, but that wasn't good enough for us: Adaline Carson had significance for my family, and Chris thought if we looked around we might find her resting place. It was worth the stop and search because a dear friend and fellow writer in New York, Mary Pope Osborne, had earlier that year published a luminous little historical novel about Carson's daughter, *Adaline Falling Star*. I had known Mary Osborne for nearly twenty years, from our times at the Bread Loaf Writers' Conference in Vermont. She had been born at Fort Sill in Oklahoma and raised on army posts, mostly in the South, an army colonel's daughter who followed the tie-dyed banner of Flower Power and explored the world's hippie capitals, from San Francisco to Katmandu. When I came to know her in Vermont in the early 1980s, she lived a dozen blocks away from my New York apartment, in Greenwich Village with her husband, Will Osborne, a successful actor, where she wrote books for young people. For years we met over lunch or dinner, encouraging each other's book projects; she was like a sister to me, and after my wife and I moved to Vermont and had our children, Mary Osborne became an honorary aunt, sending lovingly inscribed books to Mimi and Davey and entertaining us during the rare times we visited New York City; a character in one of her novels was named after Mimi. *Adaline Falling Star* came into being while Mary was researching a book on American tall tales—she was haunted by what little was known about Adaline: "I sensed," she wrote, "that Adaline had been misunderstood and that she had her own story to tell. Soon a fictional Adaline was born and she provoked this imaginative story." The novel was a tremendous hit both in our household and in the world at large. It concerned the eleven-year-old's life in her foster home in St. Louis, which must have been difficult given her "half-breed" heritage, and her flight into the Wild West looking for her people and her father. We loved Adaline's voice as much as her adventures, which reminded me of Twain's Huck Finn. "The night I was born on Horse Creek," her story begins, the sky rained fire. Dogs howled and growled. Arapaho warriors put on red war paint and did a death dance. Leastwise, that's what Pa tells me. He says my hair was as black as a crow's wing, my eyes the color of a mud pond; but my skin was the color of a half-tanned fox hide, plainly showing the mix of white and Indian blood. My ma told me Pa was so proud he shouted with joy: "Outta her way and let her ride!"

We backtracked up the highway and turned off on Cemetery Road, though Chris had earlier explored the formal burial ground and didn't recall seeing a marker there. We stood out in late afternoon sun down near the lake edge. Chris squinted and raised his chin, his eyes set on the sloping western shore, where a dry creek meandered toward a thick stand of brush and trees, and he seemed to be tuning into the terrain as if onto a radio frequency, or like a dowser sniffing for hidden water. Me, at the moment I felt like helpless Frémont trusting the impulses and superior knowledge of Kit Carson, but abruptly Chris got us piled back into our vehicles and we roared back up the dusty road toward a trailer set into the hillside. He knocked on a dark screen door and talked to someone back in the shadowy recesses, and then wheeled around and was back in the truck in an instant. "Got it!" he said. "We're to follow these ruts into the trees!" Minutes later, we dead-ended in brush and Chris was bounding toward a half-visible pile of stones. Dappled with sun, it was a crude cairn of weathered river rocks set in cement, looking oddly like an old-fashioned barbecue pit and chimney, with a bronze marker at its center: *"Adaline Carson Stilts,"* it read. *" 'Prairie Flower,' Daughter of Kit Carson. Born 1839. Died at Mono Diggings About 1860."* It was a beautiful resting place despite being untended and overgrown. Below us, Mono Lake reflected the wide sky, and the slope behind Adaline's marker rose steeply past a heedless highway toward high and wild mountain crags. We took photographs to send back to Mary Pope Osborne, who had always wanted to search for this place but had never managed the trip. As I turned to leave, Chris held my arm. "Down there over the grave," he said quietly. "See the sagebrush sprouting? Why don't you pull off a piece and send it to your friend?"

I did. Mary wrote me that when she opened the envelope with my letter and the photographs, and smelled the sharp sagebrush, and saw its origin, she cried.

We spent the night in the Owens Valley town of Bishop, California, about sixty miles south of Mono Lake; the highway slipped between vast Sierra wilderness areas named for photographer Ansel Adams and naturalist John Muir, both westward, and the Inyo National Forest to the east, passing the ashy-looking Mono Craters and the decrepit red and volcanic Aeolian Buttes. At one point, climbing toward Deadman Summit (altitude 8,041) we penetrated a mixed forest of Jeffrey pine and lodgepole pine, and Chris rolled down his window: "Smells like vanilla or bourbon, depending on your point of view," he said. We paused a few miles off the highway at Hot Creek to see people paddling in creek water at the bottom of a gorge, while, dozens of yards away on the far side of warning signs, volcanic vents warmed the immediate water to a bubbling boil. We crossed a large volcanic tableland under a heavy pall of wood smoke—somewhere in Yosemite, the radio reported, a tremendous wildfire was underway. Bishop was socked in by smoke. In the morning the wind had taken it all away, revealing to our surprise the 15,000-foot walls of the White Mountains and the 10,000-foot ridge of the East Sierra, making it suddenly clear why Bishop draws legions of Los Angeles hikers, fishers, and gawkers.

After breakfast we drove a few miles north to where Horton Creek led us between the Tungsten Hills and the high Sierra; far above us, one could find its source in a chain of high-canyon lakes in the John Muir Wilderness, but we soon stopped where the tree-fringed creek led across a hot sandy valley floor, attracting a bare-necessities state campground with just a few tents in evidence. Here for the summer was another friend of Chris Graves's. Arlene Halford was a folk artist in her early fifties, who lived with her husband, Bill. He was off somewhere, but Arlene showed us around their campsite—two tents, an old card table and two folding chairs, a picnic table, cast iron pots and pans and a coffeepot. The only shade came from two tiny trees she watered every day with buckets from the nearby creek.

Arlene was strong and earthy, with gray hair in a bun and with direct-looking eyes. She liked to keep her hands busy while she talked. Near her feet were great snarls of rusty old baling wire, heaps of barrel hoops, and ancient, rusty sardine tins, and after we all found seats she twisted the wire quickly and expertly. "I was a nurse for eighteen years in Texas," she said. "Then the AIDS crisis finally got out our way, and since state laws protected patients from telling health care workers about their illnesses, a couple of my friends contracted AIDS while caring for their patients. I decided the job was too risky."

Under her hands the wire was taking shape—four legs, a tail, a body, and she kept twisting and turning it. "I became a waitress. I always liked taking care of people, and that was another way of doing it, but my legs gave out. Then I drove an eighteen-wheeler with one husband for five years—I was married four times, no, five, actually, since I married the first one twice. Something about my taste in men, I guess. Two were womanizers, and I wouldn't stand still for that. One husband slipped a cog." A head with ears took shape in the wire. "I have three daughters. I've been with Bill—he's younger than I—for seven years. If this life was fair, he'd be making a living playing guitar and writing songs, but he mostly does it for self-satisfaction. They're great songs." She gestured at the tents. "We don't like living in a house. We had twenty acres and a house up in Fallon, Nevada, but we didn't like it."

She set the completed figure on the table in front of Mimi. It was a perfectly proportioned sculpture of a mule, standing about a foot tall. "I've been tying flies for fishermen for years on the side," she explained, grabbing a new hank of wire and starting again. "The local stores sell them. A while back, a lady said, don't you have a way to display these? Well, I thought about that. I went to art school for one year and I remembered making papier-mâché animals on wire frames. Two creeks over from here, a farmer left a lot of baling wire. Since Bishop was once known as the mule capital of the world—animals were raised here by the thousands for the freighters of the Santa Fe Trail and the California Trail, and many wagon trains, not to forget the army—I decided to make a mule to hang the tied flies on in the local stores. But then people wanted to buy the mules. So that's how I keep busy nowadays. Lately I've been branching out." She reached into her tent and extracted a sculpture of a bear rummaging in a garbage can, a common sight in the mountains.

Then she returned to the second mule, a foal to the first. I remarked that the campground seemed so arid and lonely, perhaps even dangerous—just a couple of shade trees, and the nylon walls of her tent between her and the rest of the universe, which included a hot sun. "There are lots of people here," she replied. "They're rock climbers who like the cliffs over there, or they come here to work for $5 an hour washing dishes in the Bishop restaurants or stocking cans in the groceries, and they don't have to pay any rent for the whole season. We're a nice little temporary community here."

Earlier, Chris had commissioned Arlene to make a twenty-mule team, a freight wagon, and driver, all out of baling wire; we later saw it in his California sunroom, sitting on a specially made shelf over the top of a door where its six-foot-eight-inch owner had the best view of it. Every mule on that team had its own personality. While we sat with her, Arlene worked through our commission, a family of four mules that later took up permanent residence on a living room end table, and several more for relatives. She nicely threw in a handmade picture frame—the center was a rusty antique sardine tin, surrounded by a corona of symmetrically twisted baling wire.

That morning we never got to meet her husband, Bill. The next year, he rolled his pickup while driving out in the desert and was killed instantly. Arlene, whose hardscrabble life had been borne with such grace, and that had turned so late to a vital, witty, and delicate art out of rough and raw found materials, wrote us later to say that she would endure this last blow, too. Of that we had no doubt. Our last view of her on that hot August morning was of a lovely smile and a wave as she walked through brush and white sand to look in on a neighbor who was feeling poorly and confined to a tent. "After that," Arlene said, "I think I'll just head over to our little creek across there behind the trees. The water will feel cool about now, and a swim would be nice."

Sixteen miles on Highway 395 down the Owens Valley below Bishop, we turned northeast toward the Nevada border, headed on a long arc through a chain of "ghosts and near-ghosts," as Westerners called their old mining camps. There were hundreds of them in Nevada's Nye and Esmeralda counties. Highway 168 was also called the Westgard Pass Toll Road from its late-nineteenth-century days when J. S. "Scott" Broder completed it and collected tolls, and from its early-twentieth-century days when the early good roads booster A. L. Westgard led an American Automobile Association tour across the 8,000-foot pass in 1913. At the base of the mountain road a giant sequoia had stood since it was planted that same year to honor Theodore Roosevelt. The road wound through barren foothills, occasional volcanic outcroppings, and sharp talus slopes, then steadily climbed through junipers and bristle-cones to the pass. We sped down hairpins to an ancient lake bottom below sheer cliffs and a grove of cottonwoods. At Deep Springs there were old lava flows, consisting of black, unforgivingly sharp rocks. Surmounting another high ridge through tortuous curves, we reached the bottom of a place called Fish Lake Valley, and a spot on the map called Oasis, and soon were crossing back into Nevada on Highway 266. The road passed through the Palmetto Mountains; above the road evidence of old mines and mills were as plain as new claims, which usually were marked by ramming a white plastic pipe into the hillside. Piñon trees covered the hills and mountainsides. John C. Frémont had entered California this way on his fifth expedition in 1854.

We crossed Lida Summit at 7,400 feet and saw buttes and hills beyond. The rim of Death Valley was about fifty miles southeast, while just a few miles ahead began the vast off-limits wasteland of Nellis Air Force Bombing and Gunnery Range. Beyond is the Nellis Air Force Base with its Nevada Test Site and mysterious Area 51, a favorite nexus for speculation by UFO fanatics. Looking at its gray expanse of mountains and mesas on my highly detailed DeLorme map book, I shuddered and was glad we were soon turning away from it. At Lida Junction our road dead-ended at Highway 95 and we turned north. Behind high chain link fences and a sign proclaiming "Always Open!" was the collection of trailers and low buildings known as the Cottontail Ranch, a brothel on the well-traveled tourist and trucker route between Las Vegas, 180 miles southeast, and Reno, 250 miles northwest. For us, just 15 miles away was the near–ghost town of Goldfield, center of a fabulous strike and boom in the decade and a half before World War One, when its peak population was 20,000. Between 1903 and 1940, a total of $86,765,044 in precious metals was mined and milled there.

Goldfield had been a near-ghost for a lifetime. As we sped past numerous Joshua trees and through a series of reddish hills, I was sitting next to Mary in the Durango, reading to her from the WPA guide to Nevada, which in delightfully blunt terms summed up what was left of the boomtown in the late 1930s. "Seen from the highway," I read,

> this fabulous town is drearier than a graveyard—for no one expects anything of the dead and Goldfield is not a ghost. Fifty-two city blocks of the lower part of the city are covered with brush, crumbling, windowless walls, tilted fireplugs, bits of scrap iron— the results of a fire in 1923. Above this waste on a plateau are remnants of the old city, the 200-room hotel now boarded up— the big stone crenellated courthouse, the two stone schools—one boarded up—low comfortable houses with stiff joshua-trees before their doors, and a few of the older business buildings where stock-brokers formerly did business all night long and fortunes were made and lost daily. Outside county and civic officials, most of the remaining people are householders who live on their savings, positive that Goldfield will come back again.

It hadn't, more than sixty years later. The recovering price of gold and more efficient refining techniques had transformed some once moribund communities, and some of its more stubborn residents swear that prosperity will someday return, but the place still looked as if it had been carpet-bombed. If I looked in particular directions I could have been in Dresden or Manila after VE and VJ days, 1945. The weedy empty lots seemed untouched after an additional six decades and now both of the big stone schools were shuttered and shattered. A bare, ruined Roman Catholic cathedral stood on the north end of town, blackened by the fires of 1923. The high heat of the day (it was a Saturday) might have driven everyone inside—Goldfield's population was said to be 450—but it truly seemed deserted even of survivors. We drove around city blocks of rubble alternating with blocks of intact but darkened houses and businesses. One building was built entirely of old bottles cemented into walls, but it was hard to see when it had been lit from within. Two taverns seemed to be the only enterprises inside the city limits; both had real estate agents' placards tacked on them. In front of a midtown junkyard was this sign: NOTHING IS FOR SALE.

In the center of the town rose the Goldfield Hotel, oddly and surprisingly spruced up: four stories of repointed brick with a fancy recessed entrance, and relatively clean and clear plate glass windows affording sight of broomed lobbies, Sheetrocked walls, paint-stripped decorative ornamentation, and a swept restaurant with an imposing bar. But refurbishment had obviously stopped abruptly.

After walking around town with us for a while, Mary took the children ahead to a motel up the highway in another old boomtown, Tonopah, while Chris and I dropped by the municipal building. "The fact that Goldfield is the county seat—for lack of a suitable replacement—seems to be the only thing keeping it alive," said Chris. "There are only about 1,300 souls in Esmeralda. That's a lot of empty space!" Buzzing us in was another of Chris's acquaintances, the Honorable Juanita Colvin, justice of the peace of Esmeralda Township, who had started there as clerk of the court some eighteen years before and had subsequently been elected justice of the peace a number of times, using an old-fashioned-looking pictorial WANTED poster (it showed her to great effect) in her campaign.

The license plate on her truck was LWNORDER. Later, we learned, Juanita's husband was Ben Colvin, a rancher and local-property-rights activist who, in a dispute that earned national headlines in 2001, saw sixty-two of his cattle seized and auctioned off by the federal Bureau of Land Management. BLM cited that by grazing his cattle on federal land he was trespassing and owed the government some $73,000 in back fines and fees; Colvin was also overgrazing the land held in trust by the BLM for the people of the United States. Federal land grazing was controversial back in Washington. Environmentalists contend that the practice degrades delicate ecosystems and should end; Colvin and his supporters contend that the federal government has no right to dictate how they use the land. All hearings on the case were broken by rallies and courtroom protests—this debate is growing serious in Nevada, where the federal government owns 87 percent of the state's landmass—and at the mandated cattle auction guns were evident on both sides, and tensions were high.

But this hot dispute was still in the future when we visited J.P. Juanita Colvin. Chris liked to stop in to chat when he was wandering the territory looking for old historical places and artifacts. Juanita loved talking about the history of Goldfield and the wealth of anecdotes at her fingertips in the county files. She took us on a tour of the building and showed us antique official documents she had framed as decorations up- and downstairs. "You gotta love this place," said the justice of the peace, "for the stories it contains."

After we saw her courtroom, Juanita took us down for a tour of the sheriff's offices, where we met officer Ken Aldrich, who was happy to show us around and to share some anecdotes. "Once in wintertime," he said, "a woman and her car slid into a drift off the highway south of here. She came into the police station to ask for help, dead of winter and she's wearing sandals and short-shorts. This led us to believe that she'd come up from the Cottontail Ranch. So the officers on duty took some chains and a hook down to the scene and began to pull her out. One of the officers had nothing to do, so he went back to wait inside his warm patrol car, and to pass the time he ran her plates through the computer. The car came up stolen." We laughed and he shrugged. "These people aren't the brightest pebbles on the beach."

Chris's eyes lit up when he saw the patrol room's doorstop: it was an old leg iron. "I collect iron," he told the officer. "Anything made of iron. How much do you want?" "Nah," said Ken, chuckling, "then what would we do for a doorstop?" He took us on a tour of the lockup. It looked like a ship's brig, with steel floors, four-by-six cells, tiny bunks, and a toilet out at the end of the hall. We passed four prisoners as we went up the circular stairs to see the upper level. They were playing Monopoly. "Okay Larry," said one. "You gotta go to jail."

On the sides of the highway, which ran for twenty-five miles almost arrow-straight through the desert between Goldfield and Tonopah, white crosses bloomed every few miles among the creosote bushes and sagebrush, marking vehicular deaths. We could see the Silver Peak Range, about twenty miles west, with the white-capped Sierras rising some distance beyond. We passed from Esmeralda County into Nye County. Crossing a low divide, Tonopah Summit (6,260 feet), we descended toward a mining town that had been founded in an arrested act of animal cruelty.

The story, told frequently, concerned a prospector and sometimes attorney named James Butler who in 1900 wandered into this wasteland and made an overnight camp at Tonopah Spring. The whole area was open and mysterious, the future boomtowns of Tonopah and its child, Goldfield, were not even gleams in anyone's eyes at that point. When Butler awoke in camp in the early morning he found that his burro had wandered off. It took him some time to find the beast, which instead of submitting to harness kicked up its hooves and bounded off. Butler picked up a rock to throw it—but the rock had an unusual heft to it. Instead, he put it in his pack along with a number of other samples from an outcropping. Then, with burro in hand, he went off to have the samples assayed. The ledges proved out at values of $800 per ton in gold and silver.

Prospectors, investors, miners, suppliers, saloon keepers, gamblers, and prostitutes poured in and Tonapah grew rapidly, even acquiring an urban and genteel upper crust to its society. The boom peaked in 1913 with an annual production of $9.5 million and thereafter steadily declined into the 1940s. With World War Two a nearby air base opened, keeping some vitality in town until the base was closed in the 1970s, but later there were the rising prices of precious metals, the nearby discovery of rich molybdenum ore veins, and the convenience of the town as a way stop between Las Vegas and Reno. Tonapah's onetime luxury hotel, the Mizpah, was, like Goldfield's, respectfully renovated, but the difference was that the Mizpah actually opened and seemed to enjoy a healthy occupancy rate. We had all checked into a more modest motel a few blocks away, but we strolled over to the Mizpah around dinnertime to inspect the street-level public rooms and perhaps have dinner. The restaurant was named in honor of heavyweight boxing champ Jack Dempsey, born in Colorado, who once worked as bartender and bouncer at the Mizpah. The menu was too conventional for our tastes that evening, however, so we went elsewhere. Later, I discovered a wonderful anecdote about the Mizpah concerning the legendary political figure Key Pittman, who, on the election night of his last U.S. senatorial campaign, died of natural causes at the hotel before a winner was announced. His handlers called for a lot of ice, and kept the senator cool until Pittman was called the winner. They later said that he had died during the victory party.

At breakfast the next morning we realized that this would be our last full day with Chris as our guide. From Ogden, across Utah and Nevada and with various side trips and our big circle through western Nevada, we had been in close company for ten days and some 1,700 miles; our ears rang with his booming voice and our minds with his stories. On this, our last day together, he would appear to us more like Kit Carson than ever as he led us far off the beaten track, nearly off our maps, and deep into history.

North of Tonopah we drove up the Big Smoky Valley, traversed by John C. Frémont and Kit Carson and party in November 1845 as they cut diagonally across present day Nevada, northeast to southwest over empty washboard ranges on the long traverse from Great Salt Lake to California. We drove first across a featureless desert, with the low ridge of the San Antonio Mountains just to our east, then entered the narrow valley proper, which is bordered by the abupt Toiyabe Range westward and the more gradual-rising Toquima Range eastward, both chains designated national forests. As we turned north up the Big Smoky, there stood the huge, bustling Smoky Valley Mine, operated by the Round Mountain Gold Corporation. "When gold hit $600 an ounce," said Chris, "many things just exploded in Nevada. Others just continued to rest in peace."

About thirty-five miles farther north, Chris pulled off onto gravel. Unmarked ruts led across the desert toward the Toiyabe Mountains and a cluster of greenery at the base, a couple of miles away. We bumped and lurched over there, thick dust billowing and making the second car invisible, and when we reached the vegetation we saw there was a narrow opening between perpendicular cliffs. "Ophir up ahead," pronounced Chris. "Up, up, up, and ahead!" He drove on, and Mary followed. Tracing a tiny creek and squeezing between those high walls, we climbed some 2,000 feet in the next half hour, driving across (and through) the creek three or four times, getting out to scrutinize the rough path ahead, bumping over rocks, spinning gravel, the sides of our vehicles being scraped by overhanging bushes.

Deserted hovels presented themselves along the path, then one-story stone ruins with proper door and window openings and no roofs, and then a little bank building with a large safe inside it, and then more buildings. Then there were several-story mill ruins, stair-stepping down a short slope, across the path from a smelter chimney, eight feet tall and constructed of suitcase-sized rocks. The canyon face was nearly perpendicular on the south side, but was climbable as it rose north, sporadically patched with dry weeds and rubble. Little donkey paths went up impossibly steep hillsides, terminating at collapsed mine holes. There were piles of tailings everywhere, mostly sandy-colored or russet-tinged rocks. Up the canyon the mountains rose precipitous overhead, still climbing toward a nearly 11,000-foot ridgeline two miles westward. It was deserted and eerily quiet except for the sound of wind and our heavy breathing from the altitude, and as I looked out at mysterious, long-deserted stone buildings shimmering in high heat and bright sun, I felt like a long-ago explorer discovering an ancient civilization. The name Ophir came out of antiquity, derived from the Bible—I Kings, Chapter 10, Verses 10 and 11—in which the Queen of Sheba showered King Solomon with gold from her city of Ophir, along with "great plenty of almug trees, and precious stones." Back in the nineteenth century, when so many more Americans had at least a passing familiarity with scripture, with many readily able to quote chapter and verse, Ophir would have been a recognizable reference. This explains why it became such a popular name for mining camps; there was, for instance, another Ophir in northern Nevada, near Virginia City, where Mark Twain had tramped.

We climbed out of our vehicles and spread out to investigate the place, mindful of snakes and pits and falling rocks. I skirted some fallen-in tunnel entrances and scrambled up the north slope through weeds and across tailing piles. A hundred feet above the trucks, panting from the high-altitude exertion and the 100 degree heat, I stumbled across a pile of rusted tin cans from the nineteenth century—they bore the characteristic lead solder circling on the top. Strangely, the cans lay in a line through the weeds. I curiously followed and found myself back where I started; it was a nearly perfect circle. But what was at the center? There on the hillside I found a rectangular hole, six feet long, three feet wide, and three feet deep, the side walls shored up with thick planks. I hopped down and crouched for a long time. It wasn't a grave, despite the dimensions. Then I recalled that Mark Twain had written in *Roughing It* about the hovels of desert rats. Someone had *lived* there, and covered the scraped-out hole with canvas or wood scraps and tin, and each morning and evening had made his meal from a storage of cans, and each morning and evening had tossed the cans out into the weeds into his nearly perfect circle. And after the celestial dark had fallen he would have been free to ponder the riches that would be his, as had prospector Mark Twain when he lay abroad awake, thinking, dreaming, scheming: "The floorless, tumble-down cabin was a palace, the ragged gray blankets silk, the furniture rosewood and mahogany. Each new splendor that burst out of my visions of the future whirled me bodily over in bed or jerked me to a sitting posture just as if an electric battery had been applied to me." And finally the desert rat would have fallen into exhausted sleep.

I hunted around some more before returning to the canyon base. Mary was off peering into little stone houses. Davey and Mimi were down in the ruins of the stamping mill, on the lowest level, my boy looking for lizards and my girl for gold. After Mimi clambered back upslope to show me some promising bright streaks in a handful of rocks, leaving Davey on his lizard hunt, Chris came over to my side; he called down for my boy to take a roundabout route to join us at the vehicles. "I didn't want to scare him," Chris said, "but I think there's a rattler down there . . . hear him?" I did, and we stepped carefully down along a talus slope, hoping to see the snake, but then some kind of flying locust or grasshopper sprang into the air and hovered at head level, loudly chattering, and we realized it was a false alarm. After we had our fill of Ophir, we carefully picked our way back down the canyon, realizing when facing a downward roller-coaster angle how steep the climb had been.

Finally we found ourselves down on pavement. Ophir was printed on my DeLorme map, designated as a regular city or town, connected to a gray valley floor by the thin red spider-strand of a road we had just followed. "I can just imagine some Fuller Brush Man going up there," I said. "'Oh, Ophir— might find some prospects!' And look what he finds."

North up the valley, not ten miles farther, Chris pulled off again and we all got out. He pointed to a rounded promontory in the Toiyabes with a green, brushy base, and told us that it was the entrance to Park Canyon, another abandoned diggings and mill. We were eager to find it. Mary set off first this time, piloting the high-clearance Durango, navigating across the desert flat and rounding the mountainside, and then leading us up the canyon, having the time of her life. The road was narrow and steep but in smoother shape than at Ophir, and it employed more switchbacks, gaining us more altitude in quicker time. The terrain finally opened up on a broad, grassy shoulder covered with trees and fed by a stronger spring, which burbled downhill along a

deep crack nearby. We stopped and got out, moved and thrilled by the surroundings and feeling we had to pay tribute at the sudden prospect of a three-
story wreck—a beautifully wrought stone stamping mill, only three exterior
walls remaining with many window cavities adorned with brick lintels and
granite sills. Smaller building foundations and standing chimneys rose among
high brush and saplings. A wagon path ascended around the mill walls to its
highest level before trailing off westward. There was a stirring view of the flat
Big Smoky Valley floor and the always hazy Toquima Mountains.

I had no way of knowing then, nor now, but the high shoulder may have
been a camping spot for Frémont and Carson and company in November
1845: they had looked for water at the mountain base and, not finding it, had
climbed to find a spring nearly 2,000 feet up. "A bench of the mountain near
by made a good camping ground," recalled Frémont, "for the November
nights were cool and newly-fallen snow already marked out the higher ridges
of the mountains. With grass abundant, and pine wood and cedars to keep up
the night fires, we were well provided for." One of their number, a Delaware
Indian named Sagundai, reported that he had found "fresh tracks made in
the sand by a woman's naked foot," but no other indications of life, so the
party busied itself cooking the haunches of an antelope Carson shot that
afternoon, a rarity in that empty quarter. After the meal all lounged on the
ground, smoking and talking quietly. Carson was lying on his back with his
pipe in his mouth. Suddenly he rose and pointed to the far side of the fire:
"Good God! Look there!"

"In the blaze of the fire," wrote Frémont,

> peering over her skinny, crooked hands, which shaded her eyes
> from the glare, was standing an old woman apparently eighty
> years of age, nearly naked, her grizzly hair hanging down over her
> face and shoulders. She had thought it a camp of her people and
> had already begun to talk and gesticulate, when her open mouth
> was paralyzed with fright, as she saw the faces of the whites. She
> turned to escape, but the men had gathered about her and brought
> her around to the fire. Hunger and cold soon dispelled fear and
> she made us understand that she had been left by her people at the
> spring to die, because she was very old and could gather no more
> seeds and was no longer good for anything. She told us she had
> nothing to eat and was very hungry. We gave her immediately
> about a quarter of the antelope, thinking she would roast it by our
> fire, but no sooner did she get it in her hand than she darted off
> into the darkness. Some one ran after her with a brand of fire, but
> calling after her brought no answer.

They could find only her footprints the next morning, but before they departed they left her a cache of food from their dwindling resources.

I told Chris and Mary the story while the children played hide and seek around the mill and clambered into window cavities. "It could have happened here," he agreed. "The terrain seems to match. That would have been decades before the gold seekers came through and rediscovered this place with its water and shade."

"At least," said Mary, "you can look at the canyon and imagine them being here."

Later, I stood holding hands with Mary at the upper mill walls, looking past them and the healthy greenery toward the commanding openness of the blue space above the valley. "I could spend a summer here," mused Mary, "and paint these ruins and this view." There had been certain places along our transcontinental travel that had spoken to her and given her visions like old memories: Monument Point on the Great Salt Lake had tugged at her strongest, followed by this surprising outlook over the dry, hazy valley. In my mind's eye on canvas I could see the sand-colored, resolute but now uncompleted walls, the classically proportioned window cavities, the blossoming, overtaking green trees, the rising mountains, as she would render them, and I hoped we could sometime act on that impulse and return.

Down just short of the highway, Chris abruptly pulled his pickup over. "Something's not right," he said, and got out to inspect his tires. "Just as I thought," he blurted. He could hear the air escaping from a puncture. After a quick conference, Chris decided to retrace southward to the last settlement, Carvers, where he thought he had seen a garage and gas station, hoping to reach it before his tire went entirely flat. I jumped in beside him and he peeled off onto the highway, with Mary following us. I sat next to him, white-knuckling my knees, wishing I was back with my family and hoping that his damaged tire didn't explode and send us into any oncoming traffic or into a rollover. All those white roadside crosses seemed a little brighter as we flew by them. Down at Carvers, we found the garage closed, but Chris got directions to the owner's house from a counter girl in the station store and went off to find him. In minutes he returned with the affable and accommodating Bruce Clauser of Smoky Valley Tire, who didn't mind interrupting an afternoon television show on his day off to help a stranger. His children and wife wandered by as he worked, curious to see people from far away. While Bruce patched the tire, he talked of his quiet life there, and his nostalgia for the time of his boyhood when federal clean water laws did not curtail him from grazing cattle in the mountains, or hunting, or diverting water for his own family, which had been there for generations. "Seems like the government would just prefer us to leave," he said, spinning the tire for emphasis, "but this is the only country we know." Being able to see two sides of the argument did not prevent us from feeling empathy for him, trying to make ends meet in a barren world whose boundaries, for him, were squeezing ever closer.

The northern end of the Big Smoky Valley is punctuated by two-lane U.S. 50, which winds its patient way from the Great Salt Lake Desert of Utah across a numbing but beautiful procession of high mountain ridges and deep canyons directly across the state of Nevada to the California line below Zephyr Cove on Lake Tahoe's southern shore, some 468 miles from beginning to end. Route 50 was once dubbed "The Loneliest Road in America" by the old *Life* magazine, which devastated civic boosters in Ely, Eureka, Austin, and Fallon, Nevada, and all the gas pump shacks and tourist courts in between, until finally they turned a tourism defeat into a triumph by erecting proud signs along the road's length, claiming the moniker for their own. These signs soon became riddled with bullet holes from rifles and pistols—nothing new along those corduroy stretches of sage, rabbit brush, juniper, and scrub pine. Organizers of the Lincoln Highway, which beginning in 1919 had followed Route 50 across Nevada to Carson City and then over the Sierra by the route between Lake Tahoe and Placerville, California, put up with such ballistic desecrations to their red, white, and blue signs way out on the other side of nowhere, from the time those patriotic markers were planted across the state in 1928. Had the Overland Stage troubled to mark the line followed by its many patrons, including Samuel Clemens, doubtless those markers would have been riddled, as would those of the freighting firm of Russell, Majors, & Waddell when their Pony Express route followed the same path in 1861. When we turned west onto Route 50—I was on driving duty with my family in the Dodge—we followed Chris's white pickup up into the Toiyabe Mountains, heading toward the old silver town of Austin, now a near-ghost but once capital of the Reese River district rush beginning in 1862.

Beyond the summit we plummeted down the steep and curvy Toiyabe Canyon, veering out on a high shoulder where we could see Swiss cheese holes of old mines and also the tin roofs and streetscape of Austin, far below.

In 1924, the official Lincoln Highway guide noted that Austin, population 700, had three hotels (one grandly called the International), three garages, one bank, one Nevada Central Railroad station, twenty business places, an express company and telegraph, a newspaper, public school, and free campgrounds with running water and electric lights. In 1867, Massachusetts newspaper editor Samuel Bowles visited and reported the population as 3,000, down from 8,000 in just four years.

We drove around town, finding that it seemed to be about half inhabited and half abandoned, judging by the untended or boarded-up houses scattered on streets that rose abruptly toward surrounding mountains. But Austin's two high-steepled brick churches (Episcopal and Methodist) seemed to be struggling along and there were a few stores, antiques shops, gas stations, and motels—about two each, from what I could tell—and we pulled up in front of one of the motels, an Art Deco remnant from the Lincoln Highway days. "I've been a little worried about this place all along," Chris muttered to me in the parking lot. "So far we've done all right as far as accommodations— Budget This, Super-That, Econo-You-Name-It—but I have been fearing that your fair lady is going to be appalled by this motel. But there are no alternatives." He looked over to where she was marshaling the children to transfer cans, cups, and other litter from the floor of the car into a plastic bag, and then glanced across the lot to a number of Harley motorcycles parked in front of some rooms. "Your bride is used to much better!"

I laughed and thanked him for his concern. "Mary's infinitely adventurous," I reassured him. "She may have a gentle exterior, but she's tough inside, and she's had her share of camping and exploring adventures."

"I love this place," exclaimed Mary, coming over. "Look at it—right out of the Thirties! A little battered, maybe, but who cares? Next time we follow the Lincoln Highway we're going to stay in nothing but these old relics, all the way across! 'AIR CONDITIONED! PHONE IN ROOM! TV/RADIO!' Who needs HBO and a pool without weeds coming up through the concrete? This is authenticity!"

"What did I tell you," I said to Chris. He shook his head doubtfully. After we got our rooms and situated ourselves, he called us from his room. "Oh, we're doing fine," replied Mary to his question. "The door frame has been nicely repaired after someone apparently kicked the door in, the Gideon Bible is here, the bathroom bears the scars of a home repairman improvising on a budget, and the beds sag toward the middle, but everything is scrupulously clean and well meaning. Oh—and you could read through the sheets if you wanted to! How's yours?" She laughed at whatever was his reply.

Later, while Chris watched the news, we drove around to sightsee, first going to the shoulder of the Toiyabes just outside town to stand looking across the hot and hazy Reese River Valley toward the nearby Shoshone and Desatoya mountains. The spot was dramatic and commanding, made even more so by a derelict granite tower perched squarely on the sharp edge of the mountain. Standing three stories, pierced by old girders on its southern wall that once supported tiers of balconies, it was built by an Eastern financier, Anson Phelps Stokes, in 1897. Stokes had many mining interests around Austin and the Reese River Valley, and he had constructed the Nevada Central Railroad some ninety-two miles northward to intersect with the main line Southern

Pacific tracks at Battle Mountain. He had envisioned Stokes Castle, said to be an exact replica of an ancient tower outside Rome, as a summer home for his family. The first floor had contained a kitchen and dining room, the second a large living room, and the third two bedrooms; on the roof was an outdoor living area that could be screened from the wind by raising curtains. His family members would have been able to see some sixty miles south and thirty-five miles north, but they had lasted only one summer before fleeing back to genteel and temperate-by-comparison New York City, where at least Newport, Saratoga, and the Adirondacks were reachable. The tower fell into disrepair and was victimized by vandals, and finally the interior floors and stairways and exterior balconies were removed to save sightseers and local young beer drinkers from injury, and the whole stone structure was enclosed by galvanized fencing and barbed wire.

Mine openings and silent, rust-seized machinery obtruded through scrub brush from the nearby hillside. I gathered the children and pointed southward down Reese Valley, asking them to imagine a line of nine camels trudging up the once healthy and flowing river toward Austin. Back in the years just before the Civil War, I told them, Secretary of War Jefferson Davis became convinced that camels would be of great military use to the United States in the arid Southwest, carrying supplies in the deserts. His experiments proved a success, but after Davis had left the federal government to lead the Confederacy out of the Union, the program had languished for lack of support, and the last camels were finally auctioned off in 1864 in Southern California. One of their former wranglers had bought nine—one, I think, had been originally presented to the government by the Sultan of Turkey—and they were used for a number of years transporting salt from marshes about 100 miles to the south, near Tonopah, up to Austin. "Cool!" the children exclaimed together, and ran off in the dust playing got-you-last. Mary and I walked behind them for a while, asking in our common rhetorical way, Why are we so lucky?

Still later we found ourselves back in downtown Austin while the kids and Chris pawed over dusty curios in a store. The only choice for dinner was the International Hotel, a two-story false front agglomeration fronted by a spindly balcony and creaky boardwalk below, boasting not only a bar but a grill and said to be one of the oldest buildings in Nevada—one that had been carted in pieces from Virginia City in 1863. Once, we were told, it had a ballroom and catered to Austin's upper crust as well as just-passing-through dignitaries. We were the only patrons there at the moment, and ordered from the menu of simple, hearty, and artery-clogging fare.

"We're not far from the geographical center of Nevada," announced Chris, who said it was just a half mile west of Austin Summit. He looked tired, and said he was looking forward to rejoining Carol, his wife, back in California. "I've got some chores to do out in the garden, and there's a VW Bug that must be inspected—it's a sore temptation to buy." He launched into a story about the old narrow gauge Nevada Central Railroad running northward to Battle Mountain. "The builders took their own sweet time about getting it going in the late 1870s," he said, "but finally they did it. The contractors woke to the fact that Lander County would give them a bonus of $200,000 if they finished the line to Austin by midnight on a certain deadline. Sure enough, at noon on the last day, the railroad workers were still two miles shy of Austin. Everybody in town poured out to help, men, women, and children, with horses, mules, donkeys, dogs, cats, you name it, and they were laying ties on snow and spiking those rails as best they could, but by evening it was obvious they weren't going to make it into town. But you know what happened? The town council held a special meeting that evening—and extended the city limits by a half mile! And they crossed that line with only ten minutes to spare."

As we worked on dessert and coffee, our thoughts naturally turned to the long corridor we had explored together from Ogden, Utah, to this grill out in the middle of Nevada. "I'd always wanted to see that Utah desert stretch along the old Central Pacific right-of-way," Chris mused. "That was for my education. This has been for yours—the real Old West. And it's been fun!" He got all bashful at our effusive thanks and actually said, "Pshaw!" He reminded us that he was going to steal out of town at sunup. "Well, anyway, I'll see you in a few days when we meet at the eastern base of the Sierra. We still have the Donner Trail to finish, and the original path of the Pacific Railroad over the mountains. My mission isn't over yet!"

It had been a long day, from Tonapah to Austin: the sun's glare was still back somewhere behind our eyes, and we had eaten our share of desert dust. We all turned in early, and the children were, for a change, unmindful for the late night road trip familiarity of a television set. While Mary and I slept, we both dreamed of tumbledown Ophir and stately Park Canyon, as we would for quite some time, and when we awoke, Chris's truck was gone.

The next morning a long road stretched ahead of us. We got out early, following the Lincoln Highway across the parched valleys of Reese River, Edwards Creek, Dixie Wash, and the Salt Wells Basin toward the fertile green irrigated valley surrounding Fallon, seat of Churchill County. Where Highway 50 crossed the southernmost high desert hillocks of the Stillwater Range, we pulled off the road to see Sand Mountain, a unique mile-wide sand dune some 600 feet high and two miles long. Sand Mountain is administered by the U.S. Bureau of Land Management as a rather nightmarish recreation area for desert campers and would-be cowboys riding dirt bikes and all terrain vehicles—a good number of whom are killed each year due to drinking, accidents, and excessive speed. Significant among those tens of thousands who do survive, however, are those few who tend to vandalize facilities and leave enormous piles of trash; sometimes, they start wildfires with sparks from their vehicles. The pale-colored sand dune had been built up over eons by windborne quartz particles ground off the Sierra by glaciers during the Ice Age, and a brochure picked up at a kiosk said that the wind- and shape-shifting sand rustles, sings, and booms, but all we could hear was the hot wind and the mosquito-whine of dirt bikes.

Stunned by punishing high heat and the painfully bright glare, we soon retreated to our car and pressed on the twenty-five miles to the green and pleasant land of Fallon, beneficiary of rechanneled waters from the Carson and Truckee rivers, which supported alfalfa fields, suburban-like streets, and a healthy, bustling town. Beyond Fallon and the tiny gas pump burg of Ragtown, we left the trickling Carson River and followed State Highway 95 the short distance northwest to Fernley and Wadsworth—back in the Truckee Valley where Interstate 80 rushes the last few miles to Reno, and we had completed the vast necklacelike circle across western and southwestern Nevada.

We again took advantage of the substantial mid-week discounts at the casino hotels, stowing ourselves at Circus Circus and, the next day, venturing out to post office and express facilities to mail home an accumulation of books, maps, and brochures, and to a Laundromat to do a much needed wash. I had been told about a water park in nearby Sparks by a former student who had worked there as a lifeguard, and I dropped Mary and the children at the gate for the afternoon. Then I drove into the foothills west of Reno, where elegant suburban developments sprouted with green grass, lush trees, and a complete lack of acknowledgment of the bone-dry sage and chaparral gorges and canyons they had overrun. The Sierra Nevada loomed just to the west.

I visited with Stan Paher, publisher of a series of excellent historical reprints and books about local nature, the mining industry, and ghost towns under the imprint of Nevada Publications. He ran the operation out of his pleasant split-level home, and because workers were loudly renovating some rooms we stepped outside to finish our talk in the yard. I could feel a wind coming down off the mountain heights. The barometric pressure was definitely dropping. "We're in for a thunderstorm," said Stan, looking a little worried. My first inclination was to congratulate him on the impending rain, badly needed in that harsh drought summer, but then I remembered that in the dry West, thunderstorms didn't guarantee extra moisture—just fire. It suddenly got very dark. Then lightning strikes began to walk down the mountainside a few miles west and south. I wound up our talk just as there was a gigantic flash of a strike about a mile away beyond the tidy development.

Hastening back to Sparks to pick up my family, I found they had taken shelter from the storm in the admissions building; they ran to the car in the wind while lightning struck continuously to our east on either side of the interstate. After we were all in the car but still in the parking lot, I saw a wall of dust suddenly envelop the highway, stopping all traffic. We threaded back to Reno by local streets.

That night, after dinner, the local news program led with dramatic footage of a lightning-ignited wildfire southwest of the city; it would become known as the Arrowcreek fire, blackening 5,500 acres, destroying at least six homes, threatening more than fifty others, and causing a large evacuation in the Spring Valley subdivisions. I recognized the terrain: I had seen the bolt that set the mountainside on fire. We watched live television reports as nearly 200 firefighters struggled to contain the fire and large air tanker planes dropped red chemicals on the blazes. Then we left our east-facing, lower-floor hotel room and rode an elevator to the top floor of Circus Circus, where a south-facing window at the end of a corridor gave us a panoramic but hellish sight of foothills, canyons, and mountainsides ablaze. The children leaned into us for reassurance, worried that the fires might spread and engulf the city and our high rise hotel. As we stared across tumultuous Reno toward the hot or-ange hillsides and towering smoke, I thought about Mark Twain's words when carelessness had set a fire on the northern shore of Lake Tahoe and it raced with "fierce speed" across deeply carpeting pine needles. "In a minute and a half the fire seized upon a dense growth of dry manzanita chapparal six or eight feet high," he recalled in *Roughing It*, "and then the roaring and pop-ping and crackling was something terrific. We were driven to the boat by the intense heat, and there we remained, spell-bound.

> Within half an hour all before us was a tossing, blinding tempest of flame! It went surging up adjacent ridges—surmounted them and disappeared in the canyons beyond—burst into view upon higher and farther ridges, presently—shed a grander illumination abroad, and dove again—flamed out again, directly, higher and still higher up the mountain-side—threw out skirmishing parties of fire here and there, and sent them trailing their crimson spirals away among remote ramparts and ribs and gorges, till as far as the eye could reach the lofty mountain-fronts were webbed as it were with a tangled network of red lava streams. Away across the water the crags and domes were lit with a ruddy glare, and the firma-ment above was a reflected hell!

I found this passage to read after we returned to our room and settled down. Perhaps it was thanks to Mark Twain, perhaps to that bright fire im-printed in my mind's eye: I woke a number of times in the night, anxious for the firefighters battling those flames outside Reno, and for all the people with their endangered homes built unmindfully of constant natural dangers. I was also not a little worried, though unreasonably, that the fire might actually enter the city, so I sniffed the air for smoke and made sure my wife and children were sleeping soundly. In the morning, when we rolled out of the city to face the high wall of the Sierra Nevada and rejoin our faithful guide on the Donner Trail, the air smelled of char and the skies above the tall mountains were overcast with smudges of smoke.

PART VI

15

Over the Sierra

Across the years I have traced many rivers, but because of my writing work the course that is most indelibly printed in my mind's eye is the wild and tumbling Truckee as it rushes down through the towering eastern ramparts of the Sierra Nevada. I first saw its green and foaming waters through the windshield of that hippie bus in 1973; despite my Atlantic Coast existence, I knew of the Truckee and was primed to see it, the natural route to the mountain summit for emigrants, gold rushers, and the Pacific railroad. Years later, research trips to California for *Empire Express* always included drives over the mountains and down—then back up—the Truckee, looking across at railroad grades, trestles, cuts, and fills, and pulling off at every opportunity to scramble downhill across a gravelly slope to the river's boulder-fringed edge where, sometimes, I could stoop, fill my hands with icy water, and be braced by dashing it against my face.

We drove west from Reno on I-80 along the narrowing Truckee Valley, looking southward toward the famed Truckee Meadows, whose lush grasses had strengthened ox teams for the arduous pull over the Sierra; tall buildings, urban streets, and suburban housing developments thickly coated the meadows now, with tamed grass in front lawns offering slim pickings for any oxen happening to wander through. The wall of the Sierra loomed higher as we approached. Road signs for the town of Verdi appeared. A few miles farther and we crossed the California line and entered the deep upper canyon, slipping between the Carson Range to the south and the Bald Mountain Range to the north. John C. Frémont, Kit Carson, and party had followed the river up a rocky, snowless trail for a short distance on January 24, 1844, craning their necks up toward dark-looking hills and craggy, snowy, pine-clad mountains above. Deciding that the canyon was impassable, they turned back to seek another way over.

Civilians with more grit and determination, and an earlier start in the season, proved Frémont wrong (he would follow their path himself on his next expedition in December 1845). Ten months after Frémont's first sight of the pass, in November 1844, three advance men from the Stevens-Townsend-Murphy wagon train squinted up at the mountains from the riverbank and decided that their guide and hostage, an ancient Paiute whose name they understood to be Truckee, had not been lying when he told them of the east-flowing stream from deep in the mountains, with ample grass and wood along its length. They hurried back to the meadows to get the party moving fast for there was snow in the air. It would be the first wagon train to cross the Sierra.

The emigrants had left Council Bluffs on the Iowa side of the Missouri River in mid-May, taking the Platte River route to Fort Laramie, navigating South Pass, and at Green River turning northwest to Fort Hall. After joining the Humboldt they had suffered across the Forty-Mile Desert and been suc-cored by the lower Truckee. Then, in the narrow upper canyon, guided by the eighty-one-year-old Caleb Greenwood, who had crossed the mountains by horseback, they were forced to ford the river as many as ten times in a single mile, often having to wallow upriver over slippery rocks between boulders and snags, the men hip-deep in the icy water and the oxen soon hobbling as their feet softened, split, and bled. The canyon led them upward to a high, wide tableland. Ahead, above and beyond an exquisite little blue lake, rose the nearly perpendicular, seemingly unbroken wall of mountains, barring any easy progress.

There is a 1,300-foot difference in altitude between the lakeshore and the pass, with surrounding mountains towering as high as Castle Peak (9,139 feet). In mid-morning interstate traffic we emerged from the canyon, before the resort town of Truckee (5,820 feet), and my wife and children had their first look at that high mountain valley and the mammoth dark blue wall barricading its western edge. My eyes naturally swept from south to north across solid mountains, and then way upward seeking the defining blue sky. I kept Donner Lake, the peaks, and the towering pines in my imagination—removing the town and erasing the climbing shelves of the Southern Pacific Railroad grade with its intermittent hardrock tunnels and concrete snow-sheds, and the smooth, blasted-out progress of the highway and interstate climbing the opposite northern flank. Then I could approximate what those Midwestern farmer-pilgrims saw. One other difference between August 2000 and November 1844 was the matter of the two feet of snow already clinging to the heights.

One could try to superimpose the different paths of the Stevens-Murphy Party, and the tales they told across that wild mountain panorama. There was first the horseback party of six young people—two women, two men, two boys—delegated to ride ahead of the main body and their wagons to seek help on the western slope at Sutter's Fort. This group climbed out of the canyon along the south face, following a creek to the large, mountain-ringed lake later called Tahoe, and finding their way down the western shore to an easy-riding pass over the summit and downslope to safety. There was the transit of the main party, who camped several days at Donner Lake, reconnoitered for a wagon path, and finally decided to cache six of the wagons and attempt a crossing with the remaining five. First they painstakingly unloaded and carried all their belongings to the summit through knee-deep snow, then they double-teamed the wagons and began to inch their way upward. Halfway, they were stymied by a seemingly impassable wall some twelve feet high. Thoughts of abandoning wagons and animals, of walking past the supplies and possessions of their old homes and hoping to reach civilization, paused unpleasantly in their minds.

Then old Caleb Greenwood found a faint rift in the wall. He showed the others how they could push, pull, prod, and tail-twist one ox at a time to get up over the barrier. Then all the oxen could be harnessed together; with chains lowered and fastened around each of the wagons in turn, they could raise them up the cliff. And then they could finish the climb to the summit, which they did about November 25, 1844. Pressing westward down the long, gradual slope, they had to halt after only three days' travel: the wife of Martin Murphy delivered a baby girl named Elizabeth, and the entire group halted as snow continued to fall. The group was then split—by then it was the first week of December—with women and children remaining in a hastily cobbled cabin while the men forged ahead toward Sutter's Fort; they would not be reunited until the rear party could be moved down into the Sacramento Valley on March 1.

The third path made from the lakeshore was that of the three young men who saw the main party to the summit and then volunteered to tramp back down to the cached and stuffed-with-valuables wagons and guard them over the winter. Ensconced in a log cabin they built, the men learned within a week the sobering fact that in winter the snows would never stop—and that game for hunting was almost nonexistent. What little there was evaded their guns. Faced with starvation, they improvised snowshoes and began to follow the main body toward the summit. Young Mose Schallenberger, just eighteen, could not keep up; he was soon exhausted and nearly paralyzed with muscle cramps from the strange rigors of snowshoe travel in the bitter cold. At the head of the pass his friends sadly let him turn back down toward their cabin, reasonably sure he would perish, and they pressed onward through alpine storms toward a reunion with the main party.

Schallenberger's story is a triumph of forbearance, ingenuity, and sheer luck. Today near the lake at the site of his cabin there is the Pioneer Monument, a mammoth stone block supporting a heroic bronze sculpture of a man, woman, and child. Nearby, a modest bronze marker relates the bare bones of his story: how he realized that if he could not shoot game he might trap his food, using some leg-hold traps left behind in the wagons. In that cabin alone for months, he therefore subsisted on foxes and an occasional coyote—fox meat seemed like a delicacy to his starving palate, but his stomach revolted at coyote meat no matter how he cooked it. "I never got hungry enough to eat one of them again," he recalled many years later. "There were eleven hanging there when I came away." Once he shot a crow and could not decide which meat was better.

"My life was more miserable than I can describe," he said. "The daily struggle for life and the uncertainty under which I labored were very wearing. I was always worried and anxious, not about myself alone, but in regard to the fate of those who had gone forward. I would lie awake nights and think of these things, and revolve in my mind what I would do when the supply of foxes became exhausted." There were plenty of books in the wagons; he read aloud, "for I longed for some sound to break the oppressive stillness." Then on one day at the end of February before the light had entirely failed, he was stupefied to make out the figure of a man approaching—it was a member of his emigrant party, a Canadian who was familiar with snowshoe travel and had taken advantage of a lull in the snowstorms to walk from Sutter's Fort up to the summit and then down to the Schallenberger cabin. Together, the next day, they left the cabin—and the six cached wagons and eleven hanging coyote carcasses—and set out for civilization.

Over the next two years the cabin stood empty—the wagons and their contents being finally claimed in the spring by their owners. Snow rose past the eaves and then retreated; game returned for the brief summer months molested by few interlopers. Parties led by old Caleb Greenwood or his sons rolled by—perhaps as many as fifty wagons in fall 1845—and were able to claw their way to the summit with all the wagons after tremendous toil. Frémont rode through on horseback with a select party of fifteen in early December, climbing easily to the head of the pass before much snow had fallen; several weeks later on Christmas Eve, the ambitious Ohio lawyer and recently published author of a travel guide, Lansford W. Hastings, rode by the empty little cabin, and his party of ten mounted men fortuitously crossed the mountains untroubled by storms or significant accumulation. Snow returned and quietly took possession of the Sierra, the little lake, and the Schallenberger cabin. Then in autumn 1846, there arrived a flood of emigrants. There were so many as to cause a bottleneck at the head of the pass, stirring those at the end of the line to forage for an alternative; they found a way up along the canyons later named Coldstream and Emigration, allowing all late-coming parties a passage before the Sierra was blocked—all, that is, but the hapless group now known for the brothers Jacob and George Donner.

It was late October 1846 and snow arrived a month early. Into this cabin came one Patrick Breen, Ireland-born and about fifty-four years old. He was a farmer in Keokuk, Iowa, until he sold his spread and loaded his family and supplies into three wagons and set out for California, joining the Donner-Reed

Party on the west side of the Missouri River. Breen and the rest of the group had attempted to cross the Sierra, their oxen and mules foundering in belly-deep drifts, but they had turned back some three miles before the head of the snow-choked pass when their Indian guides lost the trail. They retreated down to the lakeside, with the Breens occupying the Schallenberger cabin. Others built cabins nearby, using the hides of oxen stretched over poles for roofs, and there was another cluster of party members several miles downhill at Alder Creek—the Donner brothers and their families. The few remaining cattle wandered away and were lost in drifts. Snow began to accumulate and rose past twenty feet. Jacob Donner was among the first four to die down at Alder Creek.

The family of James Reed, deprived of husband and father after the manslaughter and banishment back at Gravelly Ford on the Humboldt, had to disperse to various cabins at the lake. Mother Margaret Reed and their daughter, Virginia, moved in with the Breens—there were fifteen living in the cabin—for a while subsisting on ox hides torn from their cabin roofs and then on the Breen's little dog—"we would take the bones and boil them 3 or 4 days at a time," Virginia Reed wrote. The snow piled up high above the cabins and they had to tunnel up to the surface for air and release of smoke; they were by then too weak to gather firewood, so they carved away pieces of the interior log walls to burn. Patrick Breen had started a diary when they were snowbound: "Thursdday 31th last of the year," he recorded,

> may we with Gods help spend the Comeing year better than the past which we purpose to do if Almighty God will deliver us from our present dredful situation which is our prayer if they will of God sees it fiting for us Amen morning fair now Cloudy wind E by S for three days past freezeing hard every night looks like another snow storm snow storms are dredful to us, snow very deep Crust on the snow.

With starvation a certainty, a breakaway group of ten men and five women set off in January to walk over the summit, following the two Indian guides. The group would henceforth be recalled as "the Forlorn Hope." Thirty-two days later the survivors of that party—five women and two men—reached a settlement and were saved. They bore horrifying tales of starvation, cannibalism, murder (the two guides were killed), and survival at last by eating moccasins, old boots, and the rawhide strings from snowshoes. Thus alerted, Californians began to prepare relief parties for those still pinned in the mountains.

We stood in shadows beneath towering Jeffrey and lodgepole pines and white firs, the ground littered with huge pinecones, at Donner Memorial State

Park. Our guide and friend Chris Graves had driven over the summit to meet us. He pointed to the base of the pioneers' monument: it was twenty-two feet high, the height the snow reached in the nightmarish winter of 1846–47. "For many years I thought about one particular family who was part of the Donner Party—the Graves family, father Franklin, mother Nancy, and son Billy—and for a long time I thought I was related through my people back in our home county in Indiana," Chris said quietly. "But finally after I researched it carefully, I saw that the Thomas Graves both sides claimed as an early ancestor was in fact two people who lived in different places and different times. Even though I know there's no blood connection now, I feel one— it's my name, too, after all, and I've walked in their footsteps across Nevada into California, and their story is my story and that of all Americans."

Behind us, blending into the landscape, was the low building of the Emigrant Trail Museum with its exhibits on local history and the Donner tragedy—telling how, as members died of exposure and starvation, their bodies were desperately consumed by their fellows. And meanwhile, over the mountains at Sutter's Fort, seven snowshoers left on a rescue mission. Member Daniel Rhoads recalled what they found when they reached the camp. "At sunset of the 16th day [of the journey] we crossed Truckee lake [now Donner Lake] on the ice and came to the spot where we had been told we should find the emigrants," he recalled many years later.

> We looked all around but no living thing except ourselves was in sight and we thought that all must have perished. We raised a loud halloo and then we saw a woman emerge from a hole in the snow. As we approached her several others made their appearance in like manner coming out of the snow. They were gaunt with famine and I never can forget the horrible, ghastly sight they presented. The first woman spoke in a hollow voice very much agitated & said "are you men from California or do you come from heaven."

"That was on February 19," said Chris, "the first rescue attempt of seven snowshoers." We were standing inside the museum's bookstore, where I had hoped to find copies of George Stewart's books on the Donners and on the California Trail, or Dale Morgan's editions of the various trail diaries. "Twenty-three of those poor, starved people," Chris went on, "most of them were kids—walked out, but then two children, younger Reeds, gave out and had to be carried back down to the cabins. Three more died on the way over the pass. Then there was the second relief expedition, which included James Frazier Reed: they arrived at the lake March 1, and took charge of seventeen. But they got trapped by a blizzard up on the heights, for a week—terrible frostbite!—and they starved, and when three more died the survivors ate them. The third relief expedition got down to the cabins around March 13 and found nine still alive though three were by then dying. The rescuers took away five people, but then dear Tamsen Donner, faithful wife that she was, said goodbye to her two little girls and went back to nurse her dying husband, George Donner."

He lowered his tone in case our children, who were trolling the bookstore for souvenirs, came closer. "By the time the fourth relief expedition got down to Donner Lake, Tamsen and a number of children were nowhere to be found. They did find Lewis Keseberg, a German, who was fat and hale and hearty, 'enjoying his breakfast,' said one of the rescuers, and lying amidst a pile of human bones, kettles filled with awful sights, and Keseberg with valuables and gold looted from the Donners. He claimed all the adults and children had died of natural causes, but the rescuers were convinced he killed them, each and all, as he needed them. Why they didn't lynch or shoot him right there has mystified folks for more than a century. And back in California no one could bear investigating him to the degree that the stories demanded."

As Dale Morgan has tabulated it, the Donner-Reed Party had been comprised of eighty-seven people when it set out from Fort Bridger, Wyoming, along the Hastings Cutoff; thirty-nine died. To this, we are reminded, one must add the lives of four California Indians—two guides killed by the Forlorn Hope group, and two who were guiding the rescuers from Sutter's Fort, and died of starvation and exposure in the snows.

"Is it any wonder," muttered Chris, as Mary and I collected the children and hurried outside the crowded museum into a bright patch of sunlight under the tall pines, "that emigrant traffic over Donner Pass never picked up again—folks were just too spooked, and who can blame them? That is, until our heroes in the Central Pacific Railroad came through, thirteen years later." He pointed upward at the formidable blue wall, and the ascending railroad grade. "Let's head up that way and look around."

In two vehicles as before we drove through bustling Truckee, once an old lumbering and railroad town that from about the 1920s on received an annual economic shot in the arm in wintertime when skiers crowded in and Hollywood crews took up residence; according to my copy of the Lincoln Highway–era *Mohawk-Hobbs Grade and Surface Guide* for auto tourists, Truckee was a favorite location for films with stories set in Alaska or Canada. Charlie Chaplin had shot *The Gold Rush* there, and Clark Gable and Loretta Young had sported fur outfits in *The Call of the Wild*. "Some companies spend up to $200,000 in making pictures here," reported the guide. "They pay $200 a day for a ten-dog team and driver. The three hotels are often filled early." Now prosperous, squeaky clean, and colorful, Truckee seemed to be the summertime destination for thousands of shiny new California SUVs— the streets were thick with them, stores were crammed with customers, and it seemed like every restaurant had a waiting list. We grabbed some snacks at a quick stop and headed west on hairpinning Route 40, passing many vacation homes as the road rose steeply above the northern and western lakeshore. Between 1926 and 1964, Route 40 ran from Atlantic City, New Jersey, to San Francisco; west of Reno the road was designated as one of two alternate routes of the Lincoln Highway (the other, from Carson City, passing Lake Tahoe, took the old Placerville–California stage route to Sacramento).

Previously the old automobile road over Donner Summit climbed along the south wall of the pass, often tracing the even older Dutch Flat Wagon Road built in the 1860s by executives of the Central Pacific Railroad to collect lucrative freight tolls until their tracks over the mountains were finished. Grades of the old road ran between 12 and 18 percent, which made the ride both difficult and hazardous. The original Lincoln Highway went that way, ducking beneath the railroad snowsheds in its final spurt to the top. When the 1926 U.S. 40 was completed, after an enormous amount of blasting, it climbed the north wall of the pass, with grades kept below 6.5 percent. It seems, though, that all roads come to an end: in California U.S. 40 was decommissioned and ceased being a signed highway in 1964; Nevada decommissioned it in 1975. By then, out West, most of U.S. 40 was buried under Interstate 80. Happily, though, there are some exceptions—such as the north-wall Lincoln Highway routing over Donner Pass, since the interstate ran parallel, two miles farther north. Therefore on Route 40 it is still possible to climb or descend the earlier grades, seeing more of the breathtaking views and being able to pull over to gawk much more often. Which is what we did, taking our time, craning our necks up from the two-lane highway toward overhanging cliffs and high Castle Peak, which the emigrant writer Edwin Bryant described after his muleback crossing as having "cyclopean magnitude" with "apparently regular and perfect . . . construction of its walls, turrets, and bastions." Donner Peak (8,315 altitude) and Lincoln Peak (8,403 altitude), bare and majestic, rose in front of us as we climbed and twisted and turned through slots and past sheer ledges. We pulled off at one point to look out across the great emptiness past soaring hawks toward the south face and the long snowsheds of the railroad grade, which my WPA guide to California likened in 1939 to a wriggling, "long black caterpillar . . . worming at intervals through granite tunnels." They had been constructed of dark old redwood timbers then, and had since been replaced by pale reinforced concrete slabs, but at this distance they still looked like caterpillars.

The climbing road approached a breathtakingly beautiful arching concrete bridge, built in 1928. Soon we pulled off at the Donner Lake overlook, stepping out into a sharp west wind that cooled our backs as it flowed around us and out into the vast, deep, empty bowl before us, which was enclosed by bare, steeply falling mountainsides, at the bottom of which was azure, kidney-shaped Donner Lake, surrounded by thick forests. It was a stunning and unforgettable view—beyond the eastern Sierra we could see Nevada. It was comparable (with a little imagination) to the sight seen in July 1860 by the first hero of the Pacific railroad, a young Connecticut-born civil engineer named Theodore Judah, who solved the horrible and hazardous puzzle of the Sierra Nevada by using the tragic Donner Party route for the Central Pacific Railroad.

During the years I worked on *Empire Express*, Theodore Dehone Judah was for me the most admirable of a large cast of historical characters. As I pored over his letters, diaries, field notes, press releases, and reports; as I traced his highly detailed hand-drawn maps from Sacramento through the forbidding mountains and past the Nevada border; as I read his widow, Anna Judah's, grief-stricken memoirs of his short but eventful life, he became emblematic of the best and noblest aspirations, joined to enterprise and inventiveness, of his nation in the era before the Civil War. Ambitious, progressive, and even some-what altruistic, he originally envisioned the Pacific Railroad as a people's rail-road, to be completed using small subscriptions of ordinary people. Altruism was replaced by an intriguing practicality when, at just twenty-eight, he ar-rived in California in spring 1854 after being recruited to build the first line on the Pacific slope, the Sacramento Valley Railroad. Already Judah had been bitten by "the Pacific railroad bug"—what civil engineer was not in the 1850s—and the sight of those dark mountains on the Sacramento horizon drew him into their heights again and again, especially after he completed the valley line and had time on his hands.

It was the summons of a sometimes prospector and druggist from the placer-mining Sierra town of Dutch Flat, one Daniel Strong, that handed Theodore Judah the key to stretching a railroad over those mountains. "Doc" Strong recognized that an ideal transit was a long and easy continuous grade, and discerned one above and below Dutch Flat, twenty-seven miles up to the summit and forty-five miles down to the Sacramento Valley; at that point the capital was only fifteen miles farther. The route followed the long, high ridge that divided the American River's North Fork from the Bear and Yuba rivers. When Strong led Judah on horseback up to Donner Pass, and they stood at the summit gazing down at the little lake, the engineer was powerfully excited—not only was this the way to gain 7,000 feet in altitude while staying within established industry guidelines for steepness of gradient and curva-ture of lines, but the Donner Pass route solved another difficult problem. The Sierra summit had two parallel ridgelines separated by a deep trough, mean-ing two expensive ascents for any locomotive. Far below them, the lake was drained by the Truckee River, which tumbled down its deep canyon through the eastern summit ridge, forming a natural passageway. He knew then that he had found the answer to an engineering problem that had troubled him for years—and that the transcontinental railroad would go through Donner Pass.

Later that day, back at Dutch Flat, they drew up incorporation papers for the Central Pacific Railroad of California on Strong's store counter. To finance a survey, over ensuing weeks and months they combed the mountain communities for small stock investors, finding only a few, and Judah went down to San Francisco to call on bankers and businessmen. Only after he was laughed out of their offices—few believed a railroad would ever conquer the Sierra Nevada, and in any case there were far more profitable and surer investments—did Judah call a meeting in Sacramento. There, at the St. Charles Hotel on K Street, after he had given his presentation and was disappointed at the modest, even halfhearted response by the assembled businessmen, Judah was taken aside by a hardware store owner by the name of Collis Potter Huntington.

Like Judah, Huntington had been born in Connecticut and spent his young adulthood in upstate New York. There the comparison must be dropped. Huntington was a shrewd shark of a businessman who had come to California to sell supplies to the gold rushers; he was as ingenious at trade as he was ruthless. Upon such an encounter rose the foundation of one of the most powerful corporations in the nineteenth-century West, although neither Judah nor Huntington could imagine such a thing when they met again, above the Huntington & Hopkins store at 54 K Street, as the engineer explained his ideas to partner Mark Hopkins. Soon the merchants drew in others, sharp like themselves: Leland Stanford, transplant from upstate New York who had failed at his law practice but prospered at wholesale groceries and soon was to be elected Republican governor of California; Charles Crocker, brawny dry goods retailer, tough enough to run an army; his brother, Edwin Bryant Crocker, inventive attorney and administrator. These men, together with a handful of others, including Daniel Strong, became the original investors in the Central Pacific Railroad, with the cash-poor but idea-rich Theodore Judah being credited with 150 shares and given a seat on the board of directors.

Later, after Judah completed his mountain survey, brilliant for the era, which proved the railroad line was possible; after he had gone East to Washington and become instrumental in engineering the 1862 Pacific Railroad bill, including a grant to the Central Pacific for the franchise to build the western portion of the great road; after he had returned to the coast to watch his dream become a reality; then the coalition holding him to the five Sacramento businessmen began to waver and then splinter. First Judah objected when fellow board member Charles Crocker was awarded construction contracts—it was an "inside deal" costly to the company's overall fortunes, he complained, to no avail. Then his partners conveniently forgot that he was to be credited company shares for survey work—and demanded cash. And then Leland Stanford arranged to submit a route map for federal approval, claiming the base of the Sierra Nevada to be twenty-one miles west of where it began in actuality—notching the railroad's federal reimbursement to a vastly higher amount per mile. Judah refused to support it. Instead, he declared he would go to New York to get other backers—and buy his partners out. He and his wife sailed by way of Panama to New York. But Judah contracted typhoid or yellow fever on the trip, and died soon after reaching New York. He was thirty-seven. The company stayed in the hands of Huntington, Hopkins, Stanford, and the Crocker brothers, each of whom would play a different role in completing Judah's dream, to their inestimable gain.

Not that it would be easy; a mammoth task awaited them and more than 12,000 laborers over the next six years. That October in 1863 in which Judah had stormed out of their boardroom heading for New York, Judah's friend and guide, Daniel Strong, had led Huntington, Stanford, and Charles Crocker up on horseback along the proposed route to Donner Summit. There, gazing down and out into the great gulf, they had pondered Judah's certitude that they could grade a trackbed down the cliffs 1,300 feet to Donner Lake. It looked like an impossibility. "I'll tell you what we'll do, Crocker," said Huntington after a thoughtful silence. "We will build an enormous elevator right here and run the trains up and down it." "Oh Lord," moaned Crocker, "it cannot be done."

Standing there at that stratospheric-seeming overlook, we turned to face a wind blowing over the Sierra Nevada from the Pacific, 200 miles away. It suddenly occurred to me with a rush that the last great physical obstacle—between our home on the far side of the continent and land's end—was now beneath our feet. "It's truly all downhill from here," I assured my exhilarated wife and children. After so many weeks on the road, that carried a great emotional import for all of us.

"Actually," said Chris, "we have a few more yards to travel before we're technically on the summit. But let's get back in our vehicles—there are a few things you've got to see."

Not far west of the overlook, under that big intense mountain sky, we turned off the highway onto an unmarked drive. Past some undistinguished little buildings, we parked and Chris led us toward a rocky outcropping where a low concrete wall formed a square, about fifteen feet on a side and roofed over with steel plate. "You're standing directly above the Sierra Summit Tunnel, old Number Six," said Chris. I could hardly contain my excitement as I took in the top of the famous vertical shaft built by the Central Pacific in 1866 to hurry construction of the 1,658-foot-long summit tunnel.

Initially the graders and tracklayers of the Central Pacific had coasted easily out of Sacramento on the valley floor and then wound through foothills, encountering a few problems but solving them, and then beginning to climb the long divide between the American and Bear rivers toward the summit. The hard work commenced—blasting out ledges, bridging creeks, gaps, and chasms, retaining overhangs and dropoffs, felling thick stands of trees and brush and grubbing out every stump, contending with horrendous mudslides on the slopes and avalanches on the heights. Tunnels proved the greatest barrier—no fewer than twelve were required in the mountains—and the longest, highest, and most vexing work was Number 6, a third of a mile through the hardest rock in North America. With the standard technology, handheld drill bits were slammed by sledge hammers and turned by hand until a hole a foot or more deep had been coaxed out of the unforgiving granite. Then black blasting powder was tapped into the hole, fused, and touched off, the explosion yielding a disappointing bit of rubble to be removed. From Europe to the United States, nitroglycerin had just begun to be imported—but calamitous accidental explosions aboard ship in Panama and at a freight office in downtown San Francisco had all but outlawed its use. With black powder, the tunnel men advanced the west and the east faces of the Summit Tunnel by only a foot a day on each side.

Hence the vertical shaft—by which more diggers could work back to back on the bottom, inching their way toward the men working from the outer ends, doubling their rate of advance.

In all my trips over these mountains I had never thought to look for it; I never suspected, for that matter, that the seventy-two-foot-deep shaft still existed. "Why fill it in?" said Chris. "That's money and potential problems. They covered the shaft top with iron rails and laid plate over it. After it was replaced by modern steel, I found the original iron rails lying over there in the weeds— almost eaten through by acids from a hundred years of locomotive smoke rising in the shaft."

He pointed to where, a few yards outside the perimeter of the shaft, three corners of a big square were marked by big bolts driven into the solid granite ground. These marked the anchors of a large, long-gone steam engine called the *Black Goose*, which straddled the shaft and was used in 1866 to hoist rock rubble from the tunnel floor. That itself was a story. The engine was cannibalized from a valley locomotive and sent up to the end-of-track on a platform car. Then, using traveling jacks, it was moved fourteen inches at a time to a gargantuan logging truck fitted with wheels two feet wide. Once it was bolted and braced, a teamster named Missouri Bill hitched ten yoke of oxen to the enterprise. It took days to get up the steep mountain slope along the wagon road, causing great consternation among mule teams and stagecoach horses that would round a bend in the road and be confronted with the terrible sight of the *Black Goose* being dragged uphill. "It finally became necessary to blind-fold teams of mules and horses to get them to pass," recalled a young surveyor, "for they would leave the road and take to the hills or the ravines, whichever looked best to them, and they weren't particular either about what they took along with them. They would endeavor to kick themselves loose of everything before starting for the bushes."

When finally hooked up over the vertical shaft, the *Black Goose* enabled the tunnel men to make four feet per day. Chris showed us the smoke soot stains still visible across the uphill rocks, deposited there more than 130 years before and unerased by time and winter snows. Through the dead of that first winter of the hissing, clanking *Black Goose*, 1866–67, the work continued with laborers living in shelters beneath gigantic snow piles, going to work through snow tunnels, seldom seeing the bright light of day but continuing the advance. Still, at that rate it would take more than a year to finish the Summit Tunnel, so general manager Edwin Bryant Crocker swallowed his caution and allowed the use of nitroglycerin. To see what the tarriers had wrought, we piled back into our vehicles, and Chris took us back to the highway, and a short way west across the summit we took another unmarked dirt lane that twisted downhill. I saw happily that we had rejoined the Central Pacific grade: rails and ties had been pried up and carted away, and a smooth narrow path perfect for hikers and cyclists curved off westward through rocks and trees. But to our left around the bend, deep in a cut, lay the western portal of long Number Six—with a tiny spot of sunlight shining at the far end. The jaggedy rough walls were identical to the stereograph taken by Alfred Hart in 1868. You could see old scores and gouges left by the workers.

It took my breath away. In years since I have hiked through, jogged through, and helped lug video equipment through; I've inched across a slick underground, 100-yard-wide, glacierlike mass of ice from wall seepage still very solid when, outside, it was a warm June day. But there is something about that first transit of the Summit Tunnel in August 2000, especially after all my years of manuscript research and my inability to actually see it with my own eyes before. It was highly illegal to drive past warning signs and enter the tunnel—though tracks had been removed and it was no longer used—but a metal barrier yawned, open. Halfway through we stopped, turned off our headlights, got out into utter blackness, and peered up at the ceiling until we thought we could discern the rectangular vertical shaft reaching upward. Resuming, we edged past a couple of squinting backpackers feeling their way through without flashlights.

Then, abruptly, we emerged into stunning daylight on a high rock shelf carved out of the sheer mountainside, and ahead of us past the cliff edge was the vertiginous sight of all that empty space with tiny Donner Lake at its bottom. Downhill to our left we could see the overlook and the Route 40 arched

Rainbow Bridge. Along the cliffs to our right the rocky railroad shelf, now just a wide dirt trail for hiking boots or sneaky truck tires, continued toward Tunnels Seven and Eight and concrete snowsheds. The eyes followed the line as it gently dropped along the mountainside, heading for the height-conquering curves of Strong's and Coldstream canyons before reappearing some miles ahead, considerably closer to the lake level. At some point downgrade it would intersect the still heavily traveled tracks used by the Union/Southern Pacific freights and Amtrak, which now drilled through the mountains in a bigger, more modern and accommodating tunnel that shortened the route and cut a few corners along the way.

A short distance from the mouth of the Summit Tunnel, we scrambled downhill to examine another landmark of the Pacific Railroad, the China Wall, built across a ravine (the original emigrants' path, many believe) to block it and retain the fill beneath the railroad grade. Standing some thirty or forty feet high, wrought of big stone blocks set there by Chinese laborers in 1867, it looked as stalwart as it had when Alfred A. Hart trained his stereo camera on it. That wall, and others nearly like it, in addition to bridge footings and sturdy culverts scattered across the mountain heights, had been doing their job beautifully—looking like works of art, for that matter—for fourteen decades. They stood as mute testimony to the solid contributions of peasant boys from Kwantung Province in China in the face of nay-sayers and outright bigots, proving them wrong.

In early 1865, the Central Pacific received a windfall in state aid and advertised widely that it needed 5,000 laborers immediately "for constant and permanent work." Despite newspaper advertisements and handbills distributed at every post office in the state, fewer than 200 new faces turned up in construction chief James Strobridge's office. With more than 4,000 jobs going unfilled, Charles Crocker suggested they hire Chinese to wield picks and shovels—many had settled in California during the Gold Rush. Crocker considered himself prejudiced, but Strobridge was even more so. "I will not boss Chinese," Strobridge retorted heatedly. "I will not be responsible for the work done on the road by Chinese labor." There the matter stalled. "Our force, I think, never went much above 800 white laborers with the shovel and the pick," Crocker would say some years later, "and after payday it would run down to six or seven hundred, then before the next payday it would get up to 800 men again, but we could not increase beyond that amount." The white workers were, Strobridge would add, "unsteady men, unreliable. Some of them would stay a few days, and some would not go to work at all. Some would stay until payday, get a little money, get drunk, and clear out." And there was the lure of gold and silver mines not so far away.

Strobridge had no choice, though, and he agreed to try fifty Chinese as fillers of dump carts—the least skilled work available, serving under white foremen. He hired them through a labor contractor at a dollar a day, or $26 per month; with this, they would board themselves. (The white workers received $30 per month, including board.) "I was very much prejudiced against Chinese labor," admitted Strobridge many years later. "I did not believe we could make a success of it." Within a short time he was proved wrong, and he hired another fifty, followed by another fifty. Eventually the railroad sent recruiters to Kwantung. The total number of Chinese would eventually exceed 12,000, more than 90 percent of the Central Pacific's labor force. "They were a great army laying siege to Nature in her strongest citadel," wrote the enthusiastic *New York Herald* reporter Albert Richardson. "The rugged mountains looked like stupendous ant-hills. They swarmed with Celestials, shoveling, wheeling, carting, drilling and blasting rocks and earth, while their dull, mooney eyes started out from under immense basket-hats, like umbrellas. At several dining camps we saw hundreds sitting on the ground, eating soft boiled rice with chop-sticks as fast as terrestrials could with soup-ladles."

Strobridge would also learn they could do considerably more than merely shovel dirt and rocks. "We tried them on the light work, thinking they would not do for heavy work," recalled Crocker. "Gradually we found that they worked well there, and as our forces spread out and we began to occupy more ground and felt more in a hurry, we put them into the softer cuts, and finally into the rock cuts. Wherever we put them we found them good, and they worked themselves into our favor to such an extent that if we found we were in a hurry for a job of work, it was better to put Chinese on at once." Labor troubles among the whites, a constant concern, continued to fan the expansion of duties. "At one time when we had a strike among our Irish brothers on masonry," Crocker reported, "we made masons out of the Chinamen." Supervisor Strobridge retorted, "Make masons out of Chinamen!" Crocker's reply brought the Irishman back to reality: "Did they not build the Chinese Wall, the biggest piece of masonry in the world?"

To run one's hands across the great stone wall just downslope from Tunnels Number Seven and Eight was a great delight, as it had been to feel the rough interior shell of the tunnel itself, fingers finding the ends of drill holes in the solid rock. Less than a year after I stood at the wall with my family, I found myself back there at the summit, leaning against sun-warmed stones answering interview questions for a Bill Moyers documentary on the Chinese in America. Cameraman Mike Chinn, an artist in the genre, took pains to capture the high-relief artistry of the wall so that viewers would understand the hallowedness of the remote place—footings laid before discouraging wintertime fell, the wall growing tier by tier by torchlight while laborers shivered in a dim snow cavern, until spring unlocked the Sierra to outside work. And the wall lasting generations.

Back on the grade and on the other side of small tunnels and a length of snowshed, we stepped through a break in the concrete wall to stand near the edge and look across toward Route 40 and the interstate and down to the lake. Chris pointed to just beyond our feet, where a tangle of huge redwood timbers from the 1860s' original snowsheds, built to shelter the tracks and keep the line open in the worst of winter, lay on the steep embankment, tangled with their equally large fir replacements, where they had been flung years before by workers replacing wood with reinforced concrete. We climbed down to the pile, examining old joinery and hardware, and for a souvenir I snapped off a foot-long splinter of that ancient Coast redwood, felled in the Santa Cruz Mountains.

Remnants of the old Dutch Flat Wagon Road climbed the south wall toward us, in places braided with a newer trail line of closely placed boulders covering a fiber optics cable—digging a trench through the solid granite of Donner Pass had proved too expensive. Imagining the slope innocent of any roads whatsoever as it had been in 1845, I thought about emigrant Sarah Ide, who remembered that it took "a long time to go about two miles over our rough, new-made road . . . over the rough rocks, in some places, and so smooth in others, that the oxen would slip and fall on their knees; the blood from their feet and knees staining the rocks they passed over. Mother and I walked (we were so sorry for the poor, faithful oxen) all those two miles—all our clothing being packed on the horses' backs. It was a trying time—the men swearing at their teams, and beating them most cruelly, all along that rugged way." They hurried over, terrified of winter. And after the winter of 1846–47 when the pass took on the horror and reputation of the Donner Party, travelers could not make the transit fast enough, perhaps as frightened of ghosts as of storms and cold.

The railroad changed all that. The thousands of laborers inhabiting the mountains in camps for several years, and the slow pace of work that allowed Californians—whether they were Central Pacific executives or San Francisco reporters or railroad passengers—to see the Sierra up close and comfortable, changed the way they viewed the mountains. In the torrid Sacramento summer of 1867 the C.P. general manager Edwin Bryant Crocker retreated to the summit for respite from the heat and his horseback party found an effervescent mineral spring only six miles from the railroad. "It will, as soon as it becomes known, be a great place of fashionable summer resort," Crocker enthused in a letter to Collis Huntington in New York. "All around that point in every direction are points of great interest—& these mineral springs the most interesting of all. There is Lake Tahoe, & Donner Lake, close by. A few miles to the North are Lakes Weber & Independence, & all around are Summit Peaks from 9,000 to 12,000 feet high, easily accessible by trails, & even by wagon roads." The tourism brochures of the future waited. In a few years C.P. moguls Stanford and Hopkins would preside over the first of such area resorts: at Hopkins's Springs (later, Soda Springs), altitude 5,975 feet, genteel

Victorian tourists could drink the healthful waters and take a short parasol-and-walking-stick stroll to view the prehistoric Indian pictographs of Painted Rock, with the once frowning mountains standing all around them.

This remarkable change did not, of course, escape the notice of Mark Twain after the railroad made Sierra travel almost commonplace. "In Sacramento it is fiery Summer always," he commented in *Roughing It*,

> and you can gather roses, and eat strawberries and ice-cream, and wear white linen clothes, and pant and perspire, at eight or nine o'clock in the morning, and then take the cars, and at noon put on your furs and your skates, and go skimming over frozen Donner Lake, seven thousand feet above the valley, among snow banks fifteen feet deep, and in the shadow of grand mountain peaks that lift their frosty crags ten thousand feet above the level of the sea. There is a transition for you! Where will you find another like it in the Western hemisphere? And some of us have swept around snow-walled curves of the Pacific Railroad in that vicinity, six thousand feet above the sea, and looked down as the birds do, upon the deathless Summer of the Sacramento Valley, with its fruitful fields, its feathery foliage, its silver streams, all slumbering in the mellow haze of its enchanted atmosphere, and all infinitely softened and spiritualized by distance—a dreamy, exquisite glimpse of fairyland, made all the more charming and striking that it was caught through a forbidden gateway of ice and snow, and savage crags and precipices.

In 1847 when James Frazier Reed led a rescue party down to Donner Lake, and the expedition was stalled by storms on its return journey just below the summit on the eastern slope, he wrote of the horrors of being stranded and beginning to starve. The "dreaded Storm is now on us," he scrawled,

> comme[nce]d snowing in the first part of the night and with the Snow Commed a perfect Hurricane in the night. A great crying with the Children with the parents praying Crying and lamentations on acct. of the Cold and the dread of death from the Howling Storm the men up nearly all night making fires, some of the men began to pray several became blind I could not see even the light of the fire when it was blazing before me.

Today near the site of that "Starved Camp," as it became known, the Donner Ski Ranch entertains thousands of winter's leisure-timers with its slopes and trails.

═══════

As water falls, so follows traffic: west of the Sierra summit, past the wild mountain scenery of Tahoe National Forest, Interstate 80 plunges and careens downhill along the Yuba River Canyon, then Canyon Creek, and then Bear River. It darts past tree-shrouded and ravine-hidden bedroom communities still claiming the names of Gold Rush camps from the 1840s and 1850s— Cisco Grove, Dutch Flat, Gold Run, New England Mills, and Auburn—then wends its way along Secret Ravine until, attaining the valley floor, it coasts into Sacramento. On the steep downslope traffic zooms at 80 miles an hour, with reasonable drivers living a chariot-race nightmare populated by crazed, veering lane-changers and barreling tractor trailer drivers whose wary eyes are peeled for runaway lanes. Along the highway, should the curious pull over to admire precipitous prospects or historic overlooks like Yuba Gap or Emigrant Gap or Blue Canyon, or to stare up at the railroad tracks gingerly picking their way down along the divide, the aroma of burning brake linings mingles with those of pine and cedar and manzanita. Nearly thirty years before, I came through in that hippie van, caffeinated out of my head and finally letting someone else drive after I had piloted all the way from Evanston, Wyoming, to Donner Summit myself. I remember sitting up front in the passenger seat, looking past nonexistent guardrails down horrific cliffs, imagining the smoking wreckage of what had once been our van, babbling some kind of nonsense that everyone in Northern California had to have one leg longer than the other to live on such a slope. Now, with Chris at the wheel of his pickup and taking the downhill curves at a furious clip, and with poor Mary and the children trailing behind us in the Durango and in danger of losing us and getting lost in cataclysmic traffic, I kept urging him to slow down a little—without wanting to distract him from eager, endless talk about explorers, emigrants, gold rushers, and railroaders, all of which was lively and fascinating. "There are no trail ruts visible across this Sierra granite," he said at one point, "but you can often locate the route by rust stains left on the rock from wagon wheel rims. And some trees still have notches cut in the bark from passing wagons." Out exploring at a place called Carpenter Flat, he said, he once kicked and freed an ox shoe from soil that had hidden it for "well nigh onto 150 years. Now that was a glorious thing." Suddenly Mary's voice came over the CB speaker. "Come on, guys," she said impatiently, "I keep losing sight of you—slow down! Now I don't know where you are or where I am!"

"Tell me the next thing you see on the roadside," Chris responded over the radio. When she did, he told her she was about two miles behind us, and he slowed into the right lane until she caught up.

I had to hand it to him: he knew his terrain, especially across Donner Pass, whether on the highway or on the railroad grade. One time, Chris and Carol were passengers on an Amtrak train that left Salt Lake City on a pleasant spring morning at its appointed time, 1:36 A.M. "and remained in that mode until Carlin Canyon, Nevada," he told me.

"At that point the train slowed, and the oh-so-well-known disembodied voice floated through the train, announcing that we were following a freight that had engine problems, and we would remain in that status until Winnemucca. That proved to be true. We left Winnemucca an hour and a half late, arrived in Reno nearly an hour late, and then bravely attempted the Sierra. We began strongly enough, reached the Summit and started down the far side, when, just east of Dutch Flat, the voice once again intruded on our reverie, and announced that a freight train had derailed in front of us, and we would be stopping for a bit, as some hazardous materials had spilled onto the tracks. I turned to my bride, and said, 'Cape Horn has struck again.' Our Amtrak speeded up, and we descended quickly to Milepost 153 where we ground to a halt. The voice told us that six cars had left the tracks at Cape Horn (I knew that, twenty-five miles back) and that somebody had put a tanker filled with ammonia between two cars filled with ammonia sulfate, the results of which presented a hazard to our train. We were on Track 2, and the derailment was on the Cape, on Track 1. It looked like we were stuck for a long time—but no: a few weeks before, I had trotted over the mountains with a Hollywood bunch, scouting locations for a Nicolas Cage movie; now, as I sat on the Amtrak staring out the window, I noted a house roof that looked familiar. I knew where we were! I called my daughter on the cell phone and told her to come to the Gold Run exit on I-80, come over the freeway, travel to Gold Run, proceed to the rail crossing, cross it, then drive along the grade easterly until she found the damned train, as we were on board. Good daughter that she is, in a half hour she and her four-wheel drive arrived, and we got off Train 5, leaving it and 179 passengers for a milder clime."

So close to the end of our transcontinental journey, we were to stay for a few nights in Auburn (altitude 1,360), an old foothill mining town that grew into a supply center and shipping transfer point for hundreds of gold rush camps, being for some time the head of wagon roads from Sacramento. We had paused for a day's R&R in Reno, but we had been actively on the road and following our usual compressed schedule for more than two weeks. The children had earned some leisure time. Chris and Carol lived in nearby Newcastle, and besides seeing them there, we would make some forays into Gold Rush country, trace the original Central Pacific right-of-way, investigate local history, but also put in some time lounging around the motel swimming pool.

From our motel, just off the interstate, we could see the elegant Renaissance Revival–style dome of the Placer County Courthouse down on Lincoln Way in Old Auburn. The courthouse was built in 1898, a classic, highly ornamented three-story granite, brick, and terra-cotta confection that now contained the county museum with its big displays on Native American art, the Gold Rush, and other aspects of local history. The old sheriff's office had been restored, and a splendid stagecoach stood to remind of its run between Auburn and Michigan Bluff; the courthouse had a gift shop and an excellent local history research center. Among the earliest visitors to the area were John C. Frémont in 1844 and the emigrant party leader John Bidwell in 1845. After the itinerant wheelwright and carpenter James W. Marshall discovered some shiny flecks of gold in the stream while building a sawmill at Coloma on January 24, 1848, excited would-be millionaires threw down their shovels, clerk's aprons, and operating instruments and fanned out into the rivers and the streams tumbling down the Sierra's western flank.

The great California Gold Rush was on. A French-born friend of Marshall's named Claude Chana led a party consisting of three other Frenchmen, twenty-five Indians, and thirty-five horses up Auburn Ravine, where, on May 16, 1848, they found some "good sized specimens" at the streamside. Today, Old Auburn stands above that spot—with its restored relic brickfront buildings, including a marvelous old-fashioned drugstore and soda fountain. The mining camp was officially named Auburn in August 1849 by a group of upstate New York prospectors, after their hometown; it thrived through the Gold Rush days and drew additional power after the Central Pacific Railroad built past it and planted a station at Auburn on May 13, 1865, being greeted by cheering crowds and a brass band or two.

With Auburn's roster of museums and historical sites—among them the Bernard House, a restored Victorian residence and winery; the Gold Country Museum, housed in a WPA building; the Joss House Museum, a Chinese temple, school, and boarding house; and the colossal statue of that founding French prospector, Claude Chana, crouching and panning for gold—an afternoon's walking tour was hardly enough time. One evening we met Chris and Carol at the former Masonic Hall, now the chapter meeting house of the Auburn Parlor of the Native Sons of the Golden West, a group of historical-minded citizens including Chris Graves. The welcome scent of charring hamburgers and foaming beer greeted us as we ushered the children inside, where a crowd of beaming members waited to welcome us. I was there to give an after-dinner talk about the Central Pacific Railroad, and to sign books, but for our entire family the evening was a joy—surrounded by people who, like many others we had encountered along the old iron road, lived with one foot in the present and one in the past.

We met so many good people that evening, but there were three who gave us particular pleasure. At one point Mary came through the crowd towing a bright-eyed, gray-haired, and vivacious woman: it was a local scholar and novelist, Jo Ann Levy, whose work on women in the California Gold Rush was nationally known. We had her book, *They Saw the Elephant*, at home, on long-term checkout from the Middlebury library. After animated talk we exchanged addresses as well as each other's inscribed books. Then another area author appeared: Meade Kibbey, whose authoritative, profusely illustrated book on the photographs of railroad photographer Alfred A. Hart was one I had wished I had owned in the last few years of working on *Empire Express*. Nearby, as we, too, exchanged books and addresses, stood a man who shared our deep interest in Hart's work: it was Dana Scanlon, whose name, in Western historical circles, always comes up when on the subject of the Central Pacific because he has the most complete private collections of original Hart stereograph views; later, during another California trip, seeing Dana again, I would be privileged to examine a stack of original, mint-condition Hart photo cards that I'd never seen anywhere else but in the pages of Meade Kibbey's book.

The evening went late, and it wasn't until we turned to see Davey and Mimi all but passed out from fatigue that we broke away. Back at the motel they were quickly asleep, and Mary and I curled up with our new books. I thumbed to the back of Meade's Hart tome, losing myself in the appendices with the miniature, chronologically arranged collection of Hart images, following the Central Pacific's construction from riverside Sacramento all the way to dry and bleak Promontory, Utah. Mary kept leaning over to see rarities I spotted; it recalled to me early days of our marriage when she worked for Dover Publications—that quirky family-owned book house famous for its art and photography books—and we would sit up, late at night, sharing a heavy book across our laps, staring into reproductions of antique photographs and paintings. Then, like other earlier times, she quietly read, to refrain from disturbing our slumbering children, a few passages from Jo Ann Levy's work on Gold Rush women, including this from one entrepreneur: "I have made about $18,000 worth of pies—about one-third of this has been clear profit," reported one anonymous woman. "One year I dragged my own wood off the mountain and chopped it, and I have never had so much as a child to take a step for me in this country. $11,000 I baked in one little iron skillet, a considerable portion by a campfire, without the shelter of a tree from the broiling sun."

"Eighteen thousand dollars of pies," Mary whispered. "How about them apples?"

We all slept late, and after lunch we followed Chris's pickup back up the steepening divide between the Bear and the North Fork of the American River, along which old Route 40, Interstate 80, and the Southern Pacific tracks have been draped in a braid past the hillside town of Colfax (altitude 2,422). As an old mining camp it had been known as Alder Grove and then Illinoistown, but after a press and publicity junket came through, in 1865, carrying the popular Republican speaker of the house, Schuyler Colfax of Indiana, the town fathers enthusiastically renamed the place after Colfax. This was later regretted. In 1872–73 after serving one term as Ulysses S. Grant's vice president, Colfax became a leading suspect in the furious Washington scandal involving congressmen buying or accepting stock in the contracting company of the Union Pacific Railroad, the Crédit Mobilier, which exploded his meteoric political career. The scandal, which all but paralyzed government for six months and captured constant headlines, helped inspire Mark Twain into his collaborative first novel, *The Gilded Age*, about the excesses of the rich and powerful after the Civil War. As he was writing it, with Schuyler Colfax and colleagues grimacing under the bright spotlight of official inquiries, Twain appeared in crowded halls lecturing about his voyage to the Sandwich Islands—Hawaii—and about the curious and amusing practices of the innocent, picturesque Hawaiians. "In the Sandwich Islands," he told chuckling audiences, "everything is done in an 'upsidedown' manner. Among other foolish things that they do is to elect the most incorruptible men to Congress."

Past Colfax, old Lincoln Highway–era tourist cabins had once perched on the sharp eastern edge of the divide, high above the canyon of the American River's North Fork, but there were only chimneys left. There, at a pull-off, we rendezvoused with another of Chris's friends, Pat Jacobsen, an expert on California fruit crate labels, that brightly figurative advertising art genre so popular with collectors all over the world. Pat had written a big, authoritative guidebook on the subject, and maintained an active Web site for purchasing from his collection of between 600,000 and 700,000 labels. Chris had told him that we were interested in obtaining fruit crate labels with railroad associations. From the back of his station wagon, filled from back bumper to driver's seat with catalogued and theme-arranged storage boxes, he produced a handful we couldn't resist: of the ones Mary later framed and hung in our house there was one, for Golden Bosc Pears, showing a gleaming yellow streamliner; another, Old Fort Yuma Selected Vegetables, featured a sunset-reddened train crossing a delicate railroad bridge and across bountiful farmland; another, showing the familiar sight of the blue Sierra lake and the summit's arched concrete bridge, was for Donner Brand Bartlett Pears.

Then we followed Route 40 only a little way before turning off onto a steep, winding paved road down almost to the canyon bottom. A little farther and we came to a dirt lane, ordinarily restricted from the curious by a metal barrier, which that day happened to be open. We switched to four-wheel drive and climbed the twisting path through heavy woods. After some time, near a tangle of felled trees and neat stacks of railroad ties, we emerged at an open, leveled area near the Southern Pacific tracks. "Welcome to Cape Horn," called Chris as we got out of our car.

In front of us was a high and jagged brown rock wall. In late 1865 the Californians had named the forbidding and obstructive cliff—which jutted into the theoretical path of the Central Pacific right-of-way—after the sea-battered southern tip of South America, around which many emigrants had sailed, pea-green, confined to their berths, and wishing for home. Cape Horn in California stood above the American River's North Fork Canyon; the river tumbled by, about 1,300 feet below and a short distance from the bottom of the slope, which was too steep (at 75 degrees) to offer any unaided access for rock cutting and blasting. If anyone lost his footing he would undoubtedly die after a quick, violent tumble. Cape Horn required a shelf. Volunteers—Chinese, of course—allowed themselves to be lowered by ropes while they hand-drilled blasting holes into the resistant rock, and had to scramble upward before their lit fuses touched off explosions and a deadly showering of debris. First a seven-foot-wide shelf was carved out, which must have been perilous and scary enough to work on as the blasters chipped away at the cliff, widening the shelf farther to accommodate a single track. Alfred Hart once clambered to the roof of a locomotive to capture the creepy, awe-inspiring drop just inches from the train; often, engineers paused their passenger trains to give people a prolonged view of the emptiness falling away toward the canyon bottom and the river.

There has been a durable story about Cape Horn: that the Chinese wove reed baskets that were then attached to the ropes and lowered with a drilling man in each. Most railroad historians, including, alas, myself, have been lulled by a few secondary sources and some 1860s journalism and travel guidebooks to repeat the baskets story. By the time *Empire Express* came out I had changed my mind about the baskets; I had no doubts about the rope lowerings, but more likely the Chinese had sat or leaned into slings—they are called bosun's chairs for their nautical use in scraping, caulking, and painting sides of ships. The Chinese would thereby have been able to maneuver using their feet to "walk" across the steep slope while using their legs to lean out into the support of the ropes. In 2001 a family descendant of the C.P. construction boss James Harvey Strobridge set out to trace the origins of the probably false baskets assertion, to credible effect. But back in late 1999, although conscious that one can't prove a negative, I preferred to back away from those durable baskets; unfortunately it was too late to change the passage in question. And sticklers, amateur and professional historians alike, will rib me about the Cape Horn baskets until we are all gone.

Today the Cape Horn shelf has been considerably widened to accommodate two tracks; one vanishes into a shortcutting tunnel built early in the twentieth century. The overhanging cliffs are covered by a multitude of horizontal electrical wires that sound a distant alarm if rocks fall from the heights and strike the wires on their way to the railroad tracks. A low wall rises at the outer edge, and we stood there, gripping stone, looking down and out to the canyon. Through the haze Mary saw either an eagle or a large hawk plummeting toward earth and lunch.

A yellow U.P. freight train rumbled slowly out of the tunnel, heading west. The engineer waved back at us, as we were leaning against the outside wall a healthy distance from the tracks, but after he had taken his train around a sharp curve and vanished, leaving only echoes, we hastened back to our vehicles in case he radioed our presence on the right-of-way to the dispatcher. As opposed to the numbers of victims in this country who stumble into the paths of speeding trains, we were being extra-careful, though mindful that we could earn anything from a stern lecture to a misdemeanor ticket and fine. And we had the expertise of someone who knew the line like it was in his backyard. "But try telling that to the sheriff. Let's go back downhill," our guide and protector said. "Carol's expecting us for dinner at five, so there's just enough time to see Bloomer Cut. We're standing here at Cape Horn thinking about those Chinese working on this place in late 1865. Let's go back further in time, a year and a half, to early 1864. And you're really going to see something."

Between Auburn and Newcastle, as we followed his pickup, Chris led us through neat suburban neighborhoods. Leaving our vehicles at a dead end, we hiked up through a narrow strip of scrub brush between parallel lines of backyard fences, passing patio equipment, swing sets, and a few alarmed and yelping penned-up dogs. Trees thickened, and then, when we were away from the housing development, we stopped short, I clutched the children, and we gaped down at a long, deep cleft in the earth—it could have been, for all I knew, a tectonic earthquake rent in the crust, but there were train tracks at the bottom. "Bloomer Cut," Chris announced.

He led us away and then down a steep, gullied path until we were down on the gravel of the right-of-way, and I gazed at a sight unchanged from the time of the Central Pacific's Alfred Hart, who photographed Bloomer Cut from where we stood. Back in 1864, the railroad stakes laid out originally by Theodore Judah's assistants ran until interrupted by a ridge of solid, cement-hard indurated gravel. According to Central Pacific contractor Charles Crocker, Bloomer Cut, 63 feet deep and more than 800 feet long, required some 500 kegs of black blasting powder each day; it was backbreaking work for the laborers hauling away rock debris in wheelbarrows and one-horse carts, and a punishing drain on the company treasury. When a delayed explosion there cost the eye of construction boss James Harvey Strobridge, work was hardly slowed. The result was a wonder to behold, whether in 1864 or 2000: it was the exact shape of a modest wedge of pie balanced on its apex, then as now only inches wider than the track.

"Come on," said Chris, "you've come this far, you should walk the length

of Bloomer Cut." Mary didn't like that, and she and the children stayed behind, as much for caution's sake as for Davey's discovery of a large blackberry
bush, heavy with fruit at the edge of the woods.

"Is this track still used?" Already I regretted stepping into the cut. The
walls climbed up and leaned out and away far above us. There was nowhere
to go if a train appeared.

"Don't be nervous," Chris replied. "You can hear the trains from far off."

We walked the length of the narrow slot, all 800 feet of it. The formation
was composed of hard, fossilized mud, thick with rounded river rocks like
raisins in pudding. From high above us on the rim vandals had recently
dropped an old television onto the tracks where it had been swiftly disassembled by a locomotive into barely recognizable fragments. One foot in front of
the other, I balanced along one of the rails for a while, then tried alternate
ties. We got to the far side, where the tracks veered sharply east, and we
turned back. When we rejoined my family on the far side I could barely get my
breath. "A nice, uneventful walk," commented Mary. I rolled my eyes.

This was not so uneventful for my friend Mark Zwonitzer, a Missouri-born
writer of books and independent writer-producer of public television documentaries for *The American Experience* show for WGBH-TV, Boston. With his
New York troupe of technicians and researchers, Hidden Hill Productions,
Mark had done films on a variety of topics, from the Irish in America to Joe
DiMaggio to the construction of Mount Rushmore. After we began working
together on an *American Experience* documentary growing out of *Empire Express*, I introduced him to a number of friends who had expertise on the Pacific railroad, including Bob Chugg and Chuck Sweet and Chris Graves. Chris
escorted Mark the length of the Central Pacific Railroad right-of-way from
Sacramento to Ogden; also along was Mike Chinn, the freelance cameraman
for the Bill Moyers shoot on Donner Summit. I think it was their first day of a
two-week journey when they got to Bloomer Cut and walked nearly its entire
length while Mark took notes and Mike shot footage. "We were standing in
the middle of the cut," Chris later wrote me, "looking up at a rock with blast
hole still visible, when I looked eastward—and beheld a freight train heading
our way, at a terrific rate of speed, coming downhill with his light flashing, his
whistle blowing.

"I yelled what I had learned during the Nixon administration was an 'ex-
pletive,' and ran like the speedster I am, for the west end of the cut—the east
end was closer, but uphill. I was surprised to find that Mark is really athletic,
as he passed me, and all I could see was the back of his head and the bottoms
of his feet. Darn guy, he stayed in the middle of the track where there was
steadier footing, while I was running on the gravel and rocks on the side of
the track. You remember: there's nowhere to avoid a train. Mark had an advantage in speed, but his disadvantage, of course, was that the train was also

in the middle of the track, and gaining on us. I pointed out the error of his way in the most polite manner I could, he casually shifted to the side, and still pulled out in front of me. Good friend Mike was bringing up the rear, still lugging his camera, in a most admirable fashion. Well, we made it to the end of the cut before the train and its most hostile engineers—they were most concerned about their hands, it seemed, as they kept pointing their middle fingers in the air as if alarmed about birds. We shuffled back to the truck, fearing the sheriff's helicopter and the trespassing charge that could be leveled at us."

It's easy to make light of a terrifying, narrow escape after the fact. When I heard the story I resolved to never put myself in similar harm's way again, as did, I am fairly sure, my friends.

That evening we visited Chris and Carol at their home in Newcastle. The ranch-style house stood on several shady acres bisected by a little stream and decorated with sculptures, heroic-sized boulders, and large iron artifacts. Inside the house, paintings of California landscapes hung opposite framed historical documents; a wide and deep solarium contained many house plants, bookcases, a more-than-lifesize RCA "His Master's Voice" figurine of Nipper, the Victor mascot, and the twenty-mule-team sculptures of Chris's friend from Bishop, Arlene Halford. Food, drink, and happy, animated talk stretched from late afternoon until late at night; it was our last evening with Chris after more or less constantly traveling in tandem for nearly three weeks across the deserts, river valleys, and mountain ranges of three states. We were road-weary and conscious of the fact that in exactly one week we would be returning to the East, but Carol and Chris made us feel supremely at home, and when it was time to leave we did so reluctantly.

Early the next morning the phone rang. It was Chris—he had arisen at dawn, and while listening to the radio heard of a horrendous multicar accident on the interstate, which was the central artery across the Sacramento valley and the route he knew we were planning to take after breakfast down to Sacramento. "The interstate's going to be plugged for hours," he warned. "You drive back here to our place in Newcastle. I'll lead you through the back roads—it's also much more scenic than I-80—and you'll get down to the capital in record time."

Elated at being saved the misery of traffic jams, we ate and packed and then soon gratefully joined our guide for one last journey. I rode up with Chris in his vintage Volkswagen Bug—I still couldn't figure out how he could fold his six-foot-eight-inch frame behind the steering wheel—and we threaded beneath oaks and cottonwoods through foothills and snaked across the valley floor, as Chris's car radio continued the reports of how long it was taking the highway department to clear the interstate. Finally, we pulled into Sacramento, a place familiar to me from stays and research at the California State Library, and we found parking spaces down at the edge of Old Sacramento, the sprawling re-created Gilded Age district at the confluence of the Sacramento and American rivers.

We got out to say goodbye to Chris. I wondered what the Pathfinder, John C. Frémont, said to the scout who actually found the paths, Kit Carson, when they parted after one of their expeditions. All I could come up with for Chris was a humble thanks and a promise to write—maybe that's all Frémont could do.

"Well, I got you here in one piece. Now remember," he drawled, pausing and grinning. "Never take no cutoffs—and hurry along as fast as you can."

That said, he folded himself into his little car, darted out into the morning traffic, and, merrily waving, vanished around a corner.

Two tiny Volkswagen beeps floated back to us.

16

From Sacramento to the Sea

The triumph and tragedy of Johann August Sutter has always spoken to me. Born in Germany in 1803, raised in Switzerland, Sutter abandoned home and family for a life of adventure, his roamings taking him to New York, St. Louis, Santa Fe, Oregon, and Hawaii, from which he sailed to San Francisco when it was still a sleepy provincial Mexican village. That was in 1839. Sutter won the confidence of the young civil governor, Juan Bautista Alvarado, and proposed that he settle California's interior. The governor gave him a baronial grant—twenty-two square leagues, a whopping 97,648 acres—and Sutter, who adopted Mexican citizenship, established himself near the confluence of the Sacramento and American rivers, constructing an armed fortress and covering adjacent lands with crops; the timber and adobe fort was hardly needed, as it turned out. Sutter named his domain New Helvetia in honor of his home in Switzerland, and actively encouraged others to settle nearby, dispensing homestead grants to almost anyone who appeared and sending aid to emigrant parties (such as the Donners) who ran into difficulties; he also maintained good political and commercial relations with the Pacific shore enclave of Russians. To his everlasting credit Sutter seems to have governed his holdings liberally, treating even Native Americans sensitively and honorably. The numbers of new settlers—most were Americans, though many were Mexican or other nationalities—grew, to the growing dismay of the Mexican provincial government. New Helvetia, with its rich soils and temperate climate, became a sort of paradise. "I found a good market for my products among the new-comers and the people in the Bay district," he recalled.

> Agriculture increased until I had several hundred men working in the harvest fields, and to feed them I had to kill four or sometimes five oxen daily. I could raise 40,000 bushels of wheat without trouble, reap the crops with sickles, thrash it with bones, and winnow it in the wind. There were thirty plows running with fresh oxen every morning. The Russians were the chief customers for

my agricultural products. I had at the time ten and fifteen thousand sheep, and a thousand hogs.

Sutter presided over his vast estates despite the tumult of 1846—the saber-rattling and brazen California ramblings of John C. Frémont and his scientific soldiers in the face of Mexican inhospitality; the Americans' Bear Flag Revolt, which captured the Sonoma presidio and established the short-lived California Republic; the U.S. Army and Navy invasion of California as soon as the Mexican War was declared, the lackluster skirmishes, and the ultimate surrender of the entire province of California to the United States.

Sutter's New Helvetia was still peaceably productive in the rainy January of 1848 when his head carpenter, James Marshall, left his labors over a new mill at Coloma and came down with urgent news he insisted had to be divulged privately, in Sutter's personal quarters; Marshall was in the act of removing a handkerchief from his pants pocket to show Sutter what he had found in a stream when one of Sutter's clerks intruded into the chamber with a question—"how quick Mr. M. put the yellow metal in his pocket again," recalled Sutter, "can hardly be described." Too late to have been kept in the dark, the clerk was shooed out and at Marshall's insistence Sutter locked the door.

It was then that Johann Sutter got a good look at the gold that would ruin him. Knowing word was spreading through his staff, he begged them to keep the discovery quiet for at least six weeks until he got some commercial work done—"I was unhappy, and could not see that it would benefit me much"—but within a fortnight increasing numbers of strangers were passing through New Helvetia with their gimlet eyes set on the foothills and mountains, and the overwhelming flood had begun. With pride and enterprise Sutter and son staked out the future city of Sacramento on farmland later in that year of 1848, but before long Sutter watched as his workers deserted his fields and tanning vats in favor of the big bonanza in the hills, watched as his crops and cattle were openly plundered by hordes of gold seekers, watched as squatters began a long pattern of defiance by perching on his property and daring him to come closer.

In disgust he sold the entire lot left to him and bitterly retreated into obscurity on a farm along the Feather River, slipping into despondency, alcoholism, and bankruptcy. California became a state in 1850. "By this sudden discovery of the gold, all my great plans were destroyed," Sutter recalled in 1857. "Had I succeeded with my mills and manufactories for a few years before the gold was discovered, I should have been the richest citizen on the Pacific shore; but it had to be different. Instead of being rich, I am ruined." He desperately appealed to the state legislature in Sacramento—the city he had founded—for some kind of aid but was rebuffed. Then Sutter went East to petition the federal government for compensation, or at least a modest sinecure recognizing his role in history and his many generosities; there, too, all turned away from him. Sutter died in 1880, penniless, in a hotel room in Pennsylvania.

Sacramento became one of the most populous cities in the state, surviving a march of catastrophes that had repeatedly trampled it flat—floods and more floods, cholera plagues, firestorms, street riots, outbursts of lawlessness and vigilantism. With each difficulty Sacramento had buried the bodies, cleaned up the mess, and erected new commercial and residential blocks. The growing city certainly had aspirations of respectability. But there was no denying that, having been born during a boom, Sacramento still had an explosive side. Every commercial block seemed to have its saloons, gambling dens, and houses of prostitution, all hoping to fleece the 100,000 or more transients who passed through in just a few years. Mark Twain called the California capital the "City of Saloons." There were a good many of them, he would tell readers of the *Territorial Enterprise*. "You can shut your eyes and march into the first door you come to and call for a drink, and the chances are that you will get it." The most elite establishment was the Tremont Hotel on J Street, a three-story brick structure housing one of the largest gambling saloons in the West, in which the minimum poker ante was $1,000, gold. "In the winter Sacramento was always crowded," recalled Mariano Guadalupe Vallejo, once baronial rancher of Sonoma, "since it was there that the gamblers assembled in droves to fleece the unwary. Gaming tables were set up everywhere; faro, monte, rouge et noire, and lasquinet were the favorite games. The amounts wagered on a single card were very great, and there were times when they reached five or six hundred ounces of gold."

After all the visits over fifteen years, my memories of Sacramento have been equally divided between the Neoclassical, colonnaded state buildings fronting on the graceful, tree-shrouded Capitol Park, and the renovated red-brick district of the old town, spread along the Sacramento riverfront. At the old State Library with its murals, statuary, and countless rare books and manuscripts (it was founded upon the donation of books by John C. Frémont), I spent many hours in the California Room, especially with the invaluable subject guide to nineteenth-century state newspapers, handwritten in ancient cursive on yellowing index cards, which was so helpful in writing *Empire Express*. Also there, in a nineteenth-century attorney's personal papers, I found—and held in my white-cotton-gloved hands—rare letters from the Central Pacific Railroad's founder and chief engineer, Theodore Judah. They detailed his survey party efforts to draw the first line over the Sierra, and, in affectionate messages to his wife, Anna, his boyish enthusiasms about camping in wild, unsettled mountain terrain. The missives were instrumental in rounding out his character beyond the traditionally consulted letters and journals down at the Bancroft Library in Berkeley. And over in the tourist-teeming Old Sacramento, where across twenty-eight acres dozens of historic buildings stand in reminder of California's raffish and exciting past, I had imbibed something of the atmosphere of Sacramento when the sidewalk throngs had been gold rushers from all over the world, or when Mark Twain had hired a hall to deliver his later-celebrated, whimsical "lecture" on the Sandwich Islands, or when it was the excitement of the age to step aboard a luxurious plush railroad coach and be drawn by a brightly painted and gilded wood-burning steam locomotive up over the Sierra and eastward alongside the toilsome wagon ruts of old, back toward Old Home.

The commercial center of Sacramento had moved away from the riverbank by the 1930s, when the Lincoln Highway–era WPA writers noted the oldest part of town and its "red brick buildings, with tall narrow windows, and tin-roofed awnings projecting over the sidewalk. Curbstones are high, recalling the times when the river flooded its banks." As with most similar historic neighborhoods, the area slid downhill from negligence until it was a ruin and a slum, a firetrap in danger of being bulldozed by urban renewal. In the mid-1960s, however, the city and private entities adopted a master plan for sensitive redevelopment that, over the years and in successive stages, has continued to grow and refine. Since my first research trip in 1985, I had been charting its growth, finding the renaissance even more exciting than those I had seen at Boston's Faneuil Hall Marketplace or Manhattan's South Street Seaport or the sprawling waterfront in Baltimore. During our summer transcontinental drive we had seen encouraging historic redevelopment efforts in Kansas City, Council Bluffs, Omaha, Cheyenne, Laramie, and Ogden, to name trails-and-rails aficionados' high points. I was especially eager to show my family Old Sacramento.

Over several days we strolled shoulder to shoulder with leisure-timers, whether they were tourists from afar or capital officeworkers over for the fine restaurants and watering holes—over five million visitors go annually, and locally, Old Sacramento has been voted the best place for a first date. One could stop at the Fratt Building, adjacent to the arched gateway footpath to the downtown district, to see a living history center and rent self-guided audio tour players that featured an actor playing Mark Twain and narrating nine stops along the way. One could pause at the Pony Express Monument, commemorating the destination of all those young saddlesore riders along the 1,966-mile route from St. Joseph, Missouri, in 1860. One could pay homage at the Theodore Judah Monument at the corner of L and Second streets (it used to stand outside the Southern Pacific Railroad Station at Fourth and I streets), which I always did when in town, feeling to be at the end of a long dotted line that ended spiritually at his gravestone back in the St. James Episcopal churchyard in Greenfield, Massachusetts. One could investigate the reconstructed Central Pacific Passenger Depot, built on the 1876 plan and best seen at mealtime since it is now a restaurant. One could board along-the-levee steam excursion trains at the Central Pacific Freight Depot, dine at a restored early-twentieth-century riverboat, the *Delta King*, or see historical armaments at the California Military Museum. Passing old hotels, theaters, fire stations, and mercantiles, one could shop in buildings erected by early civic leaders and businessmen Sam Brannan and Newton Booth, the latter of whom was an original supporter of the Central Pacific Railroad, and one could peer with great interest, as we did, at the displays of nineteenth-century tools and hardware on I Street's Huntington & Hopkins Hardware, established in 1855 on K Street by two of the Central Pacific's founding board members, future railroad moguls Collis Huntington and Mark Hopkins. A re-created Stanford Brothers Wholesale Groceries, once run by future California governor and senator Leland Stanford, another Central Pacific founder, rises next door.

We saw all of this and more, but the paramount experience was at the California State Railroad Museum on I Street, a large two-story brick structure with unparalleled exhibits, opened in 1981. The museum was rooted all the way back in 1937 in the efforts of several Bay Area railroad enthusiasts to save old rolling stock; later, as founding members of the Pacific Coast chapter of the Railway and Locomotive Historical Society, they sponsored steam excursions, raised funds, and slowly began acquiring and restoring old locomotives and cars. The Second World War stopped the railroaders' forward movement, most dramatically with the popular scrap metal collection program. At one point, it was announced that the Southern Pacific's first locomotive—the Central Pacific's valiant little woodburner, the *C. P. Huntington*—was going off for the scrap drive. Employees at the railroad's general workshop quietly kept it out of sight, the official history of the museum relates, until the war ended when it mysteriously reappeared. (It is now a centerpiece of the museum's exhibits.) Members resumed acquiring artifacts after the war with the vague hope that they could exhibit them somewhere in San Francisco, but in 1967 their efforts became known to the California Department of Parks and Recreation, as the first phase of restoring Old Sacramento had begun. An entire block backed by the I Street Bridge approach ramps was eventually reserved for four major structures, with the railroad history museum taking most of the space with a 100,000-square-foot facility, space enough for forty interpretive exhibits and twenty-one restored locomotives and cars.

I'll not forget my thrill when first seeing the *C. P. Huntington* engine, until then for me simply the subject of several faded little 1860s Hart stereographic cards I owned, suddenly transformed into a formidable black iron giant with scarlet wheels and pilot. Now it was my family's turn to see it and the other restored and cherished spotlit exhibits, such as the Central Pacific's *Governor Stanford* engine, set in a Sierra snow scene; the Virginia & Truckee's *Empire*, the engine's sharp lines making it appear that it was moving even when frozen in place; the North Pacific Coast Railroad's narrow gauge *Sonoma* (1876); the Nevada Central's ornate passenger day coach, the *Silver State* (1881); or the burly million-pound forward-cab steam locomotive—the last one ordered by Southern Pacific—known simply as No. 4294. We walked through a shining silver streamliner dining car, the Atchison, Topeka & Santa Fe's *Cochiti* (1937), where my mother or father could have dined on the route west; we peered into the Canadian National's open-section sleeping car, *St. Hyacinthe* (1929), which looked all ready for an intrigue, and admired the sumptuous wooden *Gold Coast* (1905), a private car restored and decorated by those flamboyant railroad enthusiasts and men about town, Lucius Beebe and Charles Clegg, at whose former house we had been entertained in Virginia City.

All were the products of some of the most painstaking restoration efforts in the world, a process amply explained at the museum but brought home to me personally just a few days before when Chris Graves and I had spent several hours with the museum's retired chief machinist and supervisor of restoration and maintenance, Ken Yeo, at his home. When presented with a new restoration job on an artifact, the museum obeyed three hard and fast rules: document every individual effort, be sure every change was reversible, and never compromise the future operability of any engine or car. The research and technical staff operated like detectives, removing decades of paint (and rust and corrosion), carefully replacing parts and equipment, and refinishing in original colors. Ken Yeo's adventures in scouting trips in the mountains, and his solutions to many historical and mechanical puzzles, could fill a book.

The museum easily occupied a whole day with its multitude of photographs, drawings, paintings, and advertising art, with its wall texts and exhibit cases and life-size dioramas (the best depicting Chinese workers perched on a cliff—without baskets, of course—preparing for a blast). There was a well-presented film *Evidence of a Dream*, a historical overview, and a multi-screen slide show employing a couple dozen slide projectors. To us it was not surprising that the California State Railway Museum is considered the best in the country, attended by about 600,000 people every year.

During our Sacramento days we stayed with Mary's uncle and aunt, Jim and Tish Smyth, who had emigrated to California from ancestral land on Long Island after the Second World War, and now lived in blissful teachers' retirement in the city of Stockton, their three children dispersed with their own families in various directions. One day, Mimi and Davey received a dispensation from historical tourism in favor of movies and great-aunt-and-uncle indulgences, while Mary and I drove back up to Sacramento to spend our first time away from the children in two months. Characteristically for us, having been together for twenty years, we spent much of the time talking about the children—while eating fancy adult food in Old Sacramento, and then while we went to an art museum: the Crocker, once the home of Judge Edwin Bryant Crocker, the greatest overlooked figure in the history of the first transcontinental railroad.

The museum stands on landscaped grounds on the corner of Second and O streets, a few blocks outside the Old Sacramento district, two classically ornate Italianate stone structures built on the spoils of railroad empire. It was bought and ambitiously redesigned in 1868 by Crocker, a board member of the Central Pacific Railroad and incorporator of the Western Pacific and Southern Pacific and other tentacles of the immense "Octopus" of the Far West. He was older brother to rail contractor Charles Crocker, and served both as corporate attorney and general manager. In the latter position, Edwin Bryant Crocker was crucial, absolutely pivotal, to the success of that haphazard, learn-as-you-go, seat-of-your pants operation in which four middle-aged Sacramento shopkeepers and one common-law lawyer were transformed, in five stressful years, into moguls—from whose efforts grew not only a railroad and shipping empire but treasures such as the Crocker Art Museum, the Huntington Library, and biggest of all, Stanford University.

E. B. Crocker gave his active and useful life to the Pacific Railroad. With training as an attorney and engineer (at Rensselaer Polytechnic Institute in Troy, New York), he entered the business better prepared than any of his fellow future moguls, who learned along the way, mostly through mistakes. Affectionately called "the Judge" by friends after serving a seven-month appointive term in the California Supreme Court, Crocker worked longer hours than the lazy Leland Stanford or the retiring, low-energy Mark Hopkins, keeping on top of the exhausting detail work of railroad management—scheduling, administering, publicizing, lobbying, legalizing, hiring, firing, and long-distance planning. He kept one step ahead of competitors and enemies and in relative good graces with government whether it was local, state, or federal. No detail was too insignificant, and to keep the railroad from foundering it required constant attention.

He was a dedicated letter writer, and quite a good stylist at that. Several times a week for years, he paused from his crushing duties to dip pen in inkwell and write his friend and fellow railroad director Collis P. Huntington. The latter had moved to New York to find investors, keep supplies moving steadily to the West—everything from railroad spikes to immense freight locomotives—and maintain good relations with the government in Washington no matter what it cost. Crocker's handwritten letters were ten, twelve, fourteen, or sixteen pages long, full of important details and vivid anecdotes and characterizations. As an archival resource they were invaluable in my fourteen-year project. It soon became clear to me that no one had worked very hard to read his words in 130 years. The letters had been crated up in New York after Collis Huntington's death in 1900 and stowed away among family possessions, after many years being donated, as an afterthought, to the Syracuse University Library, certainly not an obvious destination for researchers, given the fact that almost everything else was on the West Coast, at Berkeley and Stanford.

Even after the Huntington office correspondence was microfilmed it discouraged research despite wide dispersal to repositories around the nation. Each of the letters—especially E. B. Crocker's, which formed the bulk of the collection to the year 1869 and were the most informative—was microfilmed out of order: in any given handwritten letter the order on a microfilm reel might be page one, fourteen, two, twelve, three, and so on. Few had ever had the patience to follow the threads all the way through, thus missing out on E. B. Crocker's crucial role in the enterprise, and losing an extraordinary voice that had related central details and stories told by no one else. After I bought the microfilmed collection and transferred it back to paper copies there was the chore of putting those thousands of pages back into proper order, deciphering the old-fashioned handwriting (and spelling), and cataloguing the rich material therein.

The man who emerged from his letters was fascinating and tragic—Crocker was a workaholic who took his mission so seriously that he labored, knowing the physical and mental cost, until it all but killed him. Without him, Grenville Dodge, General Jack Casement, and Dr. Thomas Durant would have built the fiercely competing Union Pacific Railroad all the way from Omaha to the Pacific shore before the Californians' Central Pacific would have gotten over the Sierra. Reading his letters, I followed his exhaustion, his series of small strokes, his use of sheer will to stave off collapse until his line had built its way across California, Nevada, and Utah to the wedding of the rails at Promontory.

Just one month from the driving of the Golden Spike, Crocker would suffer a paralyzing stroke from which he would never fully recover. Initially unable to move, barely able to express himself, he would retire from business forever, spending several years with his family in Europe and with them using his railroad millions to build an extraordinary art collection that he housed in a gallery next door to the Crocker mansion in Sacramento. Because of his debilitating illness and his early death in 1875, and the willingness of Stanford and long-lived Huntington to hog credit due others, his paramount role in the enterprise was unknown for well more than a century.

Ascending the flight of steps in front of his grand stone mansion, passing beneath an arched, colonnaded entryway, I instinctively paused at the front door, feeling hesitant and reverent at entering the home of a man whose mind I'd gotten to know so well over years with his private letters. Mary seemed to understand the feeling, but she was eager to see the paintings and drawings inside and reassuringly pulled me forward. The entry hall made a statement in itself, with its highly decorative tile floor and mirror-imaged dark mahogany staircases undulating upward to the second floor. "Rather more statuary than in our mudroom back home in Vermont, Pa," Mary joked in an old-lady voice.

With the house bought in 1868 and still undergoing major renovations, including the construction of a second structure in back, the newly wealthy family had taken poor Crocker, still paralyzed from his last stroke, on a grand tour of northern Europe between 1869 and 1871. During this time as they wheeled him from grand hotel to palace, they spent freely, acquiring some 700 paintings and 1,000 Old Master drawings. The collection became one of the largest private holdings of the time and showed a wide range of taste, though undeniably heavy with major works, including Pieter Brueghel the Younger's *Peasant Wedding Dance*, Philippe de Champaigne's *Head of an Old Man*, Carlo Maratta's *Madonna and Child with St. John the Baptist*, and Sir Henry Raeburn's paired portraits of Mr. and Mrs. Charles Morris. The drawings included ones by Rembrandt, Dürer, Callot, and David, and there was a small-sized but estimable printmakers' collection, heavy on French works.

E. B. Crocker and his wife, Margaret, returned from the Grand Tour with their four young daughters in 1871. (There was another daughter, married and in her twenties; their only son had died in infancy in 1856, and they would adopt a boy, the orphaned son of Margaret's relatives, soon.) The Sacramento mansion on O Street was ready for them, although renovations would continue. The annex building in back, connected by a corridor, seemed the realization of every Victorian's whimsical dream as an entertainment and relaxation center: the ground floor not only had a billiards room but a bowling alley and skating rink; a library and a natural history museum crammed with shelves and display cases occupied the first floor; the second floor contained their art gallery. In a subsequent renovation the natural history museum became a ballroom (there were, after all, four eligible teenaged Crocker daughters) with a marvelous oval-shaped "viewing portal" in the ceiling, affording glimpses of colorful art and light in the gallery above.

Beyond the mansion's front entry hall we found the Crockers' restored family parlor, dark and well upholstered and a little forbidding in its perfection, but through hallways and galleries I steered us to the ballroom (it featured an installation of Chinese blue and white porcelain), the ceiling viewing portal, and ultimately to the galleries above. The California Gallery with its heroically sized canvasses of Western landscapes, all considered contemporary art in the Gilded Age, doubtlessly were closest to E. B. Crocker's personal taste and brought us into the presence of his sensibility. There was the masterpiece *Great Canyon of the Sierra, Yosemite* by Thomas Hill (1871) and William Keith's *Mount Tamalpais* (1872), and Norton Bush's *Soda Springs, Sierra Nevada Mountains* (1868), and Albert Bierstadt's stunning *In the Yosemite Valley*, which, though not dated, was acquired after E. B. Crocker's death in 1875.

One could easily fall into those magnificent landscapes—the vast space of the West, the defining pure light, the soaring landforms—but I was avid to also study the genre paintings of California life in the 1850s, depicting scenes personally familiar to Crocker and his contemporaries, especially Charles Christian Nahl's *Sunday Morning in the Mines* (1872), commissioned by Crocker, who became the artist's sponsor. It is probably Nahl's most famous painting, and a detail from it was reproduced on the cover of my dog-eared, travel-stained Penguin edition of Twain's *Roughing It*. Nahl, born in 1818 in Kassel, Germany, to a family of artists, had exhibited in the Paris Salon in 1846 but the Revolution of 1848 sent him and his family fleeing to the more stable United States; there, the prospect of gold lured him to California.

Luckily he was a better artist than he was a prospector. I had followed his work for years, finding paintings hanging in my old neighborhood at the Brooklyn Museum of Art, at the Smithsonian's American Art Museum, and at California's Oakland Museum and M. H. De Young Fine Arts Museum. Nahl sometimes collaborated—with his brother-in-law Frederick Wenderoth (in *Miners in the Sierra*, from 1851, now at the Smithsonian) and with his half brother, Hugo Nahl—but *Sunday Morning in the Mines* was Charles Nahl's alone. In it, a vignette of miners' sober Sunday morning scripture readings are contrasted against a scene of drunken carousing. There is something timeless about the figures—in different clothes and against another background, the miners could be German or Flemish peasants, centuries earlier—but Nahl's California work is just as valuable for its realistic depictions of the sights he (and his patron, and all the forty-niners) had witnessed in those seminal early years of the Golden State, now part of the entire nation's mythology.

Nahl established studios in Sacramento and San Francisco, painting many portraits and landscapes and resorting to commercial art or photography to tide himself over during lean times. The support he received from E. B. Crocker, though, was significant, and one wonders had illness not cut the association short, what additional masterworks might have been accomplished. The Judge finally slid past the point of no return in 1875; Charles Christian Nahl succumbed to typhoid fever in 1878.

=====

We had only six days left to us, and could hardly believe it: notions of our old, paint-peeling Vermont village house and barns, our patiently waiting, boarded dog and cat, our relatives, friends, and neighbors, began to reassert themselves in minds filled with the sights and experiences over two months and across thirteen states. Increasingly, the children talked of their playmates back home and wondered how to apportion their stories to which companions, and which of their friends would be most excited about which souvenir or historical artifact. Mary and I had been talking, too, about returning to work: I to this book, and Mary to what we expected would be full-time painting in her studio; we had decided to "retire" her from her part-time work running the women's center at Middlebury College so she could finally put art in the forefront. In a week we would be flying back to the East, leaving our car to be trucked home, so we could be in time to celebrate my mother's eightieth birthday with a large number of family and friends—even Rosemary's first cousin and best friend, Dorothy Haward Lortie, who remembered the old Kansas City days and the Haward and Donahue family picnics and my mother's mother, Rose, she who had been born in a covered wagon in 1889.

Stockton—with its riverine access to Carquinez Strait and the San Pablo and San Francisco bays making it the old shipping point for the agricultural riches of the San Joaquin Valley—had for me always been a pleasant place to stay when researching, thanks to my hosts, the Smyths. The town had been founded by Captain Charles M. Weber (it was pronounced "Wee-ber," as in the Utah river canyon); Weber had been a member of the Bidwell-Bartellson emigrant party in 1841, so we had been tracing his footsteps for quite some time. During the Gold Rush Stockton grew furiously. The travel writer Bayard Taylor, a great stylist, visited there in 1849, finding "a canvas town of a thousand inhabitants, and a port with twenty-five vessels at anchor! The mingled noises of labor around—the click of hammers and the grating of saws—the shouts of mule drivers—the jingling of spurs—the jar and jostle of wares in the tents—almost cheated me into the belief that it was some old commercial mart. . . . Four months had sufficed to make the place what it was." The town was also notable in that the last of the Pony Express riders, William Campbell, died there in June 1934. As described in my WPA guide, Campbell had worked the ninety-five-mile section from Fort Kearney to Fort McPherson in Nebraska, a stretch we knew well. "Chased once for miles by a pack of wolves," the WPA book tells us, "he left a poisoned ox on the trail for their benefit. His reward was a dozen dead wolves whose hides brought $50."

In Stockton, after freighting homeward boxes of books and clothes (the car transport company required the Durango to be empty upon consignment), we bid goodbye to Mary's aunt and uncle, exceptional hosts, knowing that we would see them soon for a few days in San Francisco. Our interim port of call would be Santa Cruz, where we were to stay with Jim and Tish Smyth's older son, Neil. For the first leg of the trip we traced the faded, asphalted-over line of the Lincoln Highway, old Route 50 straggling over from Nevada and still recovering from being called the Loneliest Road in America. This part of its run went from Sacramento through Stockton and up the hot and dry San Joaquin Valley toward the aptly named Diablo Range of the Coast Mountains. Now the way was a hodgepodge of connected numerical roads—5,205 and 580—the last of which took us into the foothills of the Coast Mountains.

For me the ensuing route was redolent with associations with the writer Bret Harte, whose earliest, locally published stories before 1870, when he was nationally discovered, were memorable; fame turned him into a parody of himself, and finally little more than a hack writer. His best stories were collected in *The Luck of Roaring Camp and Other Sketches* (1870). As we followed the modern highway it favored improved grades southward of the historic Altamount Pass (740 altitude), which the infant "Octopus" of the Southern Pacific Railroad had conquered in 1868. We drove within a few miles of one locale associated with Bret Harte—Tassajara Valley—where in 1856 the writer witnessed a fundamentalist camp revival meeting, later describing it in a story entitled "An Apostle of the Tules."

We weren't through with Harte reminders. Beyond the towns of Livermore and Pleasanton (where the old Lincoln Highway continued west, intent on getting closer to San Francisco) we turned southward on Interstate 680, passing the sprawl of Fremont and the picturesque old Mission San Jose. Then we threaded across greater San Jose proper, aiming to penetrate the deep Santa Cruz Mountains through a wild and beautiful network of creek canyons overhung with heavy foliage, along State 17. As we sped through Glen Canyon we passed within two miles of Roaring Camp, setting of Harte's most known story, published in 1868 in the *Overland Monthly*, a dark, moralistic tale about the orphaned child of a prostitute who is adopted by tough, gritty gold miners. After we finished the Santa Cruz mountain descent and settled in for a few days with our hosts there, Neil Smyth and his wife, Maria Castro, we'd go back up into the heights—not, as it turned out, to pursue Bret Harte at Roaring Camp, but to rejoin the historical but still palpable presence of the Pathfinder himself—John C. Frémont—for one more time.

Sitting on the public beach at Santa Cruz, I watched Mary and the children braving a light surf. The beach was full of sunbathers but the sea was too chilled for Californians—my wife and children were the only bathers on that beautiful, sunny August day, and as Mimi and Davey splashed in, they yelled that the cold water was nothing compared to the icy mountain water they were used to at Johnson Pond on the Bread Loaf Writers' Conference campus up in the Green Mountains back home. It had been a thrill for me to introduce them all to the Pacific Ocean—Mary and Mimi had never been farther west than Tucson, and Davey, at eight, had never left the East Coast. I remembered the last time I had seen the Pacific from Santa Cruz in 1973—invisible seals had been barking to one another from beneath and beyond the jutting pier then, too—having crossed the nation in that hippie bus in the record time of fifty-eight hours, from the George Washington to the Oakland Bay Bridge. My Berkeley friends took me down the coast for the weekend. Twenty-seven-year-old memories were very pleasant—but here was my family coming back from the ocean dip, and our delightful hosts, Neil and Maria, sitting next to me on the sand. We caught up on family gossip—Mary's mother's family has been prolific—and told them tales of our transcontinental travels.

Several days passed lazily—we were still road-weary and unused to this kind of leisure time. More back-home freighting of books, clothes, and gear continued. One day, Neil Smyth took us back up into the Santa Cruz Mountains to rendezvous with John C. Frémont's memory, at Big Trees State Park, where soft trails wound in twilight beneath towering redwoods, and locals came to jog through while listening to music or, for all I knew, self-improvement tapes, through earphones. Individual trees had been named, including "the Giant," 306 feet tall according to my WPA guide, which was merely sixty years out of date, or "Jumbo," some 250 feet high, and the "Cathedral Group."

We were not far from Roaring Camp, a fact made annoyingly clear to us by tinny loudspeakers just outside the state park limits, where a modest theme park and narrow gauge railroad purported to honor Bret Harte's memory and attracted some customers, I suppose, with loud and puerile Old West music and cartoony singing that seemed out of the era of *Romper Room*. Even Davey was unimpressed with the barrage interrupting the peace and quiet of those grand glades. "This is ridiculous," my eight-year-old blurted. "These beautiful old trees are like a church"—"a *shrine*," interjected Mimi—"and we have to listen to that yakety-yak music? I don't want to go there!" We did not, leaving "Roaring Camp" to roar for itself.

In February 1846 Frémont and party camped in that forest for several days, intoxicated by the pervasive fragrance of pine and bay laurel. They measured redwoods—9, 10, 11 feet in diameter, and even one 14 feet, soaring above 200 feet in height and with deeply furrowed and unusually thick bark. Local people had boasted of the wood's durability: "posts which had been exposed to the weather three-quarters of a century, since the foundation of the Missions, showed no marks of decay in the wood and are now converted into beams and posts for private dwellings," Frémont reported in his memoirs.

Frémont's presence with soldiers had alarmed the Mexican authorities—why was this American detachment wandering through Mexican California, while relations between the two nations were getting tenser over the American annexation of Mexican Texas, possibly moving toward war? Frémont did not help the situation, getting into a haughty correspondence with the local mayor at Pueblo de San José over a stolen horse one of the Americans had bought from a band of Tulare Indians; when a Mexican rancher who owned the mount went to Frémont to complain, he was thrown out of camp. "After having been detected in endeavoring to obtain animals under false pretences," Frémont wrote the mayor, "he should have been well satisfied to escape without a severe horsewhipping." Furthermore, said the Pathfinder, "any further communications on this subject will not, therefore, receive attention. You will readily understand that my duties will not permit me to appear before the magistrates of your towns on the complaint of every straggling vagabond who may chance to visit my camp." In a few days the Frémont band would be down the coast, camping suggestively within a day's ride of the Mexican Californians' capital at Monterey. The American consul there worked overtime to keep the peace between the authorities and Frémont, who now refused the commandant general's order to leave his country immediately. They were just weeks away from war.

There in the redwood forest, we came to the great hollow tree, a living giant with a gaping opening that is large enough to hold fifty people if they all felt their way inside and stood closely together. We crawled in and held hands, waiting for our eyes to adjust to the blackness but they never did. We whispered reverently—it was quiet enough inside the great tree that we could hear nothing from the Roaring Camp carny barkers.

Supposedly, in 1846 Frémont and his men had taken shelter inside this hollow tree to escape a heavy rainstorm. Years later, when Frémont returned here, he was asked if the legend of his sleeping in the hollow tree was correct, to which he replied to the effect, "Never let the truth get in the way of a good story."

On the day before we left Santa Cruz, Neil and I drove south to Watsonville in two cars so I could drop off our Durango at the trucking distribution terminal for its three-week journey back to Vermont on a succession of tractor trailers. I left the keys in the ignition, signed some papers, and nostalgically patted it on the fender. The vehicle that we had whimsically named *Grenville Dodge* after the Union Army's cavalry general and the Union Pacific's chief engineer apparently rode ignominiously backward the whole way; when it was deposited at the base of my driveway in Orwell, the back window and tailgate, and both exterior rearview mirrors, were coated with a squashed, baked-on stratum of insect carcasses, more than an eighth of an inch thick. What an insult and an ordeal to the proud general!

Appreciable layers of dirt and dust that subsequently flaked and washed down onto Vermont soil had been gathered in the deserts, mountains, prairies, and plains of the West over our two months; I'm sure the car was considerably lighter after losing its acquired coating.

In crossing this wide continent of ours, we had put more than 7,000 miles on our car—one way.

17

Golden Gate

I t was bright but thinly overcast when we took the last leg of our journey, northward now, up the coast road in a rented red minivan. The nearer we drew to San Francisco, the closer was California Highway 1—the westernmost American highway—pressed by fog and drizzle, but as we took that winding, stupendous narrow road along the undulating shoreline, we pulled off a number of times to find our way down to the beach. At Año Nuevo State Reserve, only ten miles above Santa Cruz, we went out on a strand populated by thousands of birds of every variety; we looked out at the sea where in the 1850s through the 1880s, whaling boats were rowed out by Portuguese settlers to kill the great mammalian passersby for their blubber; also at Punta Año Nuevo (New Year's Point), in January 1602, the Spanish crew of Captain Sebastian Vizcaíno noted the landform and its background of mountains while exploring and mapping the coastline. The children left their sneakers for us to retrieve while they ran off ahead across the sand, and Mary, whose childhood summers had always been spent on the Atlantic beaches at Northport, an old maritime village on Long Island Sound, relished her long barefoot walk on that Pacific sand, too. We sat for a while at the base of a grassy dune while the children splashed where fresh creek water flowed into salty surf, their laughter ruffling the feathers of the local populace. Mary was thinking about home now, too, I could see, but she nodded toward the shore's wildlife, the curving creek and the steeply plunging mountainside beyond the highway's narrow shelf. "This, too, is a place I want to spend some time painting," she said, "once we get a little space around ourselves. You may need to come back here, to get some details right." It sounded like a good plan to me.

Driving north along Highway 1, we were tracing the uncertain, staggering footsteps of the exploring party of Don Gaspar de Portolá, in 1769 the first white men to see Northern California by land. Sent up into unknown territory from San Diego, their job was to track down a fantasy described 172 years earlier by Don Sebastian Vizcaíno: he had looked from his ship's railing at Monterey Bay, an open gulf, and recorded it as a "fine harbor sheltered

from all winds." By the late eighteenth century this sounded like an enticing location for a fort and mission to Christianize the *indios*. Portolá's men—
officers, soldiers, friars, servants, and Indian laborers—faithfully rode north
with a mule train, suffering mightily until they had become, in Portolá's
words, "skeletons, who had been spared by scurvy, hunger and thirst." Eleven
of their number were so weakened from disease and short rations that they
were borne on improvised litters dragged behind the mules.

Monterey Bay bore so little resemblance to the official report that they
skirted its great curving length without recognizing it, and they considerably
overshot, reaching in late October a promontory now known as San Pedro
Point; from these heights they could see northward enough to realize they
had missed their goal. But when they climbed higher and turned to look east,
they glimpsed an expanse of blue water, a hitherto unknown "great arm of
the sea, extending to the southeast farther than the eye could reach." It was
San Francisco Bay. Portolá's ragged party turned southward again, doggedly
searching for the fabled harbor of Monterey, and when, weeks later, they sor-
rowfully straggled back into the San Diego mission, these discoverers of San
Francisco considered themselves failures.

We scooted up Highway 1 through thickening fog, passing Point Montara
Lighthouse, Gray Whale Cove, and then poor Don Portolá's San Pedro Point
where he, at least, could look far north toward Point Reyes. I sped us through
Pacifica and Daly City along Skyline Boulevard, and then the picturesque
blocks of San Francisco began rising perpendicular to us, now on the great
Esplanade Highway, bringing evident excitement to Mary, Mimi, and Davey,
who had never seen the city on the bay. At the intersection of Highway 1 with
California Street, we crossed over the Lincoln Highway for the last time: it ran
another mile westward and dead-ended at the base of a flagpole in Lincoln
Park overlooking the Pacific Ocean.

I came in for a landing a few blocks south of Golden Gate Park, where we
would lodge for a few days in a borrowed house on 33rd Avenue with Mary's
uncle and aunt Jim and Tish Smyth, whose San Francisco friends were at
their summer place up north.

The ground beneath the attractive neighborhood of three-story row
houses and its nearby parkland had been empty, sandy and scrub-brushy,
manicured by sea winds and moistened by fogs, when Juan Bautista de Anza
led a party of 248 colonists overland from Sonora in March 1776. (Back
in Philadelphia, delegates to the Continental Congress were still debating
whether a clean break with England was advisable.) Among the San Francisco colonists was a three-month-old boy born en route, becoming the first
white child born in California, whose name was Salvator Ignacio Linares and
who was given a proper indoor baptism once the San Francisco missions were
raised. In September 1776, Anza finished a presidio to guard the missions

and watch the entrance to San Francisco Bay; soldiers would lower the Spanish flag for the last time there in 1822 after the Mexican Revolution, and then the Mexican colors would come down after the conquest by the United States in 1846. By then, sailing ships full of settlers were beginning to feel their way up the California coastline, and emigrants had already begun to stagger over the Sierra Nevada into the new golden land; James Marshall's discovery of gold at Sutter's Mill on the American River in 1848 clarified old Calfornia's luster and with the subsequent, convulsive western tilt of population, San Francisco came to extraordinary life.

It was a life that had fascinated me as a teenaged reader of popular histories, and there certainly had been no shortage of books illustrating that gaudy era. Casting my mind back across my adult years, I could not remember how many times I'd been to San Francisco. But my mind was a montage of long visits with friends, of publishing business trips, book research, talks, and book publicity tours; afternoons at the art galleries at the Palace of the Legion of Honor and the De Young Museum; document copying at the San Francisco Public Library and the California Historical Society; long strolls down the full length of the Embarcadero past the wide waters of the bay to the misty tidal surges just west of Golden Gate; countless Chinese dinners and Mexican dinners and Japanese dinners; taking trains daily down to Stanford for research or the gleaming, brand-new BART beneath the bay to Berkeley and the Bancroft Library; and day drives to Sausalito, the Muir Woods, and Mount Tamalpais.

Only a few days were available to us, so close to the end of this summer-long expedition, and there was no way to even approach doing San Francisco justice: the city demanded its own expedition, in fact, so all I could do was plant images of hilly streets and neighborhoods, of the bay and the brown mountains beyond, and of the cultural and historical institutions, each of which requiring time we did not have. I could tell stories. And I knew that someday we would all come back.

In that rented red minivan I piloted us through Golden Gate Park, its 1,000 landscaped acres overlaid on old sand dunes in the 1870s. We drove through the Mission District and ate in Chinatown, thinking of it as it was in the 1850s and 1860s when hopeful farm boys drawn by the lure of gold or the prospect of decent labor on the railroads teemed and attempted to get themselves anchored, their welcome too often experienced in the form of intolerance and narrowing liberties. We went to Coit Tower on Telegraph Hill, the 210-foot concrete memorial raised in 1933 on the site of the first telegraph stations established in 1849 and 1853, and I steered the car down the meandering and nearly vertical Lombard Street, which left my children moaning and covering their eyes.

We rode the cable cars. We passed a number of addresses associated with the mad "Emperor" Joshua Norton, whose story captivated my children's imagination: Norton had roamed the city for several decades after the Gold Rush attired in an old military uniform and admiral's hat, sword swinging from his belt. He had declared himself Emperor of the United States and Protector of Mexico, and thanks to San Francisco's amused indulgence, ate free of charge at nearly every restaurant in town and was permitted to draw checks on any San Francisco bank, up to fifty cents in value.

Down where Market, Battery, and Bush streets intersected we looked for the Donahue Monument, or Mechanics' Fountain, a heroic bronze statue of three artisans wrestling with machinery and plate metal; it was dedicated to one Peter Donahue—not my humble great-grandfather who had coaxed his covered wagon team only as far as the eastern Kansas riverbank hills between Leavenworth and Kansas City, but a nonrelative who got all the way to California, founded an ironworks, and became a pillar of the San Francisco peninsula community.

On Nob Hill we stood at the southeast corner of California and Mason streets at the Mark Hopkins Hotel, built on the site of Central Pacific mogul Mark Hopkins's mansion—downhill critics called it a "magnificent monstrosity"—which was destroyed after the earthquake and fire of 1906; up the street when we paused outside the gray stone Gothic-style Grace Cathedral we were at the site of the baronial homes of the Central Pacific's contractor Charles Crocker and of his son, William, the banker. On the northern edge of Nob Hill we stopped before a famous apartment building with a circular driveway, and my children kept saying "This looks familiar" until I had coaxed their memories back to scenes with Kim Novak and James Stewart in Hitchcock's classic film *Vertigo*, and the children laughed in delight.

That evening we met the Smyths for dinner at the nearby Huntington Hotel, just opposite the cathedral. On the book tour for *Empire Express*, my publisher had booked me into the Huntington, which was fitting: the owners of the Huntington were railroad fanatics who had filled all the first-floor public rooms with memorabilia, a breathtaking historical collection. The restaurant was called the Big Four, after Huntington, Hopkins, Stanford, and Charles Crocker; the moguls had been popularly known as the Big Four, which unfortunately left out the equally important fifth figure, Edwin Bryant Crocker, whose early incapacitation just after the Golden Spike Ceremony had lost him a place in the memoirs of his friends and indeed in the history books. For some years I had been waging a quiet, friendly campaign to have the hotel owners change the name of their restaurant to "The Big Five," but it was a losing battle. That in no way diminished the experience of dining there with my family and the Smyths beneath framed photographs and oil portraits and railroad posters.

Talk that evening went in every direction but the children were particularly interested in the 1906 earthquake and fire that had shaken loose or burned or caused to be condemned not only the palaces atop Nob Hill but such a sickening proportion of the mostly wooden city. Later that evening, when we were back at our borrowed house, we talked more about the catastrophe: how the earthquake had struck at 5:16 A.M. on April 18, 1906, when the San Andreas Fault shook itself like a snake in spasm and settled, shaking and tumbling buildings. Water and gas mains broke, fires ignited all over town that leveled large portions of the city while citizens fled to safety at the waterside and fire brigades stood by, helpless. Out of my WPA guide I read statistics: the city "suffered a staggering blow. Casualties included 500 dead and missing; four square miles, including virtually all of the business district, were destroyed—an area of 497 blocks, some 30,000 buildings. Damage was estimated at 500 million dollars, of which 200 million remained a net loss after payment of insurance. Food and clothing for the thousands of homeless were rushed from all parts of the United States; Europe and Asia contributed millions to relieve suffering."

The next day while driving us through the Presidio, the big military reservation, national cemetery, and historic fortress site, I recalled that there was the burial place of Major General Frederick Funston, who during the chaos after the earthquake and fire had controversially assumed command of the peninsula and declared martial law, dynamiting blocks along Van Ness Avenue as a firebreak, shooting looters and arresting criminals, and organizing relief along military lines. Funston was credited with saving what was left of San Francisco. I had lived with Funston's life for nearly five years, as he was one of two principal characters in my book about the Philippine-American War and the roots of American imperialism, *Sitting in Darkness*, whose title had been inspired by Mark Twain. As we searched for his stone at the Presidio, I thought about how, at the end of that long research period, and after I had completed the marching-in-his-footsteps trek up the rugged northeast coast of Luzon in the Philippines to duplicate his historic journey to capture the leader of the Philippine nationalist movement, Emilio Aguinaldo, I had finally met the man. Funston may have died in 1917 in San Antonio, Texas, but I met him in the National Archives when I found the Personal Effects Inventory compiled by his army subordinates as they crated up his possessions to accompany his coffin on the long train ride back to his grieving widow and children at the Presidio in San Francisco. Through the inventory of books, the candlesticks and place settings, the matched pistols and framed Victorian prints and Chinese teapots, the crystal brandy snifters and cigar humidors, the sideboards and bedboards and ceremonial swords, I met the man. He may have been a braggart and a blowhard who had been accused of shooting prisoners in Luzon and who inspired Mark Twain to heights of sarcasm about Funston's ungentlemanly ruse to capture his Filipino enemy and bring a democratic nationalist movement to its knees; on the other side of the ledger sheet were his actions to save San Francisco: he may have crushed a number of toes but he had made it a redemptive act. And for me he became fully human, connected even more than through his countless journals and letters and reports, when the Personal Effects Inventory rebuilt something private and personal about the old soldier's sensibility. I would not think the thousands of Filipino-Americans who populate the Bay Area would have much of a reason to pay homage to his grave there on the Presidio in view of the Pacific Ocean he helped, by his wartime actions, to shrink. But the city can, and should.

One cannot visit San Francisco, one hopes, without connecting to its literary heritage, and for me there were many personal associations. First, I suppose, there was Richard Henry Dana, a Harvard boy who left the East in 1834 to mend his poor health working as a sailor on a California-bound hide ship. Even the long voyage around Cape Horn did not whet his appetite for what he saw in San Francisco. He was underwhelmed, and recorded his haughty impressions in his memoir, *Two Years Before the Mast* (1840), a great best-seller that has remained in print and faithfully read for more than 160 years.

Of course, Bret Harte was there in the 1860s, editing and writing for newspapers, setting down tales and sketches, founding the classic journal, *The Overland Monthly*. I remembered the early humorist Lieutenant George H. Derby, who wrote under the name of "John Phoenix" and "John P. Squibob." *Phoenixiana* (1856) was my favorite, and in his work one can sense the antecedents of Harte and Twain. Robert Louis Stevenson lived on Union Street in the early 1870s. Ambrose Bierce, sardonic journalist, poet, and storyteller, worked for many years there.

Robert Frost—whom I felt I had somehow followed to rural Vermont because of his indelible ties to the Green Mountains and the Bread Loaf Writers' Conference, my annual home for two weeks of every August for more than two decades—was the quintessential New England poet and presence, but he had been born in San Francisco in 1874. Jack London was also born in San Francisco, in 1876. Hard-boiled Dashiell Hammett had been a Pinkerton operative in the city and later set his masterful detective novels, such as *The Maltese Falcon*, there.

But paramount in that city of hills was Mark Twain, whom I had been following for what seemed my whole life and especially, this summer, on the long line that had included his boyhood home, Hannibal, on the banks of the Mississippi and his haunts described so wonderfully in *Roughing It*. He had left his reporter's post in Virginia City, Nevada, for San Francisco in late May 1864, perhaps intending to stay only a few weeks. He remained for nearly two and one half years. Twain secured a reporter's job at the *San Francisco*

Morning Call, which paid $40 a week, but he found it "fearful drudgery, soulless drudgery." As retold by Edgar M. Branch, editor of Mark Twain's early tales and sketches, by July and August 1864 Twain was strongly tempted to accept a lucrative offer as a government pilot on the Mississippi—$300 per month. He ran into an old friend, John McComb, at the corner of Clay and Montgomery streets. We purposefully drove by the spot but traffic was too fearsome to allow even a brief pause.

"Mac, I've done my last newspaper work," said Clemens, according to McComb's memory, "I'm going back East." McComb urged him to stay. "Sam," he said, "you are making the mistake of your life. There is a better place for you than a Mississippi steamboat. You have a style of writing that is fresh and original and is bound to be popular. If you don't like the treadmill work of a newspaper man, strike up higher; write sketches, write a book; you'll find a market for your stuff, and in time you'll be appreciated and get more money than you can standing alongside the wheel of a steamboat. . . . No, Sam, don't you drop your pen now, stick to it, and it will make your fortune." Clemens thought it over for a while before replying. "Now, Mac, I've taken your advice. I thought it all over last night, and finally I wrote to Washington declining the appointment, and so I'll stick to the newspaper work a while longer."

Whether or not McComb's anecdote is entirely accurate (it does cast McComb in a great, white light of literary posterity), a month later Twain's fortunes undeniably changed, for in September 1864 he was hired by Bret Harte, editor of the weekly *Californian*, to contribute four pieces per month at $12 per article. The writer accepted for the freedom it afforded him and because work in the *Californian* was avidly reprinted by many journals in the East. As a career move it was exactly the right thing to do, for the young journalist's work was noticed and remembered.

In December 1864, though, the pressure of weekly deadlines became too much, and Twain took a leave of absence and wandered up to the exhausted but still hopeful mining camps in Tuolumne and Calavaras counties, above Stockton. For three months he stayed with friends at Jackass Hill and Angel's Camp and filled notebooks with phrases, anecdotes, and sketches. One story he heard concerned a jumping frog contest in the camps. "Coleman with his jumping frog," he dashed down in his journal on February 6, "bet stranger $50—stranger had no frog, & C got him one—in the meantime stranger filled C's frog full of shot & he couldn't jump—the stranger's frog won."

Returned to San Francisco in late February 1865, Twain worked sporadically on expanding his notes into full-fledged pieces while continuing to write for the *Californian*. With the Virginia City *Territorial Enterprise* he arranged to send letters for publication—it turned into a daily telegraphed contribution, for which he was paid $100 per month. And the tastemakers back East noticed: Twain was "foremost among the merry gentlemen of the California press," said the influential literary journal, *New York Round Table*, on September 9, 1865. "He is, we believe, quite a young man, and has not written a great deal. Perhaps, if he will husband his resources and not kill with overwork the mental goose that has given us these golden eggs, he may one day take rank among the brightest of our wits." On November 18, the *New York Saturday Press* published Twain's "Jim Smiley and His Jumping Frog." It took the literary world by storm. More stories soon appeared in the *Press* and other journals and were widely reprinted. By early March 1866, Twain had contracted with the *Sacramento Union* to visit the Sandwich Islands—offer a journalist a month's Hawaiian "working vacation" and see how quickly he leaps—which stretched into five months and added twenty-five travel letters to Twain's writer's capital. Experiences in Hawaii, anecdotes from the mining camps in California and Nevada, impressions of a bumpy, dusty stagecoach journey from St. Joe to the Comstock, had begun to well up in his mind and spilled out onto paper where they were relentlessly revised. Recollections of his childhood and riverboat life on the Mississippi also began to assert themselves and he talked and wrote to friends about all these book plans. And with his sudden and surprising success as a platform lecturer and humorist, he began his tour eastward to capitalize on it all. And the long cometlike trajectory of his literary life was all in the future.

Our last full day was spent driving and tramping along the waterfront lapping the peninsula. The day began or ended on the western shore near the historic site of the Cliff House hotel (Twain had visited but was depressed and feeling peckish that day and did not enjoy himself). The misty Seal Rocks lay just offshore, populated then as now by their yipping and yammering sea lion colony. The day ended or began over on the bayside in sight of the Oakland Bay Bridge, where Mary and I pointed out Yerba Buena Island, where the bridge roadway bores through crags toward the East Bay; below the bridge on the navy reservation, Mary's cousin Kim and her husband, Captain Barron Nelson, had lived for some years, always inviting us out to visit, but we could never afford the trip until after they had moved on. The bulk of the last day we spent at touristy, crowded Fisherman's Wharf—I was glad that the children could see the famous street musicians, mimes, and "The Human Jukebox," a one-man band hidden inside what might have originally been a refrigerator carton painted and decorated to resemble a jukebox.

Pushing through crowds, I showed my family the sidewalk plaques memorializing the buried treasure found up to 100 feet below the concrete and asphalt surface: this whole solid margin of the city had once been open harbor, where ships bobbed with the tides; in the course of time many had rotted and leaked and sunk and others had been abandoned and scuttled, such as during the 1849 Gold Rush. Photographs of the harbor at the time show it so thickly filled with empty ships that one could jump from deck to deck. James L. Tyson, a physician, had arrived in San Francisco on May 18, 1849, finding little but "tents and scattered frame-tenements" and a few neat cottages. But "what commercial inducements could such a mean and insignificant-looking place as this present, to bring together such a forest of masts as the harbor disclosed? We soon learned that every vessel which had arrived since the first discovery of gold in the country, was quickly deserted by its crew, and left to idly swing at its cable's length; however anxious the captains or owners might be to depart, it was impossible to man a ship with a sufficient number to work her to the nearest port. The wages for even a common laborer in the town were higher per day than a sailor was accustomed to receive monthly." Many ended at the bottom of the harbor and in time the city's northern strand was filled in. Generations later, every new large bayside building project turned up treasures from the past. I told the children about the diggings in other areas of the city such as when the basement and foundation holes were scooped out for the Transamerica Building, or with the Bay Area Rapid Transit excavations for the deep stations and track of the Market Street line, which uncovered many historic and identifiable ships.

A few years before I had walked the length of the Embarcadero, enjoying the sunny, open views of the bay that used to be blocked by the elevated freeway; the 1989 Loma Prieta earthquake had damaged the high road so badly that it was removed. San Francisco rediscovered its waterfront and decorated it with plazas, sidewalks, landscaping, and public art, which I enjoyed as I walked. But the embossed sidewalk plaques about the buried waterfront fascinated me, as they would captivate my children's imaginations when we paused, far above the buried hulks, our minds reaching down through rubble and dark mud to grasp their images.

We ate lunch in a soaring, deco-modernistic seafood restaurant with great glass walls overlooking the harbor and Alcatraz Island—in 1843 travel writer William H. Thomas said it looked like "variegated marble, with the deposits of sea birds, and the air full of shrieking and quarreling gulls," and roaring, jousting old sea lions chilling onlookers' blood as they sailed by. Afterward, the children wanted to take a boat out to take the old federal prison tour, but something nearer and ultimately more attractive was moored at the west end of Fisherman's Wharf: the *Pampanito*, a World War Two–era submarine whose oily, claustrophobic depths we braved at Davey's behest; he was, at eight years old, as much of a World War Two scholar as I had been a Civil War scholar at the same age, and the little boy's excitement brightened not only our hearts but the faces of the many aged tourists squeezing past us in the narrow metal passageways.

Emerging onto the deck and blinking in the bright summer light, we made our way to see the rest of the San Francisco Maritime National Historical Park, a unit of the National Park Service. The centerpiece museum had been constructed to resemble a bone-white, rounded vessel complete with large round portholes, glass-brick windows, and, above, a boatlike superstructure upon which a high mast jutted with flag flapping merrily in the breeze. Artifacts, displays, and colorful maritime murals kept us occupied for an hour, after which we walked out onto the Hyde Street Pier, where the park kept its large collection of historic maritime ships—most of which we were able to board and examine.

There was the grand and graceful, three-masted, full-rigged *Balclutha* (1886), crimson and white, 301 feet long with a beam of 38.6 feet and a mainmast that soared upward to 145 feet. Built at Cardiff, Wales, the ship served for years in the Europe-to-California grain trade and then as a Pacific lumber ship carrying Puget Sound timbers for use in Australian mines; following this she went to work for the Alaskan canneries, but after retirement in the early 1930s she was featured in the film *Mutiny on the Bounty* starring Clark Gable and Charles Laughton. We saw the charcoal-colored *C. A. Thayer* (1895), a schooner built up at Humboldt Bay, California, its 219 feet of hull used first to haul lumber and then salted salmon, coal, and copra. We climbed aboard the gleaming white *Eureka* (1890), an elephantine, twin-sidewheel-paddle ferryboat nearly 300 feet long with an extreme width of 78 feet and a monster 1,500 horsepower oil-fired steam engine, which used to carry 120 automobiles and 2,300 passengers across San Francisco Bay without even breathing hard. There were several smaller workhorses, including the *Alma* (1891), a scow that once transported salt, lumber, and oyster shells; the *Hercules* (1907), a black and red ocean tugboat that reminded Mimi and Davey of Lil' Toot, hero of an ancient children's book, and Mary and me of the barge-hauling tugboats we had watched in Long Island Sound and its harbors since we were young ourselves; and the *Eppleton Hall* (1914), a sidewheeled English river tug that had begun its career on the Tyne, survived partial scrapping and fire only to be refitted as a private yacht to steam across the Atlantic and through the Panama Canal, passing through the Golden Gate in 1970, ultimately being donated to the museum in 1979.

And finally, wordlessly, having walked out to the end of the pier of that Pacific harbor, we four decided we had seen enough sights and relived enough history for one summer's journey, and we collapsed onto a park bench and stared out across San Francisco Bay, as giant oceangoing vessels glided through the Golden Gate and slid between us and Alcatraz Island, their decks stacked high with metal containers marked with Korean words, heading for port.

A little later we drove over to the Golden Gate itself, parking in a lot down from the toll plaza and walking out to admire the prospect of the familiar, iconic steel towers ascending 746 feet into the blue, supporting the webs of strong cable and the bridge's roadway. Mary and the children headed off to the gift shop, while I leaned against a railing and regarded the ocean for what seemed a very long time. Suddenly, I remembered my camera back in the van, and sprinted back to the lot to retrieve it so I could take some tourist snaps of Mary, Mimi, and Davey with the Golden Gate Bridge as their backdrop; this I would do, and they would look tired, happy, and triumphant as I did so.

Before this was accomplished, however, I was walking up the slope from the car, gesturing at my family to wait for me so we would not have to retrace their steps. Suddenly, from behind me came an enormous crash and the sound of rending metal—I instinctively stiffened, crouched, and wheeled to see if anything was hurtling toward me. Instead, I saw a beer truck entering the tunnel-like underpass beneath the toll plaza, the truck height some twelve inches taller than the underpass ceiling, and it becoming entirely wedged inside like a cork in the neck of a bottle.

When my adrenaline stopped pumping, I took the picture of my smiling wife and children. And then we went home.

It was Mimi's and Davey's first airplane ride, the weather out across the West was clear, and they could drink in the space as the pilot flew us over the steep Sierra and then eastward, exactly following the California Trail, the first transcontinental railroad, the Lincoln Highway, and today's Interstate 80. What had taken us two months we now did in a matter of hours, over all those old footpaths, wagon trails, and iron roads. Over the Sierra we thought about Chris Graves, the Chinese tunnel men, and the Donners; the length of Nevada's Humboldt Valley our minds crowded with Mark Twain, the "mule lady," Arlene Halford, Bing Crosby in Western garb, and Kit Carson's daughter Adaline; over the blue Salt Lake and the Wasatch, we thought about our desert guides Bob Chugg and Chuck Sweet, the Golden Spike engines at brilliant noonday, and the first emigrants picking their way through boulder-choked Echo Canyon. And as the plane continued eastward we picked out Green River in Wyoming with its wondrous buttes, and glimpsed a slice of the Laramie Plains, and we thought about ghost towns, outlaws, dinosaurs, rock-and-rollers, and railroaders. At some point there was lunch and a movie, but we opened the porthole shutters when the pilot announced we were nearing Scotts Bluff, and as the terrain below began greening up we thought about unmolested tribal lives, covered wagon emigrants passing millions of buffalo, of course bibulous Buffalo Bill, and young Willa Cather reading in her wall-papered room. The Missouri River was unmistakable as our pilot edged us finally northward from the path of the Pacific railroad and Lincoln Highway, but as we began crossing high above Iowa cornfields and soyfields, I leaned across where Davey was reading to press my face to the window, thoughts arcing south toward the jumping-off places for thousands of emigrants, toward river hills where my grandmother had begun her life under billowing canvas in 1889, toward bustling, jazzy Kansas City and my mother's childhood home.

I know at some point I switched seats with Mimi and sat next to Mary and we napped, touching heads, holding hands, while the children listened to their personal stereos and played hangman. We would have liked to drive all the way back—there was the need to be present at my mother's eightieth birthday party on Long Island in a day or two—but we four all agreed that we would do it again. Mary and I were certain that even looming adolesence would not break the children's connection to and affection for seeing the great landscapes and hearing the stories in the company of parents and sibling—in two months Mimi had begun that outward change from girlhood to young womanhood, and Davey seemed inches taller, too, and had changed in ineffable ways.

So we flew back to the East and to the work ahead, Mary and I promising each other to take them back "before it's too late," whatever that meant. In the weeks and months afterward, my office cascaded with papers and groaned under the weight of books; Mary's studio became colorfully cluttered with new paintings and drawings of our Vermont landscapes as she puzzled if she ever could do justice to the West without returning a few more times. The house was now filled with Western maps and photos, slices of pear-shaped iron rail and sections of wooden ties, with rusty old spikes and desert debris, and baling wire mules, and a lifetime's worth of anecdotes, stories, tales, yarns, personalities, and the processional mind's eye imagery of big, bright monumental spaces.

Daily, we found ourselves talking over and over—launching into stories with "remember when . . ."—about the people we had met and the places we had seen, shaping and buffing and polishing those stories until they shone like silver or gold and felt as solid as a mountain or as long as a river.

Epilogue

Two full years to the month after we had begun that journey, Mary became ill. It was not cancer this time—she was now five years cancer-free. It was her heart: problems that, we learned, had begun three decades before with radiation therapy that had saved her life from Hodgkin's disease at age sixteen, but left a price tag in her heart valves. Open heart surgery did not go well, and instead of a month or two of recuperation she was four months in the hospital, fighting battles daily with extraordinary courage. What we could always count on to bolster her spirits were sharing and comparing tales of our two-month, 7,000-mile adventure with our children, following those old wagon wheel ruts and vanished iron rails. Sometimes we could almost feel the hospital walls disappearing into bright desert and mountain skies. And we plotted out new journeys for the future. Toward the end of the four months, although she was still weak the doctors thought she was strong enough to go home soon. Filled with this hope and full of plans, on a September afternoon her valves clogged and Mary went to final sleep. She was only forty-six. We scattered her ashes at places significant in her life: her ancestral Long Island ground at Makamah Beach, a Hudson riverbank near Bard College, the Lake Champlain shoreline she had repeatedly painted, and in a high upland valley overlooked by Bread Loaf Mountain. Out West, Mimi and Davey and I would fly and then drive to four holy and beautiful places to which Mary intended to return, and commune, meditate, render in oils and pastels, and glorify: Vedauwoo Glen in Wyoming, Monument Point on the Salt Lake in Utah, Park Canyon in Nevada, and Año Nuevo State Reserve on the California Pacific shore.

For us who are left—for me, at the end of our twenty-two-year relationship, and of this book manuscript; for Mimi, at fourteen now grown tall and beautiful, witty and musically gifted; for Davey, at eleven growing beyond leprechaunhood but still wry and mischievous and athletic and bookish; we have kept going as we must. In our grief, the stories of our transcontinental journey—and the example of our strong, courageous, determined predecessors on that path—became a way for us to keep going, threading our way toward new vistas that we can only begin to imagine.

Orwell, Vermont
January 2003

References

Note: In several places in this narrative, I made free use of my previous writings.

1. The Odyssey Begins

Disabled Soldiers' Homes Registry, Western Branch, Leavenworth, KS.
Federal Writers' Project, Works Progress Administration, *Iowa: A Guide to the Hawkeye State*. New York: Viking, 1938.
———, *Kansas: A Guide to the Sunflower State*. New York: Viking, 1939.
———, *Missouri: A Guide to the "Show Me" State*. New York: Duell, Sloan and Pearce, 1941.
Leavenworth City Directory, R. L. Polk, 1916.
Leavenworth Convention and Visitors Bureau, *Leavenworth, Kansas Visitors Guide*, n.d.
U.S. Army, *Register of Enlistments*, 1866.
U.S. Census Bureau, 10th (1880) Census and 12th (1900) Census.

2. Jumping Off

Brackenridge, Henry Marie, *Views of Louisiana* (1814). Ann Arbor: University Microfilms, 1966.
Bradbury, John, *Travels in the Interior of America* (1817). Ann Arbor: University Microfilms, 1966.
Butler, Susan, *East to the Dawn: The Life of Amelia Earhart*. Reading, MA: Addison-Wesley, 1997.
Dary, David, *The Santa Fe Trail: Its History, Legends, and Lore*. New York: Knopf, 2000.
DeVoto, Bernard, *The Course of Empire*. Boston: Houghton Mifflin, 1952.
Federal Writers' Project, Works Progress Administration, *Iowa: A Guide to the Hawkeye State*. New York: Viking, 1938.
———, *Kansas: A Guide to the Sunflower State*. New York: Viking, 1939.
———, *Missouri: A Guide to the "Show Me" State*. New York: Duell, Sloan and Pearce, 1941.
James, Edwin, *Account of an Expedition from Pittsburgh to the Rocky Mountains*, vol. 1 (1822). Ann Arbor: University Microfilms, 1966.
Lavender, David, *The Way to the Western Sea: Lewis and Clark Across the Continent*. New York: Harper & Row, 1988.
Leavenworth Convention and Visitors Bureau, *Leavenworth, Kansas Visitors Guide*, n.d.
Mattes, Merrill J., *The Great Platte River Road*. Lincoln: University of Nebraska Press, 1969.
Morgan, Dale (ed.), *Overland in 1846: Diaries and Letters of the California-Oregon Trail*. Lincoln: University of Nebraska Press, 1963.

Morris, Richard B. (ed.), *Encyclopedia of American History*. New York: Harper & Row, 1970.

Moulton, Gary E. (ed.), *The Journals of the Lewis & Clark Expedition*, vol. 1. Lincoln: University of Nebraska Press, 1986.

Nichols, Roger L., and Patrick L. Halley, *Stephen Long and American Frontier Exploration*. Newark: University of Delaware Press, 1980.

Vestal, Stanley, *The Missouri* (Rivers of America Series). New York: Farrar & Rinehart, 1945.

3. Rails and the River

Astaire, Fred, *Steps in Time* (1959). New York: Cooper Square, 2000.

Bain, David Haward, *Empire Express: Building the First Transcontinental Railroad*. New York: Viking, 1999.

Brando, Marlon (with Robert Lindsey), *Songs My Mother Taught Me*. Toronto: Random House, 1994.

Creigh, Dorothy Weyer, *Nebraska: A Bicentennial History*. New York: Norton, 1977.

Current Biography, 1945, 1964.

Federal Writers' Project, Works Progress Administration, *Iowa: A Guide to the Hawkeye State*. New York: Viking, 1938.

———, *Nebraska: A Guide to the Cornhusker State* (1939). Lincoln: University of Nebraska Press, 1979.

Hirshson, Stanley P., *Grenville M. Dodge: Soldier, Politician, Railroad Pioneer*. Bloomington: Indiana University Press, 1967.

Little, Malcolm (with Alex Haley), *The Autobiography of Malcolm X*. New York: Grove, 1964.

Manso, Peter, *Brando: The Biography*. New York: Hyperion, 1994.

Morton, J. Sterling, *Illustrated History of Nebraska*. Lincoln: Western Publishing, 1907.

Peirce, Neal R., *The Great Plains States of America: People, Politics, and Power in the Nine Great Plains States*. New York: Norton, 1973.

Schmitz, Lou, "Omaha's Union Stations," *The Streamliner* (Union Pacific Historical Society), vol. 13, no. 4, 1999.

Simmons, Steve, "The Western Heritage Museum," *The Streamliner* (Union Pacific Historical Society), vol. 11, no. 1, 1996.

Train, George Francis, *My Life in Many States and Foreign Lands*. New York: Appleton, 1902.

Wall, Joseph Frazier, *Iowa: A Bicentennial History*. New York: Norton, 1978.

4. The Lincoln Highway

Bain, David Haward, *Empire Express: Building the First Transcontinental Railroad*. New York: Viking, 1999.

———, "Manifest Destiny": *A Newer World* by David Roberts, *New York Times Book Review*, February 27, 2000.

———, "A Timeless Passage": *Sights Once Seen* by Robert Shlaer, *New York Times Book Review*, May 7, 2000.

Current Biography, 1948.

Federal Writers' Project, Works Progress Administration, *Nebraska: A Guide to the Corn-husker State* (1939). Lincoln: University of Nebraska Press, 1979.

Fonda, Henry (with Howard Teichman), *Fonda: My Life*. New York: Orion, 1981.

Franzwa, Gregory, *Lincoln Highway: Nebraska*. Tucson: Patrice Press, 1996.

———, *The Oregon Trail Revisited*. Tucson: Patrice Press, 1997.

Frémont, John Charles, *The Expeditions of John Charles Frémont* (ed. Donald Jackson and Mary Lee Spence), vol. 1: Travels from 1838 to 1844. Urbana: University of Illinois Press, 1970.

Gladding, Effie Price, *Across the Continent by the Lincoln Highway*. New York: Brentano's, 1915.

Grinnell, George Bird, *Two Great Scouts and Their Pawnee Battalion*. Cleveland: Arthur H. Clark, 1928.

Hokanson, Drake, *The Lincoln Highway: Main Street Across America*, tenth anniversary edition. Iowa City: University of Iowa Press, 1999.

Hyde, George E., *The Pawnee Indians*. Norman: University of Oklahoma Press, 1951.

James, Edwin, *Account of an Expedition from Pittsburgh to the Rocky Mountains*, vol. 1 (1822). Ann Arbor: University Microfilms, 1966.

Lincoln Highway Association, *The Lincoln Highway: The Story of a Crusade That Made Transportation History*. New York: Dodd, Mead, 1935.

Mattes, Merrill J., *The Great Platte River Road*. Lincoln: University of Nebraska Press, 1969.

McNally, Hannah, *Nebraska: Off the Beaten Path*, second edition. Old Saybrook, CT: Globe Pequot, 1999.

National Frontier Trails Center, Independence, Missouri, *Voices from the Trails: Selected Quotes from the National Frontier Trails Center*, n.d.

Parkman, Francis, *The Oregon Trail* (ed. E. N. Feltskog). Madison: University of Wisconsin Press, 1969.

Partridge, Bellamy, *Fill'er Up*. New York: McGraw-Hill, 1952.

Post, Emily, *By Motor to the Golden Gate*. New York: Appleton, 1917.

Ramsey, Alice Clarke, *Veil, Duster and Tire Iron*. Covina, CA: Castle Press, 1961.

Richardson, Albert, *Garnered Sheaves*. Hartford, CT: Columbian Book Co., 1871.

Thomas, Tony, *The Films of Henry Fonda*. Secaucus, NJ: Citadel Press, 1983.

Van de Water, Frederic, *The Family Flivvers to Frisco*. New York: Appleton, 1927

5. The Road from Red Cloud

Andrist, Ralph K., *The Long Death: The Last Days of the Plains Indian*. New York: Macmillan, 1964.

Bain, David Haward, *Empire Express: Building the First Transcontinental Railroad*. New York: Viking, 1999.

Billesbach, Ann E., *Red Cloud, Webster County*. Red Cloud: Willa Cather Historical Center, 1988.

Bohlke, L. Brent (ed.), *Willa Cather in Person: Interviews, Speeches, and Letters*. Lincoln: University of Nebraska Press, 1986.

Cather, Willa, *Alexander's Bridge* (1912).

———, *My Antonia* (1918).

———, *One of Ours* (1922).

———, *O Pioneers!* (1913).

————, *The Song of the Lark* (1915).

————, *Willa Cather's Collected Short Fiction, 1892–1912*, Introduction by Mildred R. Bennett (1965).

Creigh, Dorothy Weyer, "Edwin Perkins and the Kool-Aid Story," *Adams County Historical Society News*, vol. 21, no. 4, 1998.

Federal Writers' Project, Works Progress Administration, *Nebraska: A Guide to the Cornhusker State* (1939). Lincoln: University of Nebraska Press, 1979.

Great Platte River Road Memorial Foundation, Opening release packet, June 2000.

Hastings Museum, *Yester News*. Monthly Newsletter of the House of Yesterday/Hastings Museum, Hastings, Nebraska.

————, "The Hastings Museum Story," November 1984.

————, "Home Sweet Home," March 1975.

————, "Museum's 30th Anniversary," October 1956.

————, "Nineteenth Century Household," March 1983.

————, "People of the Plains: Birth of the Project," July 1983.

————, "Sixty-five Years of Growth," January 1992.

Hyde, George E., *The Pawnee Indians*. Denver: University of Denver Press, 1951.

————, *Red Cloud's Folk*. Norman: University of Oklahoma Press, 1937.

Larson, Robert W., *Red Cloud: Warrior-Statesman of the Lakota Sioux*. Norman: University of Oklahoma Press, 1997.

Ronning, Kari, and Elizabeth Turner, *Willa Cather's University Days: The University of Nebraska, 1890–1895*. Lincoln: Center for Great Plains Studies, 1995.

6. Hell on Wheels

Bain, David Haward, *Empire Express: Building the First Transcontinental Railroad*. New York: Viking, 1999.

Cody, William Frederick, *Life and Adventures of "Buffalo Bill."* New York: Willey, 1927.

Davies, Henry E., *Ten Days on the Plains* (1871) (ed., with introduction, Paul Andrew Hutton). Dallas: Southern Methodist University Press, 1985.

Federal Writers' Project, Works Progress Administration, *Nebraska: A Guide to the Cornhusker State* (1939). Lincoln: University of Nebraska Press, 1979.

Hutton, Paul Andrew, *Phil Sheridan and His Army*. Lincoln: University of Nebraska Press, 1985.

Kasson, Joy S., *Buffalo Bill's Wild West: Celebrity, Memory, and Popular History*. New York: Hill & Wang, 2000.

Kenfield, Harvey, and Howard Kenfield, *Petrified Wood Gallery*. Self-published pamphlet, 1991.

Mercoid Control Corporation, "Commemorating Ira Emmett McCabe, Chief Engineer, 1921–1938," June 1968. Courtesy Dawson County Historical Society.

Morgan, Dale (ed.), *Overland in 1846: Diaries and Letters of the California-Oregon Trail*. Lincoln: University of Nebraska Press, 1963.

Morton, Nancy Jane, *Captive of the Cheyenne* (ed. Russ Czaplewski). Lexington, NE: Dawson County Historical Society/Baby Biplane Books, 1993.

Parry, Henry C., "Observations on the Prairies: 1867," *Montana Magazine of History*, Autumn 1959.

Reisdorff, James, "North Platte Canteen," *The Streamliner*, Union Pacific Historical Society, vol. 6, no. 2, 1990.

Russell, Don, *The Lives and Legends of Buffalo Bill*. Norman: University of Oklahoma Press, 1960.

Sandoz, Mari, *The Buffalo Hunters: The Story of the Hide Men*. New York: Hastings House, 1954.

Stanley, Henry M., *My Early Travels and Adventures in America* (1895) (introduction, Dee Brown). Lincoln: University of Nebraska Press, 1982.

Twain, Mark, *Roughing It* (1872). New York: Penguin, 1985.

Young, Geraldine, "Lincoln County Western Heritage Museum and Village: History and Tour." North Platte, 1990 (revised 1999, Phyllis Shavlik).

7. The View from the Bluffs

Bain, David Haward, *Empire Express: Building the First Transcontinental Railroad*. New York: Viking, 1999.

Federal Writers' Project, Works Progress Administration, *Nebraska: A Guide to the Cornhusker State* (1939). Lincoln: University of Nebraska Press, 1979.

Franzwa, Gregory M., *The Oregon Trail Revisited*. Tucson: Patrice Press, 1997.

Mattes, Merrill J., "Chimney Rock on the Oregon Trail," *Nebraska History*, vol. 26, no. 1, 1955.

————, *The Great Platte River Road*. Lincoln: University of Nebraska Press, 1969.

————, *Scotts Bluff*. National Park Service Historical Handbook No. 28. Washington, D.C.: National Park Service, 1958 (revised 1992).

Morgan, Dale (ed.), *Overland in 1846: Diaries and Letters of the California-Oregon Trail*. Lincoln: University of Nebraska Press, 1963.

National Park Service, Scotts Bluff National Monument, "The Rebecca Winters Story," n.d.

————, "Scotts Bluff: Official Map and Guide," n.d.

Stanley, Henry M., *My Early Travels and Adventures in America* (1895) (introduction, Dee Brown). Lincoln: University of Nebraska Press, 1982.

Twain, Mark, *Roughing It* (1872). New York: Penguin, 1985.

Waitley, Douglas, *William Henry Jackson: Framing the Frontier*. Missoula, MT: Mountain Press, 1998.

8. Magic City

Ames, Charles Edgar, *Pioneering the Union Pacific*. New York: Appleton, 1969.

Bain, David Haward, *Empire Express: Building the First Transcontinental Railroad*. New York: Viking, 1999.

Chaffin, Lorah B., *Sons of the West: Biographical Account of Early-Day Wyoming*. Caldwell, ID: Caxton, 1941.

Federal Writers' Project, Works Progress Administration, *Wyoming: A Guide to Its History, Highways, and People* (1941). Lincoln: University of Nebraska Press, 1981.

McCoy, Tim (with Ronald McCoy), *Tim McCoy Remembers the West*. New York: Doubleday, 1977.

Mueller, Ellen Crago, *Calamity Jane*. Cheyenne: Self-published monograph, 1981.

Schmitz, Lou, "Union Station Memories," *The Streamliner*, Union Pacific Historical Society, vol. 14, no. 3, 2000.

Sherr, Lynn, and Jurate Kazickas, *Susan B. Anthony Slept Here: A Guide to American Women's Landmarks*. New York: Times Books, 1994.

Simmons, Steve, "Union Pacific's Steam Locomotive Program," *The Streamliner*, Union Pacific Historical Society, vol. 10, no. 4, 1996.

Spring, Agnes Wright, *The Cheyenne and Black Hills Stage and Express Routes*. Glendale, CA: Arthur H. Clark, 1949.

Waite, Thornton, "The Ames Monument," *The Streamliner*, Union Pacific Historical Society, vol. 14, no. 3, 2000.

Williams, Susan E., "The Great West Illustrated: A Journey Across the Continent with Andrew J. Russell," *The Streamliner*, Union Pacific Historical Society, vol. 10, no. 3, 1996.

Wister, Fanny Kemble (ed.), *Owen Wister Out West: His Journals and Letters*. Chicago: University of Chicago Press, 1958.

9. Road Tested on the Red Plains

Bain, David Haward, *Empire Express: Building the First Transcontinental Railroad*. New York: Viking, 1999.

"'Big Al' Helps Link Prehistoric Past and Present," University of Wyoming Public Information Office, 1996.

Breithaupt, Brent H., "Como Bluff, Wyoming Territory, 1868–1877: An Initial Glimpse of One of the World's Premier Dinosaur Sites," *Dinofest International Proceedings* (1977).

———, "From Dinosaurs to Lawmen to the Model T: The Dinosaur Museums of Albany County Fieldtrip," August 1993. Courtesy University of Wyoming Library.

Complete Official Road Guide of the Lincoln Highway, fifth edition (1924). Tucson: Patrice Press, 1993.

"Cooperative Effort Saves 'Big Al' for Public, Scientists," University of Wyoming Public Information Office, 1996.

Federal Writers' Project, Works Progress Administration, *Wyoming: A Guide to Its History, Highways, and People* (1941). Lincoln: University of Nebraska Press, 1981.

Jaffe, Mark, *The Gilded Dinosaur: The Fossil War Between E. D. Cope and O. C. Marsh and the Rise of American Science*. New York: Crown, 2000.

Lamb, F. Bruce, *The Wild Bunch: A Selected Critical Annotated Bibliography of the Literature*. Worland, WY: High Plains, 1993.

McCoy, Tim (with Ronald McCoy), *Tim McCoy Remembers the West*. New York: Doubleday, 1977.

McPhee, John, *Rising from the Plains*. New York: Farrar, Straus & Giroux, 1986.

Mitchell, W. J. T., *The Last Dinosaur Book: The Life and Times of a Cultural Icon*. Chicago: University of Chicago Press, 1998.

Patterson, Richard, *Butch Cassidy: A Biography*. Lincoln: University of Nebraska Press, 1998.

Payne, Darwin, *Owen Wister: Chronicler of the West, Gentleman of the East*. Dallas: Southern Methodist University Press, 1985.

"Prehistoric Mammal Named in Honor of UW Paleontologist," University of Wyoming Public Information Office, 2000.

Preston, Douglas J., *Dinosaurs in the Attic: An Excursion into the American Museum of Natural History*. New York: St. Martin's, 1986.

Ramsey, Alice Huyler, *Veil, Duster, and Tire Iron*. Covina, CA: Castle Press, 1961.

Seelye, John, Introduction to Owen Wister, *The Virginian*. New York: Penguin, 1988.

"Significant Dinosaur Tracksite Discovered in Wyoming," University of Wyoming Public Information Office, 1998.

Smith, Henry Nash, *Virgin Land* (1950).

Stegner, Wallace, Foreword, Ben Merchant Vorpahl, *My Dear Wister: The Frederick Remington–Owen Wister Letters*. Palo Alto: American West, 1972.

———, *The Sound of Mountain Water: The Changing American West* (1959). New York: Penguin, 1997.

Wister, Fanny Kemble (ed.), *Owen Wister Out West: His Journals and Letters*. Chicago: University of Chicago Press, 1958.

Wister, Owen, *Roosevelt: The Story of a Friendship, 1880–1919*. New York: Macmillan, 1930.

———, *The Virginian: A Horseman of the Plains* (1902). New York: Penguin, 1988.

10. Crossing the Divide

Bain, David Haward, *Empire Express: Building the First Transcontinental Railroad*. New York: Viking, 1999.

Beadle, J. H., *The Undeveloped West, or Five Years in the Territories*. Philadelphia: National, 1873.

"Big Nose George Turns Up," *Rawlins Daily Times*, May 12, 1950. Courtesy Carbon County Museum, Rawlins.

Biographical Dictionary of the American Congress.

Butler, Susan, *East to the Dawn: The Life of Amelia Earhart*. Reading, MA: Addison-Wesley, 1997.

Carlson, Susan, *Wyoming Historical Markers at 55 MPH: A Guide to Historical Markers and Monuments on Wyoming Highways*. Cheyenne: Beartooth Corral, 1994.

DeVoto, Bernard, *Across the Wide Missouri*. Boston: Houghton Mifflin, 1947.

Earhart, Amelia, "Your Next Garage May House an Autogiro," *Cosmopolitan*, August 1931.

Federal Writers' Project, Works Progress Administration, *Wyoming: A Guide to Its History, Highways, and People* (1941). Lincoln: University of Nebraska Press, 1981.

Franzwa, Gregory M., *The Oregon Trail Revisited*. Tucson: Patrice Press, 1997.

Frémont, John C., "A Report of the Exploring Expedition to Oregon and North California in the Years 1843–44," in *The Expeditions of John Charles Frémont*, vol. 1 (ed. Donald Dale Jackson and Mary Lee Spence). Urbana: University of Illinois Press, 1970.

Harpending, Asbury, *The Great Diamond Hoax* (1915). Norman: University of Oklahoma Press, 1958.

Lovell, Mary S., *The Sound of Wings: The Life of Amelia Earhart*. New York: St. Martin's, 1989.

Morgan, Dale (ed.), *Overland in 1846: Diaries and Letters of the California-Oregon Trail*. Lincoln: University of Nebraska Press, 1963.

New York Times, New York Herald, San Francisco Bulletin, 1872 accounts of Table Rock diamond swindle.

Peters, Arthur King, *Seven Trails West*. New York: Abbeville, 1996.

Ramsey, Alice Huyler, *Veil, Duster, and Tire Iron*. Covina, CA: Castle Press, 1961.

Rawlins Landmark Committee, *Walking Through History: Historic Downtown Rawlins Walking Tour*, 1994.

Reflections: A Pictorial History of Carbon County. Carbon County Historical Society, 1990.

Robins, Norma Jean, *Rock Springs Historic Downtown Walking Tour*. Rock Springs Historical Board, 1996.

Stansbury, Howard, *An Expedition to the Valley of the Great Salt Lake* (London, 1852). Ann Arbor: University Microfilms, 1966.

Storti, Craig, *Incident at Bitter Creek: The Story of the Chinese Massacre*. Ames: Iowa State University Press, 1991.

Twain, Mark, *Roughing It* (1872). New York: Penguin, 1985.

Union Pacific Coal Company, "Rock Springs, That Grew into a Great City," *History of the Union Pacific Coal Mines, 1868 to 1940*. Omaha: Colonial Press, 1940. Sweetwater County Historical Society reprint.

Wilkins, Thurman, *Clarence King: A Biography*. New York: Macmillan, 1958.

11. Green River to the Rim

Bain, David Haward, *Empire Express: Building the First Transcontinental Railroad*. New York: Viking, 1999.

Carlson, Susan, *Wyoming Historical Markers at 55 MPH: A Guide to Historical Markers and Monuments on Wyoming Highways*. Cheyenne: Beartooth Corral, 1994.

Carr, Stephen, "Union Pacific Tunnels on the Transcontinental Route," *The Streamliner*, Union Pacific Historical Society, vol. 10, no. 3, 1996.

Cassity, Michael, "Washakie," in Frederick E. Hoxie (ed.), *Encyclopedia of North American Indians*. Boston: Houghton Mifflin, 1996.

Cooley, John, *The Great Unknown: The Journals of the Historic First Expedition Down the Colorado River*. Flagstaff, AZ: Northland, 1988.

DeVoto, Bernard, *Across the Wide Missouri*. Boston: Houghton Mifflin, 1947.

Federal Writers' Project, Works Progress Administration, *Wyoming: A Guide to Its History, Highways, and People* (1941). Lincoln: University of Nebraska Press, 1981.

Frémont, John C., "A Report of the Exploring Expedition to Oregon and North California in the Years 1843–44," in *The Expeditions of John Charles Frémont*, vol. 1 (ed. Donald Dale Jackson and Mary Lee Spence). Urbana: University of Illinois Press, 1970.

Green River Historic Preservation Commission, *Echoes from the Bluffs*, vol. 1. Colorado Springs, CO: Pioneer Printing, 1998.

"Green River History and Facilities," *The Streamliner*, Union Pacific Historical Society, vol. 15, no. 3, 2001.

Green River: Nature's Art Shop. Green River Historic Preservation Commission, 1997.

Hebard, Grace Raymond, *Washakie: An Account of Indian Resistance of the Covered Wagon and Union Pacific Railroad Invasion of Their Territory*. Cleveland: Arthur H. Clark, 1930.

Holliday, J. S., *The World Rushed In*. New York: Simon & Schuster, 1981.

Lester, Margaret Moore, *History of Piedmont*, 1995 reprint from *The Visitors' Guide* of the *Bridger Valley Pioneer*, Summer 1995.

Morgan, Dale (ed.), *Overland in 1846: Diaries and Letters of the California-Oregon Trail*. Lincoln: University of Nebraska Press, 1963.

Peters, Arthur King, *Seven Trails West*. New York: Abbeville, 1996.

Powell, John Wesley, *The Exploration of the Colorado River and Its Canyons* (1875). New York: Penguin, 1987.

Ramsey, Alice Huyler, *Veil, Duster, and Tire Iron*. Covina, CA: Castle Press, 1961.

Self-Guided Tour of Historic Green River. Green River Historic Preservation Commission, 1993.

Stegner, Wallace, *Beyond the Hundredth Meridian*. Boston: Houghton Mifflin, 1953.

Stansbury, Howard, *An Expedition to the Valley of the Great Salt Lake* (London, 1852). Ann Arbor: University Microfilms, 1966.

Tippets, Susan Thomas, *Piedmont Ghost Town, Uinta County, Wyoming*, 1995 reprint from *The Visitors' Guide* of the *Bridger Valley Pioneer*, Summer 1995.

Twain, Mark, *Roughing It* (1872). New York: Penguin, 1985.

Williams, Susan E., "The Great West Illustrated: A Journey Across the Continent with Andrew J. Russell," *The Streamliner*, Union Pacific Historical Society, vol. 10, no. 3, 1996.

Zwinger, Ann, *Run, River, Run: A Naturalist's Journey Down One of the Great Rivers of the American West*. Tucson: University of Arizona Press, 1975.

12. Through the Canyons to Paradise

Andrews, Thomas F., "The Ambitions of Lansford W. Hastings: A Study in Western Myth-Making," *Pacific Historical Review*, vol. 39, 1970.

——, "The Controversial Hastings Overland Guide: A Reassessment," *Pacific Historical Review*, vol. 37, 1968.

Bain, David Haward, *Empire Express: Building the First Transcontinental Railroad*. New York: Viking, 1999.

——, "Golden Spike Keynote Address, May 10, 2002, Golden Spike National Historic Site, Promontory, Utah." Unpublished manuscript.

Beadle, J. H., *The Undeveloped West, or Five Years in the Territories*. Philadelphia: National, 1873.

Bowles, Samuel, *The Pacific Railroad—Open: How to Go, What to See*. Boston: Fields, Osgood, 1869.

Campbell, Eugene E., "The Mormon Migrations to Utah," in Richard D. Poll (ed.), *Utah's History*. Provo: Brigham Young University Press, 1978.

Carr, Stephen L., "The Zig-Zag in Echo Canyon," *The Streamliner*, Union Pacific Historical Society, vol. 13, no. 3, 1999.

Crawford, D. Bolyd, *History of Ogden, Utah, in Old Post Cards*. Ogden: Maury Grimm, 1996.

Daughters of Utah Pioneers (ed. Milton R. Hunter), *Beneath Ben Lomond's Peak: A History of Weber County*. Salt Lake City: Publishers Press, 1966.

DeLafosse, Peter H. (ed.), *Trailing the Pioneers: A Guide to Utah's Emigrant Trails, 1829–1869*. Logan: Utah State University Press, 1994.

DeVoto, Bernard, *The Easy Chair*. Boston: Houghton Mifflin, 1955.

——, *Forays and Rebuttals*. Boston: Little, Brown, 1936.

Dowty, Robert R., *Rebirth of the Jupiter and the 119: Building the Replica Locomotives at Golden Spike*. Tucson: Southwest Parks and Monuments Association, 1994.

Eldredge, John, *A Tour Guide of Echo Canyon, Utah's Forgotten Natural Wonder*. Self-published pamphlet, 1997.

Federal Writers' Project, Works Progress Administration, *Nevada: A Guide to the Silver State*. Portland, OR: Binfords & Mort, 1940.

———, *Utah: A Guide to the State*. New York: Hastings House, 1941.

Frémont, John C., "A Report of the Exploring Expedition to Oregon and North California in the Years 1843–44," in *The Expeditions of John Charles Frémont*, vol. 1 (ed. Donald Dale Jackson and Mary Lee Spence). Urbana: University of Illinois Press, 1970.

Homstad, Carla, Janene Caywood, and Peggy Nelson, *Cultural Landscape Report: Golden Spike National Historic Site, Box Elder County, Utah*. Denver: National Park Service, Intermountain Region, 2000.

Miller, David E., "Explorers and Trail Blazers," in Richard D. Poll (ed.), *Utah's History*. Provo: Brigham Young University Press, 1978.

Morgan, Dale L. (ed.), *Overland in 1846: Diaries and Letters of the California-Oregon Trail* (1963), 2 vols. Lincoln: University of Nebraska Press, 1993.

———, and J. Roderic Korns (revised and updated by Will Bagley and Harold Schindler), *West from Fort Bridger: The Pioneering of Immigrant Trails Across Utah, 1846–1850* (1950). Logan: Utah State University Press, 1994.

Raymond, Anan S., and Richard E. Fike, *Rails East to Promontory: The Utah Stations*, second edition. Livingston, TX: Utah Bureau of Land Management/Pioneer Enterprises, 1994.

Richardson, Albert D., *Garnered Sheaves*. Hartford, CT: Columbian Book Co., 1871.

Stansbury, Howard, *An Expedition to the Valley of the Great Salt Lake* (London, 1852). Ann Arbor: University Microfilms, 1966.

Stegner, Wallace, *The Uneasy Chair: A Biography of Bernard DeVoto*. New York: Doubleday, 1974.

Strack, Don, *Ogden Rails: A History of Railroads in Ogden, Utah, from 1869 to Today*. Ogden: Golden Spike Chapter, Railway and Locomotive History Society, 1997.

Twain, Mark, *Roughing It* (1872). New York: Penguin, 1985.

Utah Historical Records Survey, Works Progress Administration, *A History of Ogden*. Ogden City Commission, 1940.

Williams, Terry Tempest, *Refuge: An Unnatural History of Family and Place*. New York: Pantheon, 1991.

———, *An Unspoken Hunger: Stories from the Field*. New York: Pantheon, 1994.

13. Following the Humboldt

Bain, David Haward, *Empire Express: Building the First Transcontinental Railroad*. New York: Viking, 1999.

Baker, Pearl, *The Wild Bunch at Robbers Roost* (1971). Lincoln: University of Nebraska Press, 1989.

Bell, Mike, "Interview with the Sundance Kid," *Journal of the Western Outlaw-Lawman History Association*, Summer 1995.

Curran, Harold, *Fearful Crossing*. Las Vegas: Nevada Publications, 1982.

Delano, Alonzo, *Life on the Plains and Among the Diggings* (1854). Ann Arbor: University Microfilms, 1966.

Earl, Phillip I., "Bing Crosby in Nevada," paper of the Nevada Historical Society, reprinted in *Las Vegas Review-Journal*, June 7, 1998.

Federal Writers' Project, Works Progress Administration, *Nevada: A Guide to the Silver State*. Portland, OR: Binfords & Mort, 1940.

Glass, Mary Ellen, and Al Glass, *Touring Nevada: A Historic and Scenic Guide*. Reno: University of Nevada Press, 1983.

Helfrich, Devere, Helen Hunt, and Thomas Hunt, *Emigrant Trails West*. Reno: Trails West, 1984.

Hersh, Lawrence K., *The Central Pacific Railroad Across Nevada, 1868 and 1997: Photographic Comparatives*. Self-published, 2000. (LKH, P.O. Box 5199, North Hollywood, CA 91616-5199.)

Hunt, Thomas, *Ghost Trails to California*. Las Vegas: Las Vegas Publications, 1974.

————, (ed.), "James Wilkins Diary," *Overland Journal*, vol. 8, no. 3, 1990.

Kibbey, Mead B., *The Railroad Photographs of Alfred A. Hart, Artist*. Sacramento: California State Library Foundation, 1996.

MacGregor, Greg, *Overland: The California Emigrant Trail of 1841–1870*. Albuquerque: University of New Mexico Press, 1996.

Marcy, Randolf, *The Prairie Traveler: A Handbook for Overland Expeditions*. New York: Harper, 1859.

McGlashan, C. F., *History of the Donner Party* (1880).

Morgan, Dale (ed.), *Overland in 1846: Diaries and Letters of the California-Oregon Trail*. Lincoln: University of Nebraska Press, 1963.

————, and J. Roderic Korns (revised and updated by Will Bagley and Harold Schindler), *West from Fort Bridger: The Pioneering of Immigrant Trails Across Utah, 1846–1850* (1950). Logan: Utah State University Press, 1994.

Mullen, Fran, and Marilyn Newton, *Donner Party Chronicles: A Day by Day Account of a Doomed Wagon Train, 1846–47*. Reno: University of Nevada Press, 1999.

Olch, Peter D., "Treading the Elephant's Tail: Medical Problems on the Overland Trails." *Overland Journal*, vol. 6, no. 1, 1988.

Patterson, Edna B., and Louise A. Beebe, *Halleck Country: The Story of the Land and Its People*. Fleischmann College of Agriculture, University of Nevada, Reno, 1982.

Patterson, Richard, *Butch Cassidy: A Biography*. Lincoln: University of Nebraska Press, 1998.

Shaw, Reuben (ed. Milton Quaife), *Across the Plains in Forty Nine*. Chicago: R. R. Donnelley, 1948.

Stewart, George R., *The California Trail* (1962). Lincoln: Bison, 1983.

14. Silver State

Bain, David Haward, *Empire Express: Building the First Transcontinental Railroad*. New York: Viking, 1999.

Carter, Harvey L. (ed.) *"Dear Old Kit": The Historial Christopher Carson*. Norman: University of Oklahoma Press, 1968.

————, and Thelma S. Guild, *Kit Carson: A Pattern for Heroes*. Lincoln: University of Nebraska Press, 1984.

De Quille, Dan, and William Wright, *The History of the Big Bonanza*. Hartford, CT: American Publishing, 1876.

DeVoto, Bernard, *Mark Twain's America*. Boston: Little, Brown, 1932.

Federal Writers' Project, Works Progress Administration, *California: A Guide to the Golden State*. New York: Hastings House, 1939.

———, *Nevada: A Guide to the Silver State*. Portland, OR: Binfords & Mort, 1940.

Glass, Mary Ellen, and Al Glass, *Touring Nevada: A Historic and Scenic Guide*. Reno: University of Nevada Press, 1983.

Nevins, Allen, *Frémont: Pathmarker of the West*. Lincoln: University of Nebraska Press, 1992.

Osborne, Mary Pope, *Adaline Falling Star*. New York: Scholastic, 2000.

Roberts, David, *A Newer World: Kit Carson, John C. Frémont, and the Claiming of the American West*. New York: Simon & Schuster, 2000.

Rogers, Franklin R., *The Pattern for Mark Twain's* Roughing It: *Letters from Nevada by Samuel and Orion Clemens, 1861–1862*. Berkeley: University of California Press, 1961.

Shamberger, Hugh A., *Goldfield: Early History, Development, Water Supply*. Carson City: Nevada Historical Press, 1982.

Spence, Mary Lee, and Donald Jackson (eds.), *The Expeditions of John Charles Frémont*, vol. 2. Urbana: University of Illinois Press, 1973.

Twain, Mark, *Early Tales and Sketches*, vol. 1 (ed. Edgar Marquess Branch and Robert H. Hirst). Berkeley: University of California Press, 1979.

———, *Mark Twain of the Enterprise: Newspaper Articles and Other Documents, 1862–1864* (ed. Henry Nash Smith). Berkeley: University of California Press, 1957.

———, *Mark Twain's Letters*, vol. 1 (ed. Edgar Marquess Branch, Michael B. Frank, and Kenneth M. Sanderson). Berkeley: University of California Press, 1988.

———, *Roughing It* (1872). New York: Penguin, 1985.

Williams III, George, *Rosa May: The Search for a Mining Camp Legend*. Carson City: Tree by the River Publishing, 1979.

15. Over the Sierra

Bain, David Haward, *Empire Express: Building the First Transcontinental Railroad*. New York: Viking, 1999.

Brands, H. W., *The Age of Gold: The California Gold Rush and the New American Dream*. New York: Doubleday, 2002.

California State Railroad Museum Guidebook. Sacramento, 1999.

Delano, Alonzo, *Life on the Plains and Among the Diggings* (1854). Ann Arbor: University Microfilms, 1966.

Driesbach, Jan, Harvey L. Jones, and Katherine Holland, *Art of the Gold Rush*. Berkeley: University of California Press, 1998.

Duncan, Jack E., *To Donner Pass from the Pacific: A Map History, 1852–2002*. Newcastle, CA: Privately printed, 2002.

Federal Writers' Project, Work Progress Administration, *California: A Guide to the Golden State*. New York: Hastings House, 1939.

Helfrich, Devere, Helen Hunt, and Thomas Hunt, *Emigrant Trails West*. Reno: Trails West, 1984.

Holliday, J. S., *The World Rushed In: The California Gold Rush Experience*. New York: Simon & Schuster, 1981.

Hunt, Thomas, *Ghost Trails to California*. Las Vegas: Las Vegas Publications, 1974.

——— (ed.), "James Wilkins Diary," *Overland Journal*, vol. 8, no. 3, 1990.

Johnson, Susan Lee, *Roaring Camp: The Social World of the California Gold Rush*. New York: Norton, 2000.

Kibbey, Mead B., *The Railroad Photographs of Alfred A. Hart, Artist*. Sacramento: California State Library Foundation, 1996.

Kowalewski, Michael (ed.), *Gold Rush: A Literary Exploration*. Berkeley: Heyday Books, 1997.

Kurutz, Gary F., *The California Gold Rush: A Descriptive Bibliography of Books and Pamphlets Covering the Years 1848–1853*. San Francisco: Book Club of California, 2000.

Levy, Jo Ann, *They Saw the Elephant: Women in the California Gold Rush*. Norman: University of Oklahoma Press, 1992.

———, "We Were Forty-Niners Too: Women in the California Gold Rush," *Overland Journal*, vol. 6, no. 3, 1988.

Lincoln Highway Association, *Complete Official Road Guide of the Lincoln Highway*, fifth edition (1924), facsimile edition. Tucson: Patrice Press, 1993.

———, *The Lincoln Highway: The Story of a Crusade That Made Transportation History*. New York: Dodd, Mead, 1935.

MacGregor, Greg, *Overland: The California Emigrant Trail of 1841–1870*. Albuquerque: University of New Mexico Press, 1996.

McGlashan, C. F., *History of the Donner Party* (1880).

McPhee, John, *Assembling California*. New York: Farrar, Straus & Giroux, 1993.

Marcy, Randolf, *The Prairie Traveler: A Handbook for Overland Expeditions*. New York: Harper, 1859.

Morgan, Dale (ed.), *Overland in 1846: Diaries and Letters of the California-Oregon Trail*. Lincoln: University of Nebraska Press, 1963.

Mullen, Frank, and Marilyn Newton, *Donner Party Chronicles: A Day by Day Account of a Doomed Wagon Train, 1846–1847*. Reno: University of Nevada Press, 1999.

Placer County Heritage Travel: Gold Country One-Day Ride. Auburn, CA: Golden Triangle Publications, 2000.

Rohrbough, Malcolm J., *Days of Gold: The California Gold Rush and the American Nation*. Berkeley: University of California Press, 1997.

Shaw, Reuben (ed. Milton Quaife), *Across the Plains in Forty Nine*. Chicago: R. R. Donnelley, 1948.

Spence, Mary Lee (ed.), *The Expeditions of John Charles Frémont*, vol. 3. Chicago: University of Illinois Press, 1984.

———, and Donald Jackson (eds.), *The Expeditions of John Charles Frémont*, vols. 1 and 2. Chicago: University of Illinois Press, 1970, 1973.

Starr, Kevin, *Americans and the California Dream*. New York: Oxford University Press, 1973.

Stewart, George R., *The California Trail* (1962). Lincoln, NE: Bison, 1983.

———, *The Opening of the California Trail*. Berkeley: University of California Press, 1953.

16. From Sacramento to the Sea

Bain, David Haward, *Empire Express: Building the First Transcontinental Railroad*. New York: Viking, 1999.

Federal Writers' Project, Works Progress Administration, *California: A Guide to the Golden State*. New York: Hastings House, 1939.

Harte, Bret, *The Luck of Roaring Camp and Other Sketches* (1870). Boston: Houghton Mifflin, 1903.

Kowalewski, Michael (ed.), *Gold Rush: A Literary Exploration*. Berkeley: Heyday Books, 1997.

Kurutz, Gary F., *The California Gold Rush: A Descriptive Bibliography of Books and Pamphlets Covering the Years 1848–1853*. San Francisco: Book Club of California, 2000.

Old Sacramento Walking Tour, Old Sacramento Management, 1998.

Spence, Mary Lee (ed.), *The Expeditions of John Charles Frémont*, vol. 3. Chicago: University of Illinois Press, 1984.

———, and Donald Jackson (eds.), *The Expeditions of John Charles Frémont*, vols. 1 and 2. Chicago: University of Illinois Press, 1970, 1973.

Starr, Kevin, *Americans and the California Dream*. New York: Oxford University Press, 1973.

Steinheimer, Richard, *California State Railroad Museum: Railroading in California and the West*. Sacramento: California Department of Parks and Recreation/Albion Publishing Group, 1991.

Twain, Mark, *Early Tales and Sketches*, vol. 1 (1851–1864) and vol. 2 (1864–1865) (eds. Edgar Marquess Branch and Robert H. Hirst). Berkeley: University of California Press, 1979, 1981.

17. Golden Gate

Bain, David Haward, *Empire Express: Building the First Transcontinental Railroad*. New York: Viking, 1999.

———, *Sitting in Darkness: Americans in the Philippines*. Boston: Houghton Mifflin, 1984.

Dana, Richard Henry, Jr., *Two Years Before the Mast*, (1840). New York: Collier, 1937.

Federal Writers' Project, Works Progress Administration, *California: A Guide to the Golden State*. New York: Hastings House, 1939.

Kowalewski, Michael (ed.), *Gold Rush: A Literary Exploration*. Berkeley: Heyday Books, 1997.

Kurutz, Gary F., *The California Gold Rush: A Descriptive Bibliography of Books and Pamphlets Covering the Years 1848–1853*. San Francisco: Book Club of California, 2000.

Lincoln Highway Association, *Complete Official Road Guide of the Lincoln Highway*, fifth edition (1924), facsimile edition. Tucson: Patrice Press, 1993.

———, *The Lincoln Highway: The Story of a Crusade That Made Transportation History*. New York: Dodd, Mead, 1935.

Starr, Kevin, *Americans and the California Dream*. New York: Oxford University Press, 1973.

Twain, Mark, *Early Tales and Sketches*, vol. 1 (1851–1864) and vol. 2 (1864–1865) (eds. Edgar Marquess Branch and Robert H. Hirst). Berkeley: University of California Press, 1979, 1981.

Further Sources

The following listings are addresses and Web sites for many places that are mentioned throughout this book.

ILLINOIS

Abraham Lincoln Home National Historic Site
413 South Eighth Street
Springfield, IL 62701-1905
217-492-4241 (ex. 221)
www.nps.gov/liho/

Lincoln Highway Association
P.O. Box 308
Franklin Grove, IL 61031
815-456-3030
www.lincolnhighwayassoc.org/

MISSOURI

Mark Twain Museum
208 Hill Street
Hannibal, MO 63401-3316
573-221-9010
www.marktwainmuseum.org

Lewis and Clark Education Center
www.lewisandclarkeducationcenter.com

Kansas City American Jazz Museum
1616 East Eighteenth Street
Kansas City, MO 64108
816-474-8463
www.americanjazzmuseum.com

National Frontier Trails Center
318 West Pacific
Independence, MO 64050
816-325-7575
www.frontiertrailscenter.com

IOWA

Union Pacific Railroad Museum
200 Pearl Street
Council Bluffs, IA 51503
712-329-8307
www.uprr.com/aboutup/history/museum/
    index.shtml

Western Historic Trails Center
Council Bluffs, IA
www.iowahistory.org/sites/western_trails/
    western_trails.html

NEBRASKA

Durham Western Heritage Museum
801 South Tenth Street
Omaha, NE 68108
402-444-5071
www.dwhm.org

Joslyn Museum
2200 Dodge Street
Omaha, NE 68102-1292
402-342-3300
www.joslyn.org

Malcolm X Birthsite
2448 Pinkney Street
Omaha, NE

Stuhr Museum of the Prairie Pioneer
3133 West Highway 34
Grand Island, NE 68801
308-385-5028
www.stuhrmuseum.org

Willa Cather Pioneer Memorial and
    Educational Foundation
413 North Webster Street
Red Cloud, NE 68970
402-746-2653
www.willacather.org

Hastings Museum
1330 North Burlington Avenue
Hastings, NE 68902
800-508-4629
www.hastingsmuseum.com

Great Platte River Road Archway
    Monument
One Archway Parkway
Kearney, NE 68847
877-511-ARCH
www.archway.org

Buffalo Bill State Historic Site
Buffalo Bill Avenue
North Platte, NE 69101
308-535-8035

Scotts Bluff National Monument
190276 Highway 92 West
Gering, NE 69341
308-436-4340
www.nps.gov/scbl/

Chimney Rock National Monument
Bayard, NE 69334
308-586-2981
www.nps.gov/chro/

WYOMING

Wyoming State Museum
2301 Central Avenue
Cheyenne, WY 82002
307-777-7022
wyomuseum.state.wy.us/index.asp

Cheyenne Frontier Days Old West Museum
4610 North Carey Avenue
P.O. Box 2720
Cheyenne, WY 82003
307-778-7290
www.oldwestmuseum.org

University of Wyoming Geological Museum
S.H. Knight Geology Building
University of Wyoming
Laramie, WY 82071
www.uwyo.edu/geomuseum/

American Heritage Center
University of Wyoming
1000 East University Avenue
Laramie, WY 82071
307-766-4295
ahc.uwyo.edu/default.htm

Laramie Plains Museum
603 Ivinson Avenue
Laramie, WY 82070
307-742-4448
www.laramiemuseum.org/

Wyoming Territorial Park
975 Snowy Range Road
Laramie, WY 82070
800-845-2287
www.wyoprisonpark.org/

Rock River Museum
P.O. Box 14
Rock River, WY 82083
307-378-2205

Fort Fred Steele
wyoparks.state.wy.us/steele1.htm

Carbon County Museum
Ninth and Walnut
Rawlins, WY 82301
307-328-2740
www.wyshs.org/mus-carboncty.htm

Rock Springs Historical Museum
212 D Street
Rock Springs, WY 82901
307-362-3138
www.wyshs.org/mus-rockspringshist.htm

Sweetwater County Historical Museum
3 East Flaming Gorge Way
Green River, WY 82935
307-872-6435
www.sweetwatermuseum.org/

Fort Bridger State Historic Site
P.O. Box 35
Fort Bridger, WY 82933
307-782-3842
wyoparks.state.wy.us/bridger1.htm

UTAH

Ogden Railroad Museum
2501 Wall Avenue
Ogden, UT 84401
801-629-8446
www.theunionstation.org/

Golden Spike National Historic Site
Promontory, UT
www.nps.gov/gosp/

NEVADA

Northeastern Nevada Museum
1515 Idaho Street
Elko, NV 89801
775-738-3418
www.nenv-museum.org/

Trails West, Inc.
P.O. Box 12045
Reno, NV 89510
www.emigranttrailswest.org

Nevada State Railroad Museum
2180 South Carson Street
Carson City, NV 89701
775-687-6953
www.nsrm-friends.org/htm

CALIFORNIA

Emigrant Trail Museum
Donner Memorial State Park
Truckee, CA 96161
530-582-7982
www.ceres.ca.gov/sierradsp/donner.html

Bodie State Park
State Route 270
Bodie, CA
www.ceres.ca.gov/sierradsp/bodie.html

Mono Lake State Reserve
Highway 395
Lee Vining, CA
www.ceres.ca.gov/sierradsp/mono.html

California State Railroad Museum
111 "I" Street
Old Sacramento, CA 95814
916-323-9280
www.csrmf.org

Crocker Art Museum
216 "O" Street
Sacramento, CA 95814
916-264-5423
www.crockerartmuseum.org/

San Francisco Maritime Museum
Building E, Fort Mason Center
San Francisco, CA 94123
415-561-7006
www.maritime.org/index.htm

Index